Ecology on Campus

Robert W. Kingsolver

Bellarmine University

PEARSON

Benjamin
Cummings

San Francisco   Boston   New York
Cape Town   Hong Kong   London   Madrid   Mexico City
Montreal   Munich   Paris   Singapore   Sydney   Tokyo   Toronto

Senior Acquisitions Editor: Chalon Bridges
Project Editor: Nora Lally-Graves
Managing Editor, Production: Mike Early
Marketing Manager: Jeff Hester
Production Services: Elm Street Publishing Services, Inc.
Cover Design: Jana Anderson
Cover Image: Corbis/Pete Seaward

This book is printed on recycled paper containing at least 30% post-consumer waste.

ISBN 0-8053-8214-3

PEARSON
Benjamin Cummings

5 6 7 8—QWD—09 08
www.aw-bc.com

Contents

Preface

Most ecology instructors learned their discipline through field work in a natural setting, and want to share the same kinds of experiences with their students. This is a commendable goal, to be pursued whenever possible, but it is not always a realistic one. Constraints imposed by the academic calendar, laboratory schedules, large class sizes, inclement weather, and transportation challenges make field trips very difficult for ecology classes in the prevailing academic environment. Logistical problems have forced an increasing number of ecology professors to minimize or completely abandon field exercises for undergraduates. Substitution of second-hand experience, such as analysis of on-line data sets or interpretation of research papers, provides reasonably good preparation for the few students who maintain interest in ecology long enough to enter graduate school. Unless instructors facilitate early and active engagement in science, however, we know that most students will memorize and then forget the course contents without developing any lasting interest or practical expertise.

The premise of this book is that engaging and instructive exercises involving small-scale ecological systems can be conducted within a weekly laboratory period on a typical North American college or university campus. Although less biologically interesting than a lake or forest, most campuses have lawns, a variety of trees and landscape plantings, a greenhouse or growth chamber, biological collections, laboratories, and a dense population of students to observe. With careful and creative selection of model systems within this environment, ecology classes can gain immediate experience to complement readings and lectures about natural systems. For the same reasons geneticists selected *Drosophila* to stand in for larger species in laboratory studies, ecologists can investigate population growth in yeast cultures, spatial organization of turf grasses, invertebrate communities in landscaping mulch, or behavioral ecology of birds at a feeder. The book's aim is not to confine students' thinking to the artificial environment of a campus, but to use the resources at hand to maximally engage students in experimental design, ecological measurement, observation, modeling, and an initiation into kinds of statistical analysis they will need to extend their investigations of the natural world beyond the confines of academe.

The manual emphasizes universal ecological principles, with an emphasis on classic theories and simple mathematical models. Though admittedly inadequate to explain all the complexity we see in nature, these well-established ideas provide a conceptual basis for a more sophisticated understanding of ecology as students gain experience in the discipline. Few ecologists, for example, would say the logistic growth curve accurately describes population dynamics we typically see in nature, but most would agree it is a concept students should master as a first step toward understanding density-dependent population phenomena. Key examples of foundation literature, along with selected references to current studies, are provided at the end of each chapter to help students advance beyond the introductory concepts demonstrated here. Selected descriptions of ecologists' work, both historical and current, are included to highlight ecology as an ongoing pursuit, and not simply a body of information. To reinforce the open-ended nature of scientific inquiry, "For Further Investigation" ideas are suggested at the end of each chapter, either for use as discussion topics, or to stimulate development of student projects extending the laboratory method in a more detailed study.

Ecologists will recognize chapter titles as standard ecological topics, with the possible exception of Chapter 6, on population genetics. My rationale for inclusion is that basic concepts of polymorphism, genotypic frequency balance, and mechanisms of allelic substitution are vital ecological concepts, needed to understand evolutionary mechanisms and conservation biology applications. These fundamentals were traditionally covered in undergraduate genetics courses, but in recent years, have been squeezed out of most introductory genetics syllabi by the proliferation of molecular topics. This is a serious omission, especially given the widespread public misunderstanding of evolutionary processes. If ecologists do not include a discussion of genetics at the population level in their courses, it is unlikely that biology majors will encounter these important ideas elsewhere in the undergraduate curriculum.

Each chapter provides explanation of a fundamental concept, and then offers several options for demonstrating the idea with hands-on exercises. Because laboratory courses are synchronized with lecture courses to varying degrees, the introduction to each chapter in this book is intended to provide sufficient background for students to understand the context of the exercises, regardless of other texts or reprints used in their course. I have intentionally chosen an informal and accessible writing style as a complement to the more formal structure of most texts and published literature. "Check your progress" boxes ask questions at frequent intervals to elicit a more active engagement with challenging concepts. Students may benefit from reading chapters in this book for a thumbnail introduction to a topic before they attempt to read published literature or advanced texts. Although terms are explained as they are introduced, a glossary is also included for convenience. This feature, along with cross-references within the text, facilitates assignment of these chapters in any order.

After the chapter introduction, three different exercises are typically presented as alternative choices. Options include both outdoor and indoor activities to allow for unpredictable weather and seasonal strategies in spring vs. fall courses. These activities are drawn from many years of my own experience teaching ecology as a laboratory course. I am personally familiar with the limitations on time and resources faced by most ecology instructors. These exercises therefore make use of inexpensive and readily available equipment and materials wherever possible. At the beginning of each exercise, the research question is clearly stated, and a "Preparation" section explains everything that needs to be done before the laboratory begins. This is intentionally included at the beginning of each exercise, rather than in a separate instructor's manual, because students need to understand the entire procedure in which they participate. Inclusion of these instructions also makes the text more useful as a guide for independent student projects.

Within each set of exercises, "calculation pages" walk students through standard mathematical modeling or statistical methods. The overriding objective is to give students with limited applied math experience a series of concrete applications to demystify and enliven the quantitative thinking central to ecological inquiry. Mathematical manipulations of the student's own data are designed to reinforce an intuitive understanding of the quantitative ideas that so often stymie aspiring biologists. In most of the activities, students will be able to perform all calculations and complete the statistical analyses using calculators. For a student's introduction to simple models or univariate statistics on a small data set, calculation by hand is not too onerous, and has greater heuristic value than using standard statistical packages. Where hand calculation would take too long, an explanation of the purpose and assumptions of the computer-assisted calculation is provided. Directions for constructing spreadsheet models are included in two of the exercises, but students are shown how to build their own programs rather than to plug values into prefabricated computer models. Any numerical parameters needed, such as random numbers or critical values of a statistical test, are included as tables or appendices. For students wishing to extend the analysis to a larger project, the calculation pages serve as templates for designing computer spreadsheets of their own.

The "here and now" approach of this book emphasizes that human beings and their artificially constructed habitats conform to basic ecological principles. Although this is not an environmental science text, students will be challenged to apply ecological theory to the world they see every day. The underlying philosophy is that ecology, like all of science, informs every aspect of life, and not just a "natural world" confined to parks and preserves. Habitats and systems relatively unaffected by human beings will be constantly referenced in the book, but the concept that we are a part of nature, and thus bound by its rules, is reinforced by the on-campus focus.

I would like to thank the following ecologists, who reviewed the manuscript and helped make it more generally applicable:

Kama Almasi, *University of Wisconsin at Stevens Point*

Claude Baker, *Indiana State University*

Romi Burks, *Southwestern University*

Rob Channell, *Fort Hays State University*

David Corey, *Midlands Technical College*

Lloyd Fitzpatrick, *University of North Texas*

Frank Gilliam, *Marshall University*

Mark Gustafson, *Texas Lutheran University*

William Hallahan, *Nazareth College*

Lauraine Hawkins, *Penn State University*

Floyd Hayes, *Pacific Union College*

John Jahoda, *Bridgewater State College*

Amiel Jarstfer, *LeTourneau University*

Stephen Johnson, *William Penn University*

Chad Jones, *University of Washington*

John Korstad, *Oral Roberts University*

John Kasmer, *Northeastern Illinois University*

Kimberly Kolb, *California State University at Bakersfield*

Tom Langen, *Clarkson University*

Susan Lewis, *Carroll College*

Chris Migliaccio, *Miami Dade Community College*

John O'Brien, *University of North Carolina at Greensboro*

Rowan Sage, *University of Toronto*

Maynard Schaus, *Virginia Wesleyan College*

James Thorne, *University of Pennsylvania*

Max Terman, *Tabor College*

Neal Voelz, *St. Cloud State University*

Maxine Watson, *Indiana University*

Phillip Watson, *Ferris State University*

I would also like to thank the Benjamin Cummings editorial team for their advice, the generations of students who tested previous versions of these exercises, my family for considerable patience during the writing, and my mentor and friend Chuck Rodell, who encouraged my earliest professional interest in ecology.

Chapter 1

Describing a Population

Figure 1.1   A fish population is sampled by seining.

INTRODUCTION

Ecology is the ambitious attempt to understand life on a grand scale. We know that the mechanics of the living world are too vast to see from a single vantage point, too gradual to observe in a single lifetime, and too complex to capture in a single narrative. This is why ecology has always been a quantitative discipline. Measurement empowers ecologists because our measuring instruments extend our senses, and numerical records extend our capacity for observation. With measurement data, we can compare the growth rates of trees across a continent, through a series of environmental conditions, or over a period of years. Imagine trying to compare from memory the water clarity of two streams located on different continents visited in separate years, and you can easily appreciate the value of measurement.

Numerical data extend our capacity for judgment too. Since a stream runs muddier after a rain and clearer in periods of drought, how could you possibly wade into two streams in different seasons and hope to develop a valid comparison? In a world characterized by change, data sets provide reliability unrealized by single observations. Quantitative concepts such as averages, ratios, variances, and probabilities reveal ecological patterns that would otherwise remain unseen and unknowable. Mathematics, more than any cleverly crafted lens or detector, has opened our window on the universe. It is not the intention of this text to showcase math for its own sake, but we will take measurements and make calculations because this is the simplest and most powerful way to examine populations, communities, and ecosystems.

Sampling

To demonstrate the power of quantitative description in ecology, you will use a series of measurements and calculations to characterize a **population**. In biology, a population is defined as a group of individuals of the same species living in the same place and time. Statisticians have a more general definition of a population, that is, all of the members of any group of people, organisms, or things under investigation. For the ecologist, the biological population is frequently the subject of investigation, so our biological population can be a statistical population as well.

Think about a population of red-ear sunfish in a freshwater lake. Since the population's members may vary in age, physical condition, or genetic characteristics, we must observe more than one representative before we can say much about the sunfish population as a group. When the population is too large to catch every fish in the lake, we must settle for a **sample** of individuals to represent the whole. This poses an interesting challenge for the ecologist: how many individuals must we observe to ensure that we have adequately addressed the variation that exists in the entire population? How can this sample be collected as a fair representation of the whole? Ecologists try to avoid **bias**, or sampling flaws that overrepresent individuals of one type and underrepresent others. If we caught our sample of sunfish with baited hooks, for example, we might selectively capture individuals large enough to take the bait, while leaving out smaller fish. Any estimates of size or age we made from this biased sample would poorly represent the population we are trying to study.

After collecting our sample, we can measure each individual and then use these measurements to develop an idea about the population. If we are interested in fish size, we could measure each of our captured sunfish from snout to tail (Figure 1.2). Reporting every single measurement in a data table would be truthful, but not very useful, because the human mind cannot easily take in long lists of numbers. A more fruitful approach is to take all the measurements of our fish and systematically construct a composite numerical description, or **statistic**, which conveys information about the population in a more concise form. The average length (also called the **mean** length) is a familiar way to represent the size of a typical individual. We might find, for instance, that the mean length of sunfish in this lake is 12.07 centimeters, based on a sample of 80 netted fish. The symbol μ is used for the mean of all fish in the population, which we are trying to estimate in our study. The symbol $\bar{x}$ is used for the mean of our sample, which we hope to be close to μ.

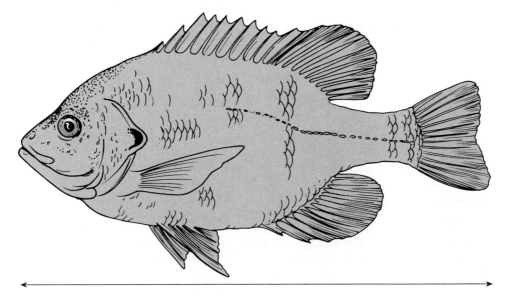

Figure 1.2   Measuring length of red-ear sunfish.

Means are useful, but they can be misleading. If a population were made up of small one-year-old fish and much larger two-year-old fish, the mean we calculate may fall somewhere between the large and small size classes—above any of the small fish, but below any of the large ones. A mean evokes a concept of the "typical" fish, but the "typical" fish may not actually exist in the population (Figure 1.3).

Figure 1.3   The calculated mean describes a "typical" fish that does not actually exist in a population composed of two size classes.

For this reason, it is often helpful to use more than one statistic in our description of the typical member of a population. One useful alternative is the **median**, which is the individual ranked at the 50th percentile when all data are arranged in numerical order. Another is the **mode**, which is the most commonly observed length of all fish in the sample.

Picturing Variation

After calculating statistics to represent the typical individual, it is still necessary to consider variation among members of the population. A **histogram** is a simple graphic representation of the way individuals in the population vary.

Check your progress:

If you wanted to determine the average height of students on your campus, how would you select a sample of students to measure?

Hint: Avoid statistical bias by making sure
every student on campus has an equal chance
of inclusion in the sample.

To produce a histogram:

1. Choose a measurement variable, such as length in our red-ear sunfish. Assume we have collected 80 sunfish and measured each fish to the nearest millimeter.

2. On a number line, mark the longest and shortest measurements taken from the population (Figure 1.4). The distance on the number line between these points, determined by subtracting the smallest from the largest, is called the **range**. In our example, the longest fish measures 16.3 cm, and the shortest fish measures 8.5 cm, so the range is 7.8 cm.

$$\text{Range} = 16.3 \text{ cm} - 8.5 \text{ cm} = 7.8 \text{ cm}$$

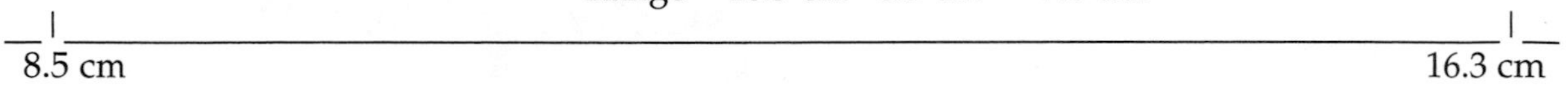

```
_|_____|_
8.5 cm                                                16.3 cm
```

Figure 1.4   Number line indicating range.

3. Next, divide the range into evenly spaced divisions (Figure 1.5). In our example, each division of the number line represents a **size class**. It is customary to use between 10 and 20 size classes in a histogram. For our example, we will divide the range of sunfish sizes into 16 units of 0.5 cm each. The first size class includes fish of sizes 8.5 through 8.9 cm. The next size class includes fish of sizes 9.0 through 9.4 cm, and so forth to the last size class, which includes sizes 16.0–16.4.

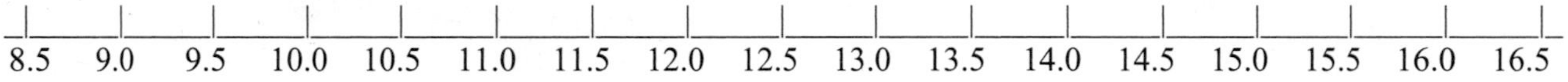

```
_|___|___|___|___|___|___|___|___|___|___|___|___|___|___|___|_
8.5  9.0  9.5  10.0 10.5 11.0 11.5 12.0 12.5 13.0 13.5 14.0 14.5 15.0 15.5 16.0 16.5
```

Figure 1.5   Number line divided into size classes of 0.5 cm.

4. Having established these size classes, it is possible to look back at our measurement data and count how many fish fall into each class. The number of individuals falling within a class is the **frequency** of that class in the population. By representing each measurement with an X, as shown in Figure 1.6, we can illustrate frequencies on the number line.

5. On the completed histogram illustrated in Figure 1.7, the stack of X-marks is filled in as a vertical bar. The height of each bar represents the proportion of the entire population that falls within a given size class.

Check your progress:

In the sample described by the histogram (Figure 1.7), how many fish measured between 10.0 and 10.4 cm?

Answer: 4

Figure 1.6 Counting frequencies.

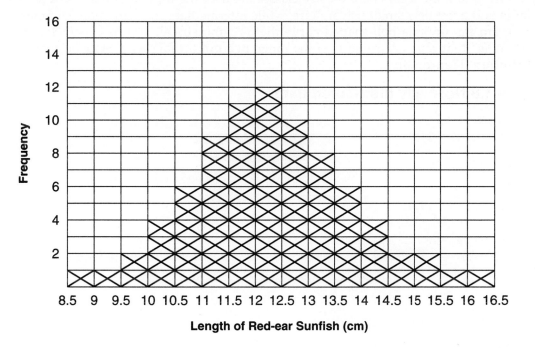

Length of Red-ear Sunfish (cm)

Figure 1.7 Histogram of fish lengths.

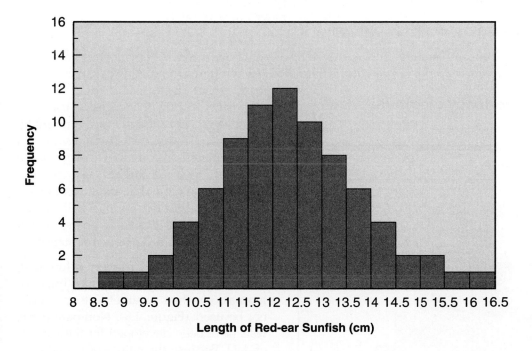

Length of Red-ear Sunfish (cm)

Describing the Pattern of Variation

Notice that in our sample histogram, the most common size classes are near the middle of the distribution (Figure 1.7). Extremely large and extremely small fish are rare, while intermediate sizes are more common. The "skyline" of the histogram fits under a bell-shaped curve that is symmetrical, and has characteristic rounded "shoulders" and "tails" that taper to the ends of the range in a predictable way. Statisticians call this shape a **normal distribution** (Figure 1.8).

Figure 1.8 Normally distributed data.

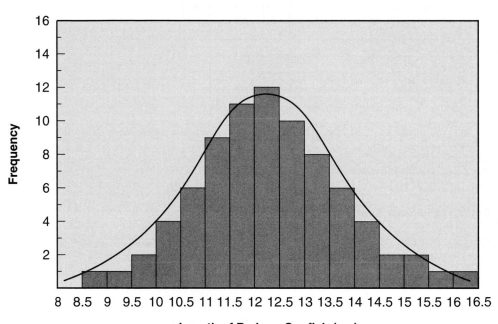

Length of Red-ear Sunfish (cm)

This pattern of variation is common in nature, and is encountered quite often when one effect is influenced by many independently acting causes. Since size in fish is influenced by temperature, diet, water purity, and

Figure 1.9a Normally distributed data.

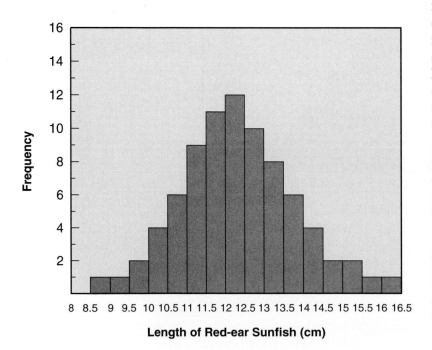

Length of Red-ear Sunfish (cm)

many other factors, it would not be surprising to find that sizes of fish in a mixed-age population are normally distributed. Because the normal distribution is encountered so commonly, many of the statistical tools ecologists use to test hypotheses assume that variations in their data are distributed in this bell-shaped form. Models and tests based on this kind of distribution are called **parametric statistics**.

If the histogram of variation is lopsided, has more than one peak, or is too broad or too narrow, then parametric tests should not be used (Figure 1.9). **Non-parametric tests** have been developed for these kinds of data. Because the nature of variation in your measurements is critical to further analysis, it is always a good idea to draw a histogram and compare your data to a normal distribution before taking your analysis any farther.

Measuring Variation

How trustworthy is the mean that we calculate from our sample of fish? Two factors come into play. First, the **sample size** is critical. If fish in the lake vary a lot in size, a mean calculated from a small sample (say 10 fish)

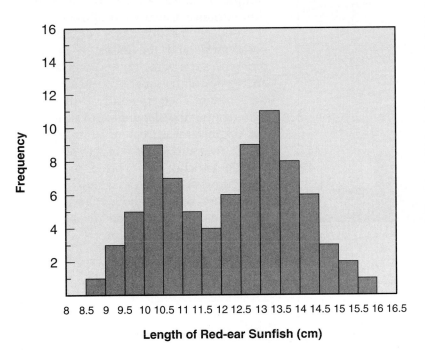

Figure 1.9b Bimodally distributed data.

might be significantly off the mark. By chance, the 10 fish you caught might be larger or smaller than the average population size you are trying to describe. If the sample is expanded to 1000, it is much more likely that your calculated mean will accurately reflect the population average. *A fundamental principle of data collection is that the sample size must be large enough to eliminate sampling errors due to chance departures from the population mean.* To keep our thoughts straight, we use **n = size of the sample**, and **N = size of the entire population**. N is usually unknown, but can be estimated in a number of ways. (See Chapter 4.)

How large, then, must a sample be? This depends on the amount of variation in the population. Samples taken from a fish farm where all the fish are nearly the same size will give reliable estimates, even if the sample is small. In a natural population with a great range of sizes, the sample has to be expanded to ensure that the larger variation is accounted for. Thus, *the more variable the population, the larger the sample must be* to achieve the same level of reliability. It becomes obvious that we need a statistic to measure variation.

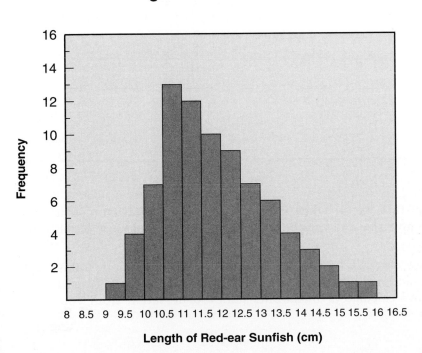

Figure 1.9c Skewed data.

To measure the amount of variation around the mean, we use a statistic called the **standard deviation (abbreviated s.d.)**. To distinguish between our sample and the entire population, we define **s = the standard deviation of the sample**, and **σ = the standard deviation of the whole population**. The standard deviation is expressed in the same units as the original measurements, which would be cm in our hypothetical fish study. A standard deviation can thus be shown as a portion of the range on a number line. In normally distributed populations, 95% of all individuals fall within 1.96 standard deviations from the mean. This means that the X-axis of a histogram can be marked off in four standard deviation units (two above the mean, and two below), and roughly 95% of the observations will fall within that region (Figure 1.10).

Figure 1.10 Standard deviations.

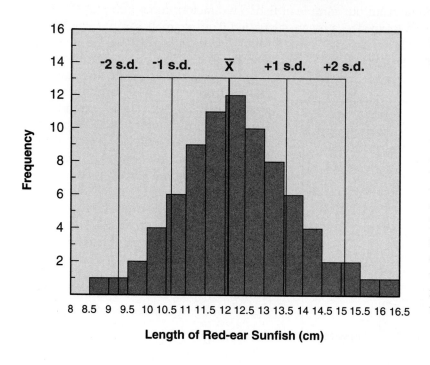

Length of Red-ear Sunfish (cm)

To calculate the size of a standard deviation, it is actually easier first to calculate a related statistic called the **variance**. The variance is the square of the standard deviation, so we use s^2 = **the sample variance**, and σ^2 = **the population variance**. Calculation of the variance is based on the difference between each observation and the mean. If all these differences are squared, and we calculate an average of the squared values, we have the variance. (See Appendix 1.) It is good to remember that the units on variance are the original measurement units squared. If we measure length in cm, then the sample variance is reported in cm^2. In calculating standard deviations, we take the square root of the variance, which returns us to our original measurement units, which is length in cm.

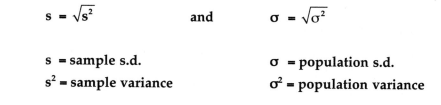

$$s = \sqrt{s^2} \qquad \text{and} \qquad \sigma = \sqrt{\sigma^2}$$

s = sample s.d. $\qquad\qquad$ σ = population s.d.

s^2 = sample variance $\qquad$ σ^2 = population variance

Check your progress:

If the standard deviation of a population is 9.5, what is the population variance?

Answer: 90.25

Confidence Intervals

There is one more statistic that you will find useful when characterizing the typical sunfish in your population with a mean. Incorporating sample size and variation, you can develop a measure of reliability of the mean called the **standard error (S.E.)**.

Assume there are 25 people in your ecology class. Each of you goes to the same pond sometime this week, nets a sample of sunfish in the same way, and measures 100 randomly selected individuals. Releasing the fish unharmed, you return to the lab and calculate a mean and standard deviation from your data.

Everyone else does the same. Would your 25 sample means be identical? No, but the variation in means would be considerably smaller than the total variation among the fish in the pond. Repeating a sampling program 25 times is usually impractical. Fortunately, statistics gives us a way to measure reliability when we have only one mean developed from one sample. The variation among all possible sample means can be predicted from the sample size and the variation in the pond's sunfish with the following formula:

Looking at the formula, you can see the relationship between error in our estimate, the variability of sunfish, and the sample size. The smaller the S.E. is, the more trustworthy your calculated mean. Note that the sample standard deviation is in the numerator of the calculation, so the more variable the size of the fish, the less accurate the estimate you made from a random sample. Sample size, on the other hand, is in the denominator. This implies that a large sample makes your mean more reliable. The formula shows that the more variable the population, the larger our sample must be to hold S.E. to an acceptably small margin of error.

$$\text{S.E.} = s/\sqrt{n}$$

S.E. = standard error of the mean

s = standard deviation of sample

n = sample size

Since standard errors tend to be normally distributed, it is a safe assumption that 95% of the variation in all possible means will fall within 1.96 S.E. of the actual mean. This fact can be used to calculate a 95% confidence interval as follows:

95% Confidence interval = $\bar{x} \pm 1.96$ S.E. $\bar{x}$ = sample mean

S.E. = standard error

Check your progress:

Calculate the 95% confidence interval for a mean of 14.3, derived from a sample of 25, where the standard deviation is 4.2. What are the upper and lower limits?

Answer: 12.65 to 15.95

To go back to our number line, the confidence intervals can be represented by brackets around the sample mean. A 95% confidence interval implies that the actual population mean (μ) will fall within the brackets you have placed around your estimate ($\bar{x}$) 95% of the time under these experimental conditions (Figure 1.11).

The importance of confidence limits cannot be overstated. Scientifically, it is dishonest to report a sample mean by itself. Without sharing with your readers the sampling methods, sample size, and the variability of the population, there is no way for them to know how accurately the sample mean represents the population. *Always report sample size with a mean, and add some measure of variation.* Any of the statistics representing variation (s, s², or S.E.) can be reported, since any one of these can be used to calculate the other two.

Figure 1.11 95% Confidence interval.

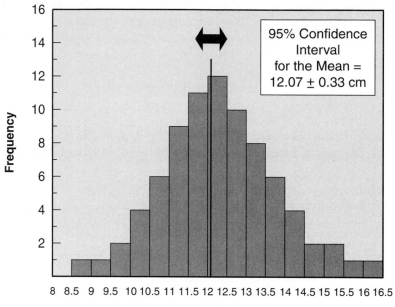

METHOD A: SEED WEIGHTS IN LEGUMES

[Laboratory activity]

Research Question
How does a population of bean seeds vary in weight around the typical individual?

Preparation
At a grocery or health food store, find uncooked dried beans of several types. Bags of approximately one pound are ideal. Lima beans, pinto beans, navy beans, and great northern beans are good varieties to choose from. If possible, buy unsorted beans; these best represent the population in the field.

Materials (per laboratory team)
1-pound bag of beans (Different teams can use different varieties.)

Analytical balance or electronic balance sensitive to 0.01 g (.001 g is preferable).

Plastic weighing tray, as large as the balance will accommodate.

Electronic calculator

Procedure

1. *Recognize that each seed in your bag is a living organism*, harvested from a mature population of annual plants. If sprouted and allowed to mature, these seeds would produce a population of bean plants. Seed weight is of obvious interest in cultivated plants like beans, but is also biologically important in wild plants, because seed weight affects the distribution, growth rate, and survivorship of the seedling. (For example, see Rees, 1995.) Since the maternal plant must expend more resources to produce larger seeds, plants making larger seeds cannot produce as many. In most plant species, the tradeoff between larger seeds vs. more numerous seeds is influenced by genetics, subject to **natural selection**, and varying from one plant type to another. Identifying mean and variance for seed weights is therefore a biologically important description of **reproductive strategy** in a plant population.

2. *Develop a sampling plan.* Your sample size will be 80 beans. If you were to cut the bag open and take only the beans on top, would your "grab sample" represent the population fairly? If larger and smaller beans settled differently during shipment, this approach might result in a biased sample. Spreading beans out on a lab bench, mixing them to randomize your sample, and selecting a sample of 80 closest to your side of the bench is a much better way to ensure that your sample is random.

3. *Weigh and record observations.* Weigh each of the 80 beans in your sample. How accurate is your balance? To simplify your data analysis, record weights in milligrams. (For example, rather than 0.087 g, record 87 mg. If your balance measures only to two decimal places, record 0.09 g as 90 mg. Enter your measurements on the calculation page at the end of this chapter.

4. *Produce a histogram.* Following the example in the introduction, produce a histogram of seed weights.

5. *Calculate descriptive statistics.* Calculate a mean, standard deviation, standard error, and confidence limits for this population, following the directions in the introduction and appendices.

6. *Interpret your data.* What does the histogram show about the variation of seed weights in this species? Answer the Questions for Discussion that follow the calculation pages. Considering the way natural selection works, would you always expect symmetrical frequency distributions of seed weights?

7. *Check your accuracy.* Working with your lab partners, count every bean in the bag. Then weigh the whole population by weighing the beans in a plastic weighing tray and subtracting the weight of the tray. If your balance cannot accommodate all the beans at once, divide the population into parts, weigh each part, and then sum the separate measurements. Divide the weight of the whole population by the number of beans in the population to calculate the population mean. How does the population mean compare with the sample mean you calculated in step 3? Does the population mean fall within the 95% confidence intervals you calculated?

METHOD B: NEEDLE LENGTH IN CONIFERS

[Outdoor/indoor activity]

Research Question

How do pine needles vary in length, within and among individual trees?

Preparation

Locate several pine trees of the same species. Needles of other conifers may be used if they have needles long enough to be measured with a mm ruler. If students cannot collect needles themselves, pine branches can be collected elsewhere, or "pine straw" can be obtained from garden supply firms. Laboratory teams can be larger if greater effort is required to collect and measure 80 needles.

Materials (per laboratory team)

Metric ruler, marked in mm

Electronic calculator

Procedure

1. *Recognize that each needle is a plant organ, developing according to a genetic program influenced by local conditions.* A pine needle performs the critically important job of photosynthesis, producing chemical energy for the tree. A needle's length may affect how well it functions. If the needle is too short, it may lack sufficient photosynthetic tissue to produce an adequate supply of food. On the other hand, needles that are too long may fail to transport fluids adequately to the tip, or may accumulate too much ice and break limbs in the winter. Limits on size and shape affecting the performance of a biological trait are called **functional constraints**, and they help explain why many species' characteristics remain within predictable ranges.

    Although needles on the same tree might be expected to conform to a genetically determined size, differences in leaf age, sun or shade, exposure to wind, temperature, or moisture supplied through a particular branch could influence needle length. Within a tree, we can recognize many sources of variation. Among a population of trees, the variation is probably even larger because different trees have different genes, and probably experience a broader range of environmental conditions.

2. *Develop a sampling plan.* Your sample size will be 80 needles. If you have access to a grove or row of pine trees, spread out your needle collection to include roughly equal numbers of needles from each of the trees. A sampling plan that includes the same number of needles from each tree in your research area is a **stratified sample**. Decide whether you will pull a needle from a live branch, or pick a fallen needle from the ground. Since pine needles decay slowly, they can be collected long after they fall from the tree. If you collect live needles from the tree, will you always collect from a low branch, or will you try to collect equal numbers of needles from high, mid-height, and low branches?

    When collecting needles from pines (genus *Pinus*) you will discover that their needles come in bunches. The brown collar of tissue holding the bunch of needles together is actually a dwarf branch, called a fascicle. The number of needles in a bunch is fairly consistent, and is useful for identification. For instance, the Eastern White Pine (*Pinus strobus*) typically has five needles per fascicle, while Red Pine (*Pinus resinosa*) has two.

Make a decision about which needle in the bunch you will measure. The longest one? A randomly selected one? What will you do if you encounter a broken needle? Whatever your method, it would be best to measure only one needle per bunch, so that your data are not clustered into subgroups. Pull the needles apart carefully, so as not to introduce error by breaking the base, and measure one needle according to your predetermined sampling plan.

If the population of pines on campus is not that large, collect all your needles from the same tree. Recognize that this collection is not a population in the biological sense, since only one individual produced all the needles. However, your collection is a population in the statistical sense, because you are measuring a sample of a much larger number of needles. Different lab groups can sample different trees so that you can compare your results.

3. *Measure lengths and record observations.* Measure each of the 80 needles in your sample. How accurate is your ruler? Record needle lengths in mm on the data pages at the end of this chapter. Follow directions on the calculation page to produce a histogram of needle lengths and to calculate a mean, standard deviation, standard error, and confidence limits for this population.

4. *Interpret your data.* What does the histogram show about the variation of needle lengths in this species? Answer the Questions for Discussion that follow the calculation pages. Think about the functional constraints on evergreen needles, and try to explain the distribution of sizes in biological terms.

5. *Check your accuracy.* Compare your results with those of another lab group. Does the other group's calculated mean fall within the 95% confidence limits you calculated for your own mean? If not, how do you interpret the difference between the two estimates? If two groups sampled from the same tree, then significant differences in your calculated means might result from sampling bias, measurement error, or calculation mistakes. If two lab groups sampled different trees, then the data may reflect real biological differences between the two trees.

CALCULATIONS (METHOD A OR B)

1. *Enter your 80 measurements* (x_i) in the second column of the table, recording 20 measurements per page on each of the next four pages. You may wish to split up this task, with each member of your team completing a page.

2. *Sum the measured values* (seed weight or needle length) for each page, and then complete calculation of the mean ($\bar{x}$) in the calculation box at the end of the tables by adding totals from all four pages and dividing by the sample size (n = 80).

3. Subtract the mean from each of the 80 measurements to *obtain the deviation* above or below the average. (Deviation from mean for sample i = d_i.)

4. *Square each deviation* (d_i^2).

5. *Add up all the squared deviations* on each page, then sum the totals for the four pages to compute the sum of squared deviations (Σd^2).

6. Divide the sum of squared deviations by (sample size − 1) to *calculate the sample variance* (s^2).

7. Take the square root of the variance to *calculate the standard deviation* (s).

Data (Methods A or B)—Page 1

Sample number (i)	Measurement for sample i (x_i)		Deviation (d_i)		Squared deviations (d_i^2)
1		$- (\overline{x}) =$		$^\wedge 2 =$	
2		$- (\overline{x}) =$		$^\wedge 2 =$	
3		$- (\overline{x}) =$		$^\wedge 2 =$	
4		$- (\overline{x}) =$		$^\wedge 2 =$	
5		$- (\overline{x}) =$		$^\wedge 2 =$	
6		$- (\overline{x}) =$		$^\wedge 2 =$	
7		$- (\overline{x}) =$		$^\wedge 2 =$	
8		$- (\overline{x}) =$		$^\wedge 2 =$	
9		$- (\overline{x}) =$		$^\wedge 2 =$	
10		$- (\overline{x}) =$		$^\wedge 2 =$	
11		$- (\overline{x}) =$		$^\wedge 2 =$	
12		$- (\overline{x}) =$		$^\wedge 2 =$	
13		$- (\overline{x}) =$		$^\wedge 2 =$	
14		$- (\overline{x}) =$		$^\wedge 2 =$	
15		$- (\overline{x}) =$		$^\wedge 2 =$	
16		$- (\overline{x}) =$		$^\wedge 2 =$	
17		$- (\overline{x}) =$		$^\wedge 2 =$	
18		$- (\overline{x}) =$		$^\wedge 2 =$	
19		$- (\overline{x}) =$		$^\wedge 2 =$	
20		$- (\overline{x}) =$		$^\wedge 2 =$	
Page 1 Sum $\Sigma(x_i) =$		**Page 1 Sum of Squared Deviations**		$\Sigma(d_i^2) =$	

Data (Methods A or B)—Page 2

Sample number (i)	Measurement for sample i (x_i)		Deviation (d_i)		Squared deviations (d_i^2)
21		$-(\overline{x}) =$		$^\wedge 2 =$	
22		$-(\overline{x}) =$		$^\wedge 2 =$	
23		$-(\overline{x}) =$		$^\wedge 2 =$	
24		$-(\overline{x}) =$		$^\wedge 2 =$	
25		$-(\overline{x}) =$		$^\wedge 2 =$	
26		$-(\overline{x}) =$		$^\wedge 2 =$	
27		$-(\overline{x}) =$		$^\wedge 2 =$	
28		$-(\overline{x}) =$		$^\wedge 2 =$	
29		$-(\overline{x}) =$		$^\wedge 2 =$	
30		$-(\overline{x}) =$		$^\wedge 2 =$	
31		$-(\overline{x}) =$		$^\wedge 2 =$	
32		$-(\overline{x}) =$		$^\wedge 2 =$	
33		$-(\overline{x}) =$		$^\wedge 2 =$	
34		$-(\overline{x}) =$		$^\wedge 2 =$	
35		$-(\overline{x}) =$		$^\wedge 2 =$	
36		$-(\overline{x}) =$		$^\wedge 2 =$	
37		$-(\overline{x}) =$		$^\wedge 2 =$	
38		$-(\overline{x}) =$		$^\wedge 2 =$	
39		$-(\overline{x}) =$		$^\wedge 2 =$	
40		$-(\overline{x}) =$		$^\wedge 2 =$	
Page 2 Sum $\Sigma(x_i) =$		**Page 2 Sum of Squared Deviations** $\Sigma(d_i^2) =$			

Data (Methods A or B)—Page 3

Sample number (i)	Measurement for sample i (x_i)		Deviation (d_i)		Squared deviations (d_i^2)
41		$- (\overline{x}) =$		$\wedge 2 =$	
42		$- (\overline{x}) =$		$\wedge 2 =$	
43		$- (\overline{x}) =$		$\wedge 2 =$	
44		$- (\overline{x}) =$		$\wedge 2 =$	
45		$- (\overline{x}) =$		$\wedge 2 =$	
46		$- (\overline{x}) =$		$\wedge 2 =$	
47		$- (\overline{x}) =$		$\wedge 2 =$	
48		$- (\overline{x}) =$		$\wedge 2 =$	
49		$- (\overline{x}) =$		$\wedge 2 =$	
50		$- (\overline{x}) =$		$\wedge 2 =$	
51		$- (\overline{x}) =$		$\wedge 2 =$	
52		$- (\overline{x}) =$		$\wedge 2 =$	
53		$- (\overline{x}) =$		$\wedge 2 =$	
54		$- (\overline{x}) =$		$\wedge 2 =$	
55		$- (\overline{x}) =$		$\wedge 2 =$	
56		$- (\overline{x}) =$		$\wedge 2 =$	
57		$- (\overline{x}) =$		$\wedge 2 =$	
58		$- (\overline{x}) =$		$\wedge 2 =$	
59		$- (\overline{x}) =$		$\wedge 2 =$	
60		$- (\overline{x}) =$		$\wedge 2 =$	
Page 3 Sum $\Sigma(x_i) =$		**Page 3 Sum of Squared Deviations** $\Sigma(d_i^2) =$			

Data (Methods A or B)—Page 4

Sample number (i)	Measurement for sample i (x_i)		Deviation (d_i)		Squared deviations (d_i^2)
61		$-(\overline{x}) =$		$^\wedge 2 =$	
62		$-(\overline{x}) =$		$^\wedge 2 =$	
63		$-(\overline{x}) =$		$^\wedge 2 =$	
64		$-(\overline{x}) =$		$^\wedge 2 =$	
65		$-(\overline{x}) =$		$^\wedge 2 =$	
66		$-(\overline{x}) =$		$^\wedge 2 =$	
67		$-(\overline{x}) =$		$^\wedge 2 =$	
68		$-(\overline{x}) =$		$^\wedge 2 =$	
69		$-(\overline{x}) =$		$^\wedge 2 =$	
70		$-(\overline{x}) =$		$^\wedge 2 =$	
71		$-(\overline{x}) =$		$^\wedge 2 =$	
72		$-(\overline{x}) =$		$^\wedge 2 =$	
73		$-(\overline{x}) =$		$^\wedge 2 =$	
74		$-(\overline{x}) =$		$^\wedge 2 =$	
75		$-(\overline{x}) =$		$^\wedge 2 =$	
76		$-(\overline{x}) =$		$^\wedge 2 =$	
77		$-(\overline{x}) =$		$^\wedge 2 =$	
78		$-(\overline{x}) =$		$^\wedge 2 =$	
79		$-(\overline{x}) =$		$^\wedge 2 =$	
80		$-(\overline{x}) =$		$^\wedge 2 =$	
Page 4 Sum $\Sigma(x_i) =$		**Page 4 Sum of Squared Deviations** $\Sigma(d_i^2) =$			

Calculation of Variance and Standard Deviation

Page 1 Sum $\Sigma(x_i) =$		**Page 1 Sum of Squared Deviations** $\Rightarrow$ $\Sigma(d_i^2) =$	
Page 2 Sum $\Sigma(x_i) =$		**Page 2 Sum of Squared Deviations** $\Rightarrow$ $\Sigma(d_i^2) =$	
Page 3 Sum $\Sigma(x_i) =$		**Page 3 Sum of Squared Deviations** $\Rightarrow$ $\Sigma(d_i^2) =$	
Page 4 Sum $\Sigma(x_i) =$		**Page 4 Sum of Squared Deviations** $\Rightarrow$ $\Sigma(d_i^2) =$	
Grand Total $\Sigma(x_i) =$		**Grand Total Sum of Squared Deviations** $\Rightarrow$ $\Sigma(d_i^2) =$	
Mean $\Sigma(x_i) / n =$		**Sample Variance** $\Rightarrow$ $s^2 = \Sigma(d_i^2)/(n-1) =$	
Standard Deviation $\Rightarrow$ $\sqrt{s^2} =$			

(Methods A or B)

Summary of Results (Method A or B)

Sample Size (n) = _____

Sample Mean ($\bar{x}$) = _____

Sample Variance (s^2) = _____

Standard Deviation (s) = _____

Standard Error (S.E.) = _____

95% Confidence Interval for Mean = _____

Histogram of Group Data

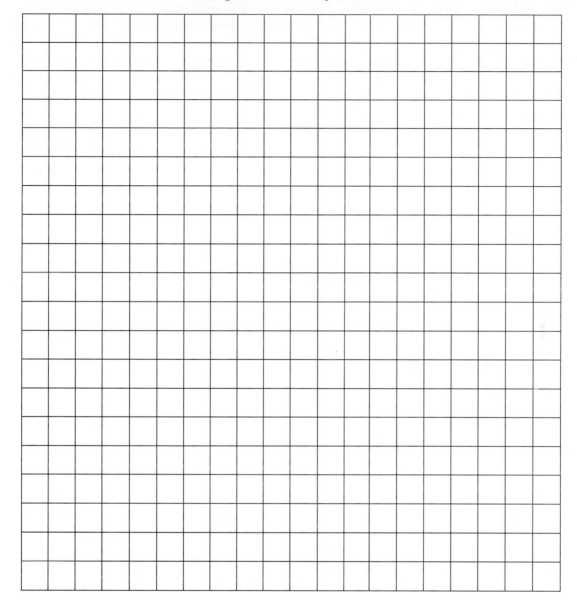

FREQUENCY

MEASUREMENT CLASSES

Questions (Method A or B)

1. In comparing your histogram to the description of a normal distribution, did you seem to get a fairly good fit to the bell-shaped curve, or did you notice a different pattern? Describe these differences: was the histogram bimodal? skewed? flattened or sharply peaked? If you performed statistical tests using these data, would you be comfortable using parametric statistics, or would you seek a non-parametric alternative? Do you think you made enough measurements to make a certain judgment on this question, or do you think more data may be needed?

    _____

    _____

    _____

    _____

2. The mean, median, and mode are three different statistical approaches to describe the "typical" individual in a population. Recall that the median is the middle observation when all data are arranged in numerical order, and the mode is the most commonly observed measurement. Based on your data, does it matter very much which of these three statistics is used? Explain how this answer is related to your answer for the previous question.

    _____

    _____

    _____

    _____

3. What do the variance, S.E., or standard deviation estimates tell you about your population that the mean does not tell you? Why is it important to report some measure of variation, along with the sample size, whenever you report a calculated mean?

    _____

    _____

    _____

    _____

4. Variation among members of a population can lead to natural selection, but only if two conditions are met: First, the trait must be relevant to an individual's survival and/or reproductive rate. Second, variation in this trait must be **heritable**, that is, at least partly controlled by genes. How would you design experiments to determine the importance of this trait in determining survival and reproduction? How would you test the extent to which this trait is heritable?

    _____

    _____

    _____

FOR FURTHER INVESTIGATION

1. In a field guide to the trees in your region, identify the species of evergreen you sampled in this exercise. What does the book say about the needle length of this species? Does the description include a range of needle lengths, or just a mean length? If keys are included in the guide, is needle length used to distinguish this species from others? Based on your data, how often might a single measurement of a randomly selected needle lead to an incorrect species identification?

2. Are some varieties of beans inherently more variable than others? Calculate variances for different types of beans, or from the same variety purchased from different suppliers. Is genetic variety a good predictor of seed weight, or are other considerations such as growing conditions in a given crop year more significant?

FOR FURTHER READING

Howe, H.F. and J. Smallwood. 1982. Ecology of seed dispersal. *Annual Reviews of Ecology and Systematics* 13:201–228.

Preston, Richard J. Jr. 1989. *North American Trees.* Iowa State University Press, Ames.

Rees, Mark. 1995. Community structure in sand dune annuals: Is seed weight a key quantity? *The Journal of Ecology* 83(5):857–863.

Chapter 2

Allometric Relationships

Figure 2.1 Water striders supported by the surface tension of water depend on a high surface area in proportion to their body weight.

INTRODUCTION

The way a plant or animal interacts with its physical environment depends a lot on body size. Physical processes such as gravitational pull, surface adhesion, evaporation, and heat exchange do not act the same way on large organisms as on small ones. For example, a water strider is an insect that can move across the surface film of water in a pond because at the size and weight of an insect, surface tension is a stronger force than gravitational pull (Figure 2.1). We cannot imagine a dog-sized animal accomplishing the same feat. An amoeba does not need specialized gas exchange organs, because at its size, diffusion through the exterior membrane is sufficient to get rid of carbon dioxide and to take in oxygen. A human being attempting to breathe only through the skin would quickly die of asphyxia, because at our size, too many of our cells are too far from the surface for diffusion to supply adequate oxygen for life.

To understand more precisely how body size affects interactions with the outside world, let's look at the way an organism's dimensions affect its length, volume, and surface area. Imagine a bacterial cell floating in a river, and imagine it is shaped like a cube with each side of the cube measuring one micron (1 μ) (Figure 2.2). If the length of the cube is 1 μ, the surface of any side of the cube must be $1 \, \mu \times 1 \, \mu = 1 \, \mu^2$. Since a cube has six sides, the **surface area** of the cell is $6(1 \, \mu^2)$, or **$6 \, \mu^2$**. Calculating the volume is just as easy. Since volume is equal to length times width times height, the **volume** of our cell is $1 \, \mu \times 1 \, \mu \times 1 \, \mu = \mathbf{1 \, \mu^3}$.

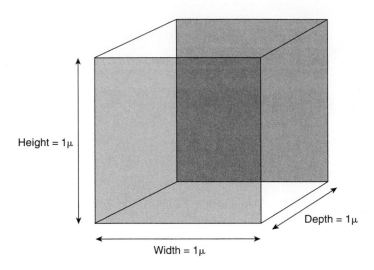

Figure 2.2   Dimensions of a small cube-shaped cell.

Now we are prepared to think about oxygen consumption in this simple organism. Since oxygen is consumed by the cytoplasm, oxygen demand should be roughly proportional to the cell volume. The more volume of cytoplasm, the more oxygen is needed. Oxygen intake, however, is limited by the surface area of the cell, since oxygen can only diffuse in through the cell membrane. The more surface, the faster oxygen can diffuse in.

To examine the oxygen balance in the cell, we would want to know how quickly oxygen can move in (related to surface area) in comparison to the rate oxygen is consumed (related to volume). The important measure, then, is the ratio of surface area to volume. This fraction is expressed as **surface/volume**, and measured in μ^2/μ^3, which simplifies to units of μ^{-1}. In our example, the surface to volume ratio is $6\mu^2/1\mu^3$, which means six units of surface area for every one unit of volume. The greater the surface/volume ratio, the easier it is for our bacterium to absorb oxygen as fast as oxygen is consumed.

What happens if the organism grows to a larger size (Figure 2.3)? Let's assume it doubles in every dimension so that its length is now 2μ. Its surface area is six times the length times the width, or $6(2\mu \times 2\mu) = 24\mu^2$. The volume would now be length times width times height, or $2\mu \times 2\mu \times 2\mu = 8\ \mu^3$. What does this do to the surface/volume ratio? We can calculate the new ratio as $24\mu^2/8\ \mu^3$, which simplifies to $3\mu^2/1\mu^3$. We now have only three units of surface area for each unit of volume—exactly half of what we had before! Consulting the following table should convince you of a very important relationship in the physics of survival for organisms: *Doubling the size of an organism cuts its surface to volume ratio in half.*

Small Cell 1μ x 1μ x 1μ

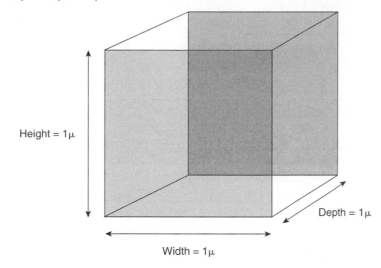

Height = 1μ

Depth = 1μ

Width = 1μ

Doubled Cell 2μ x 2μ x 2μ

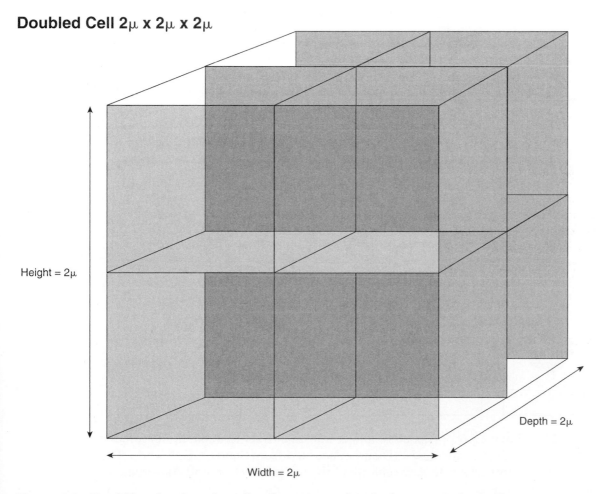

Height = 2μ

Depth = 2μ

Width = 2μ

Figure 2.3   Doubling the size of a cell reduces its surface/volume ratio by half.

LENGTH	SURFACE AREA	VOLUME	SURFACE/VOLUME RATIO
1μ	$6\mu^2$	$1\,\mu^3$	$6\mu^2/1\mu^3 = \mathbf{6}\,\mu^{-1}$
2μ	$24\mu^2$	$8\,\mu^3$	$24\mu^2/8\mu^3 = \mathbf{3}\,\mu^{-1}$
4μ	$96\mu^2$	$64\,\mu^3$	$96\mu^2/64\mu^3 = \mathbf{1.5}\,\mu^{-1}$
8μ	$384\mu^2$	$512\,\mu^3$	$384\mu^2/512\mu^3 = \mathbf{0.75}\,\mu^{-1}$

The table shows that volume (a cubic function) grows faster than surface area (a squared function), so the bigger the organism, the smaller its surface area in relation to its volume. This has a number of implications in ecology. Any living process that involves movement through a surface will be faster in small creatures and slower in large ones. For example, digestion involves movement of molecules across the intestinal tract, so a large animal has a greater challenge than a small one moving digested food through the surface area of the small intestine. Not surprisingly, large animals have compensated for this disadvantage by changing their anatomy, with intestines tending to be longer in comparison to body length (Figure 2.4), and with surface-multiplying structures such as ridges and fingerlike projections of the intestinal lining. In biology, this type of change in the relative proportions of body parts is called **allometry**.

Ruffed Grouse
(Bonasa umbellus)
Body Length: 29 cm

Chicken
(Gallus domesticus)
Body Length: 46 cm

Emu
(Dromiceius novaehollandiae)
Body Length: 130 cm

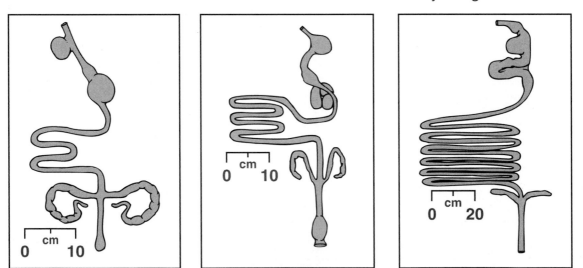

Figure 2.4 Larger birds have a proportionately longer intestinal tract to make up for a reduced surface/volume ratio as body size increases. (Diagram adapted from Stevens, C. E. and I. D. Hume. 1998. Contributions of microbes in vertebrate gastrointestinal tract to production and conservation of nutrients. *Physiological Reviews* 78(2):393–427.)

Although a large animal may require a longer intestine in proportion to its body size to compensate for its small surface/volume ratio, large body size can be an advantage. Body heat in mammals and birds, for example, is generated in proportion to the volume of body tissues, but escapes through the surface of the skin. In a cold climate, the rate of heat escape in comparison to heat generated is therefore proportional to the surface/volume ratio. A moose with a relatively small surface area in proportion to its huge volume keeps warm without much difficulty on a snowy day, but a chipmunk in the same habitat struggles to keep warm. In this case, it is the smaller animal that requires special adaptations to compensate for size, including a dense fur coat, an underground burrow, and hibernation to cope with the cold.

Within a species, populations from colder climates (farther north in the Northern Hemisphere) tend to have a larger mean body size. Adult white-tailed deer from Florida, for example, are smaller on average than adult white-tailed deer from Maine. This effect is called Bergmann's rule, after its discoverer, German biologist Karl Bergmann. You may notice a relationship between body size and heat balance among your classmates as well. If two athletes of different sizes have about the same metabolic rate and similarly low body fat, the 100-kg football player may wear shorts on the same day that the 50-kg tennis player wears a sweater. In the cold, the larger athlete enjoys the advantage of a smaller surface/volume ratio.

When the weather is hot, large mammals suffer a disadvantage because excess heat must be shed through a relatively small surface area. Small animals do not have difficulty losing excess heat, but large ones may require special adaptations. African elephants, for example, use their large and highly vascular ears to magnify their surface area and shed heat. Elephants may even bathe in a river to cool themselves on hot days. In humans, you may have noticed that small children run around outside playing actively on a summer day that seems uncomfortably hot for resting adults. A relatively high surface area in small individuals is an advantage when heat transfer to the environment is needed to keep cool.

Mammals and birds are called **endotherms** because they regulate body temperature internally. Reptiles, amphibians, and invertebrates are called **ectotherms**, because they rely on external sources of heat or cooling from the environment. A lizard may bask on a rock in the sun to elevate its temperature in the morning, but seek a cool spot under the rock when afternoon air temperatures exceed its optimum. Surface to volume ratio affects ectotherms too. A small reptile has a relatively large surface area, and can raise the temperature of its body volume by sunning itself fairly quickly. A large reptile, having a lower surface/volume ratio, would need to bask for a longer time to achieve the same temperature change. This explains why very large reptiles are found in the tropics, where it is seldom necessary to raise the body temperature. As one travels north from Central America toward the Arctic, the upper limit on reptile size gets smaller and smaller, until reptiles disappear altogether from animal communities in the far north.

Finally, the design of an animal or plant's support system must take surface and volume into account. For legs of vertebrate animals, the strength of the skeletal structure is proportional to the cross-sectional area of the supporting bone. The weight that must be supported, however, is proportional to volume, which increases faster than surface area as body size increases. It is not surprising, given the surface/volume consideration, that small antelope such as the dikdik can stand on delicate legs, while larger animals such as a rhinoceros need legs that are shorter and thicker in proportion to body size. If we apply the same reasoning to measurements of trees, we would expect the width of the trunk in a sapling to be smaller in proportion to its height than we would expect to see in a mature tree. Whether plant or animal, gravity poses more serious challenges to larger terrestrial organisms because their volume (and therefore their weight) increases more quickly than the surfaces (of feet, for example) that support them.

METHOD A: MODELING SIZE AND SHAPE RELATIONS

[Laboratory activity]

Research Question

How does body size affect surface/volume ratio and body proportions?

Materials (per laboratory team)

1 box of sugar cubes

Calculator

1 plastic Petri dish, 4-inch diameter

1 plastic Petri dish, 2-inch diameter

5 rolls of pennies

Modeling clay

Paring knife or single-edged razor blade

Metric ruler

Procedure 1: Surface/Volume Ratio

1. Use about 10 of the sugar cubes to build a model of a small animal. Imagine something simple to construct, like a worm or a lizard lying flat on the ground. Vertical structures and elaborate body parts will be hard to replicate as this modeling exercise proceeds. We will use one side of a sugar cube as our unit of length, which we will call L. The number of cubes is equal to the volume of the organism in cubic units (L^3). After you have built your organism, count the number of sugar cubes you have used, and enter the volume in the box below.

> Number of cubes in small animal =   = Volume in (L^3) units.

Now count the number of sugar cube surfaces facing the outside of the organism. Do not forget to count the surfaces on the bottom side. Do not count surfaces facing one another, since these are internal to the organism. The surface area of your organism is the number of surfaces facing the outside multiplied by the surface of one side, which is (L^2).

> Number of surfaces facing the outside of small animal =   = Surface Area in (L^2) units.

Calculate the surface/volume ratio by division. The units are (1/L), which is the same as (L^{-1}), but you can think of this fraction as the units of outside surface per average cube of volume.

> Surface Area in (L^2) units/Volume in (L^3) units =   = Surface/Volume in (L^{-1}) units.

The surface/volume ratio you have just calculated is significant in physiology, because it conveys how much surface area of the skin is available, on average, for each cube of body tissue on the inside.

2. Now use more cubes to construct an animal twice as large. Make sure you double every dimension—twice the length, twice the breadth, and twice the height. If you had legs on your organism, make sure the legs are twice as long, twice as wide, and twice as tall. In your large organism, recalculate the volume, surface, and surface/volume ratio as before:

Number of cubes in large animal =   = Volume in (L^3) units.

Number of surfaces facing the outside of large animal =   = Surface Area in (L^2) units.

Surface Area in (L^2) units/Volume in (L^3) units =   = Surface/Volume in (L^{-1}) units.

---

3. Sketch your small and large models in the space below, and analyze your data by answering Questions for Method A, Procedure 1.

Questions (Method A, Procedure 1)

1. How does the surface area per unit volume of your large animal compare to the surface area per unit volume of the small one?

   _____

   _____

   _____

   _____

   _____

2. Which animal will retain its body heat more easily in a cold environment? Why?

   _____

   _____

   _____

   _____

   _____

3. If their respiratory systems are similarly designed and proportioned, which animal will be able to acquire oxygen from the outside air more easily? Explain.

   _____

   _____

   _____

   _____

   _____

4. List some anatomical features in large animals, such as human beings or tuna fish, which amplify their surface area through out-pocketing or folding of surfaces. Are these structures needed in small animals such as amoebas or roundworms? Explain.

   _____

   _____

   _____

   _____

   _____

5. Ancestors of the modern horse were much smaller than the animal we know today. As a lineage of animals becomes larger over evolutionary time, how must the shape of the body change to accommodate the demands of a larger size?

   _____

   _____

   _____

   _____

Procedure 2: Allometric Design of the Animal Skeleton

In this exercise, modeling clay will be used to make legs for a model animal. Lids or bottoms of plastic Petri dishes will be used to simulate the rest of the skeleton, and pennies will be added to the skeleton to represent the body mass of the animal.

1. Make legs by rolling modeling clay out into a long cylinder on the surface of the lab bench. Keep rolling until the cylinder is 6 mm in diameter. Make the diameter of the cylinder as uniform as possible, using a book or other flat surface to provide even pressure as you roll out the clay. Measure the diameter with a ruler to check your progress.

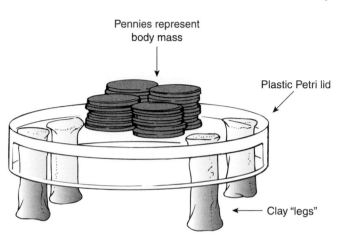

2. When you have made the cylinder the right thickness, use the paring knife to cut four 6-mm diameter legs from your clay cylinder, each one 4 cm long.
3. Roll out the remaining clay to a diameter of 3 mm. Cut four legs from this stock, each 2 cm long.
4. Use the bottom (or top) of the larger Petri dish as the frame of the large animal. Place the rim of the dish down on the bench top, with the flat side facing up. Stick the four larger legs to the bottom of the dish.
5. Carefully invert the dish to stand the model animal on its legs (Figure 2.5).

Figure 2.5   Making a model animal from a Petri dish.

6. Repeat the construction procedure with the smaller legs and the smaller dish to make an animal half the size of the first. Note that its legs are half the length and half the diameter of the first, so the two animals have *equal proportions*. The large animal is twice the size of the smaller one in every dimension.
7. Stand the two animals on their "feet" side by side. Simulate body weight by placing pennies on the dish representing the skeleton. First put one penny on the small animal frame. For the small animal, this amounts to one unit of volume (1 penny wide × 1 penny long × 1 penny high = 1 penny of volume).
8. Then on the large animal frame, place eight pennies in four closely spaced stacks of two (2 pennies wide × 2 pennies long × 2 pennies high = 8 pennies of volume). Note that you have doubled the body size in every dimension, so the large animal gets eight units of volume for every one unit added to the small animal. This demonstrates clearly that volume is cubed when length is doubled.

9. Take turns, adding one penny to the small animal and eight pennies to the large animal per round. You will have one column of pennies on the small animal, and four columns twice as high on the large animal, as shown in Figure 2.5. Count the total number of pennies you are able to stack until the legs buckle and the body frame collapses. The last penny could be considered the "straw that broke the camel's back." The number the frame supported just before the last penny is the maximum sustainable load. Record the maximum sustainable load for the small animal and the large animal in the table below. If data from the entire class are available, record the mean maximum load for small animals and large animals.

ANIMAL SIZE	YOUR MAXIMUM LOAD	CLASS MEAN MAX. LOAD
LARGE		
SMALL		

10. Comment on your results in the space below. Does allometry explain any difference you observed between the two models?

Questions (Method A, Procedure 2)

1. Is the smaller animal able to bear a larger weight in proportion to its size? Why?

_____

_____

_____

_____

_____

2. Examine the pictures in Figure 2.6 of different kinds of mammals. Note their differences in body mass, indicated in kg. How do the design of the legs differ in larger vs. smaller animals? Explain this design difference in terms of surface/volume ratio.

_____

_____

_____

_____

_____

3. In theory, what would constitute the maximum sustainable load of an animal twice the length, twice the width, and twice the height of your larger animal? Explain how your estimate was calculated.

_____

_____

_____

_____

_____

4. Why does body size affect proportions more significantly in terrestrial mammals than in marine mammals such as dolphins and whales?

_____

_____

_____

_____

_____

a

b

c

d

Figure 2.6 Comparison of animal forms. a) African elephant, 7000 kg. b) White rhinoceros, 3000 kg. c) Common zebra, 250 kg. d) Dikdik, 4 kg.

METHOD B: HEAT EXCHANGE IN BEETLES

[Laboratory activity]

Research Question
How does surface/volume ratio affect rates of body temperature change in beetles?

Preparation
The beetles used in this exercise are all available from supply houses, but collected species can be used as well. Make clamshell beetle cages (5 for each group) by joining two plastic weighing trays together with pieces of tape on two sides, putting the concave sides together (Figure 2.7). Put five beetles of each species in a clamshell cage, tape it closed, and label it with the species name. Five cages, one for each species, will be needed by each laboratory group. Place all beetles in a refrigerator (not a freezer) overnight before the laboratory.

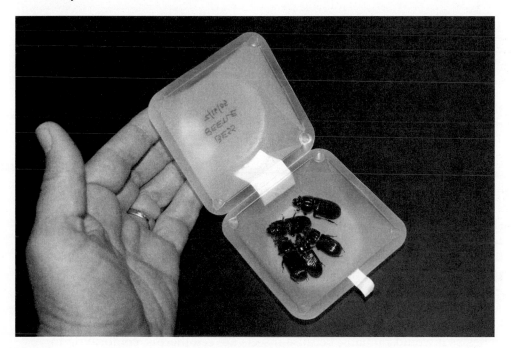

Figure 2.7   A beetle cage, made from two plastic weight boats.

Materials (per laboratory team)

5 live flour beetles (*Tribolium*)

5 live lady beetles (*Harmonia*)

5 live mealworm adults (*Tenebrio*)

5 live dermestid beetles (*Dermestes*)

5 live bess beetles (*Populus*)

Analytical balance, sensitive to 0.01 g

10 plastic weighing trays

Transparent tape

Calculator

Incandescent lamp

Small paint brush

3 × 5" index card

1 sheet of typing paper

Clock or timer, to record time in seconds

5 fruit fly culture vials, with sponge tops

Procedure

1. Retrieve five clamshell beetle cages, labeled with five different species names, from the refrigerator. Complete the next step as quickly as possible, before the beetles warm up too much. Place a beetle cage (with beetles inside) on the balance. Record "beetles plus cage" weight in the data table under the appropriate species heading. Now open the cage by peeling tape off one side, and use your paintbrush to move the beetles out of the cage and onto the index card. Reweigh the cage without the beetles and record this weight. Confirm that the cage weight is less than when it contained beetles. Place the beetles back in the clamshell cage, making sure it is taped closed. Repeat for all five beetle species, and then return all beetle cages to the refrigerator.

2. Cool the beetles back down to 5 degrees C (refrigerator temperature) for at least 15 minutes. This does not stress the beetles as it would humans, because insects are ectotherms, conforming to environmental temperature normally.

3. While your beetles are cooling, subtract the cage weight from the beetle + cage weight to determine the weight of the five beetles. Then divide each group weight by five to determine an average beetle weight for each species, and record your results in the data table. Set up the x-axis of the graph below with appropriate units to include all five average weights. Mark the axis with arrows, labeling the mean weight for each species.

4. On a plain white sheet of paper, make an arena for your experiment. Use an empty fly vial as a template to draw 25 circles on the paper. Arrange the circles in five rows and five columns, spread them out across the page with at least an inch of space between each circle. Now label the circles with letters representing the four species. Mark 5 of the circles "F" for flour beetle, 5 of the circles "L" for lady beetle, 5 of the circles "M" for mealworm adult, 5 of the circles "D" for dermestid beetle, and 5 of the circles "B" for bess beetle. Use the following pattern, called a **Latin square**, to avoid clustering beetles of a given species in one part of your arena. (In a Latin square design, each species appears only once in each row and each column. This avoids clustering a species in a warmer or cooler spot within the arena.)

F	L	M	D	B
L	M	D	B	F
M	D	B	F	L
D	B	F	L	M
B	F	L	M	D

5. Position a desk lamp or other incandescent light source over the lab bench so that it will shine on your arena fairly evenly, and at a distance of 20 cm or so. Turn the light off.

6. Remove the clamshell cages with your chilled beetles from the refrigerator, open them quickly, and use the paintbrush to position one beetle in each of the circles on your arena. Make sure the kind of beetle matches the letters on the circles in a Latin square design. Complete this step as quickly as you can, before the beetles warm up again.

7. Turn the light on and begin timing. As the beetles absorb radiant heat and warm up, they will begin to move. When any beetle begins crawling and leaves its circle, note the elapsed time and use your brush to "sweep" the beetle out of the arena without disturbing the other insects. Place the beetle back in a fly vial, and record its time in seconds on the data table.

8. Give the beetles 30 minutes to begin moving. If any have not aroused by that time, eliminate them from the data set. Make sure your average time for each body size is based on the number of beetles that actually moved.

9. Plot mean weight vs. mean time required to warm up to crawling temperature on the graph at the end of this exercise. Your graph should contain five points, one for each species.

Questions (Method B)

1. In a sunning beetle, is the amount of muscle tissue that must be warmed in order to crawl proportional to surface area, or to body volume?

   _____

   _____

   _____

   _____

2. How was time to warm up related to body size in your experiment? Is this result consistent with your understanding of surface/volume relations?

   _____

   _____

   _____

   _____

3. One of the world's largest insects is the Goliath beetle, which can be over 100 mm in length. Its natural range is confined to equatorial Africa (Figure 2.8). What problems would a Goliath beetle face in a temperate climate? Based on an extrapolation of your data, how long would you guess it would take for a Goliath beetle to warm up and become active under your experimental conditions?

   _____

   _____

   _____

   _____

4. Do the points on your graph approximate a linear or curvilinear relationship? Based on the geometry of surface and volume, would you expect this line to be straight? Explain.

   _____

   _____

   _____

   _____

5. Classic science fiction films often portrayed giant insects as monstrous threats to humanity. Based on surface/volume considerations, why is it unlikely that an ant larger than a person could be a significant threat to anyone?

   _____

   _____

   _____

   _____

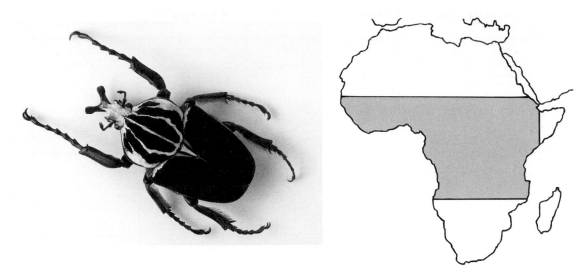

Figure 2.8   The Goliath beetle (*Goliath cacicus*) can be over 100 mm in length and 100 grams in weight. Its range is equatorial Africa, generally within 10 degrees of the equator.

Data Table for Method B: Heat Exchange in Beetles

Beetle type:	Flour beetles	Lady beetles	Adult mealworms	Dermestid beetles	Bess beetles
Weight of Cage + Beetles:					
Weight of empty cage:					
Weight of 5 beetles (by subtraction)					
Mean Body Weight (g):					
Time to move (sec): 1					
2					
3					
4					
5					
Mean time to move (sec):					

Results: Mean Body Weight vs. Time to Warm to Crawling Temperature

Mean time required to warm up to crawling temperature (sec)

Mean beetle weight (g)

METHOD C: SIZE AND SHAPE OF TREES

[Outdoor activity, any time of year]

Research Question
How does surface/volume ratio affect the morphology of trees?

Preparation
Locate 25 trees on campus of the same **taxon**. (Example: all of the oaks on campus, or all of the pines along a drive. The older trees in your sample must be very large. Dwarf trees such as crab apples or Bradford pears are not adequate subjects for this exercise.) It is important to represent trees of many sizes, from newly planted saplings to trees near the end of their life span. The class may manufacture inclinometers ahead of time in order to leave more time for measuring trees during the laboratory time.

Materials (per laboratory team)
Large semicircular protractor

30-cm length of string or fishing line

Metal washer or fishing weight that can be tied to the string

Large-diameter plastic soda straw

Scissors

Transparent tape

50 m survey tape

Procedure
1. First assemble your instrument for measuring angles, which is called an **inclinometer** (Figure 2.9). First thread the string through the hole at the center of the protractor (or tie the string around the base and tape it in place if there is no hole). Trim the string to about 10 cm beyond the edge of the protractor and tie on the weight.
2. To make a sight for your inclinometer, use tape to attach the straw to the protractor at right angles to the straight edge. Make sure the bottom of the straw touches the center of the protractor circle and also touches the 90-degree mark at the top of the circle, as shown in the picture. Also take care that the straw does not stick out past the curved edge of the protractor, since you will be placing that end of the sight close to your eye.
3. When you sight through the straw at an object on a level with your eye, the string should hang beside the zero mark on the protractor. This means zero elevation from your position to the object you are looking at. As you elevate your sight to an object above your head, you should be able to read an angle between zero and 90 degrees, which indicates the **angle of inclination**. You are now ready to take your inclinometer outside to measure the heights of trees.
4. Measurements of trunk diameter and height are needed from a sample of 25 taxonomically similar trees on campus. For each tree in your sample, first measure the tree's circumference by wrapping the metric measuring tape around the trunk at chest height. *Record the circumference of the tree* in the data table at the end of this exercise.
5. Next stretch the measuring tape to a distance roughly equal to the tree's height away from the trunk of the tree. *Record your distance from the tree as the "length of baseline."*
6. Stand at the end of the tape, and use the inclinometer to sight a place on the trunk level with your eye (Figure 2.10). Have a classmate measure the distance from the base of the tree to this level point, which you can call "distance to baseline" in meters. *Record the "distance to baseline"* in the data table.

Figure 2.9 A simple inclinometer can be made from a protractor, a soda straw, a piece of string, and a washer or fishing weight.

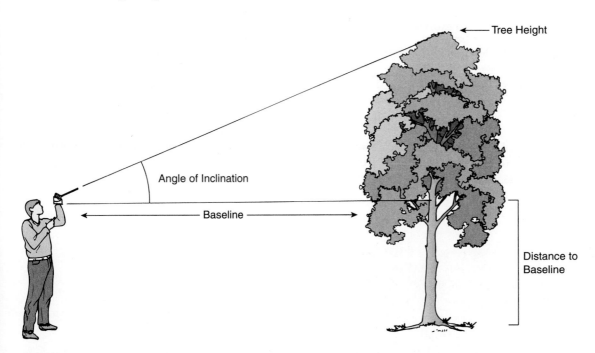

Figure 2.10 Measuring the height of a tree using an inclinometer.

7. Now measure the degree of inclination from your vantage point to the very top of the tree. *Record the angle of inclination* in your data table.
8. Repeat steps 4 through 7 for each of the trees in your sample, recording circumference, distance to baseline, length of baseline, and angle of inclination for each tree.

CALCULATIONS

After returning to the lab or classroom, calculate the height of the tree from the baseline to the top of the tree from the trigonometric formula:

$$\textbf{Tangent (A) = H / L}$$

A = Angle of inclination
H = Height of tree
L = Length of baseline
D = Distance to baseline

This may be rearranged as (Tangent A) ∗ L = H

To calculate the total height of the tree, you need to add the small increment you measured from the base of the tree to the level of the baseline. The entire formula, then, is:

$$[\text{ (Tangent A) } \ast \text{ L }] + D = H$$

You can calculate the tangent of the angle of inclination from the TAN key on many calculators, or by looking up the tangent of the angle in the table at the end of the chapter.

Finally, calculate the diameter of the tree trunk from the formula

Circumference = π ∗ diameter.   **π ≅ 3.14**

This is conveniently reorganized in the following formula:

$$\text{Diameter = Circumference} / 3.14$$

Enter your calculations, including the calculated height of the tree and diameter of the trunk, in your data table. When your calculations are completed, create a graph with a point for each tree in your sample. Plot trunk diameter on the x-axis and tree height on the y-axis. Choose scales of each variable to use as much of the graph area as possible.

Data Table for Tree Trunk Dimensions

Sample Number	Circum-ference		Trunk Diam.		Angle of Inclination		Length of Baseline		Distance to Baseline		Tree Height
		÷3.14=		Tan (		) ×		+		=	
		÷3.14=		Tan (		) ×		+		=	
		÷3.14=		Tan (		) ×		+		=	
		÷3.14=		Tan (		) ×		+		=	
		÷3.14=		Tan (		) ×		+		=	
		÷3.14=		Tan (		) ×		+		=	
		÷3.14=		Tan (		) ×		+		=	
		÷3.14=		Tan (		) ×		+		=	
		÷3.14=		Tan (		) ×		+		=	
		÷3.14=		Tan (		) ×		+		=	
		÷3.14=		Tan (		) ×		+		=	
		÷3.14=		Tan (		) ×		+		=	
		÷3.14=		Tan (		) ×		+		=	
		÷3.14=		Tan (		) ×		+		=	
		÷3.14=		Tan (		) ×		+		=	
		÷3.14=		Tan (		) ×		+		=	
		÷3.14=		Tan (		) ×		+		=	
		÷3.14=		Tan (		) ×		+		=	
		÷3.14=		Tan (		) ×		+		=	
		÷3.14=		Tan (		) ×		+		=	
		÷3.14=		Tan (		) ×		+		=	
		÷3.14=		Tan (		) ×		+		=	
		÷3.14=		Tan (		) ×		+		=	
		÷3.14=		Tan (		) ×		+		=	
		÷3.14=		Tan (		) ×		+		=	
		÷3.14=		Tan (		) ×		+		=	

Results: Tree Diameter vs. Tree Height

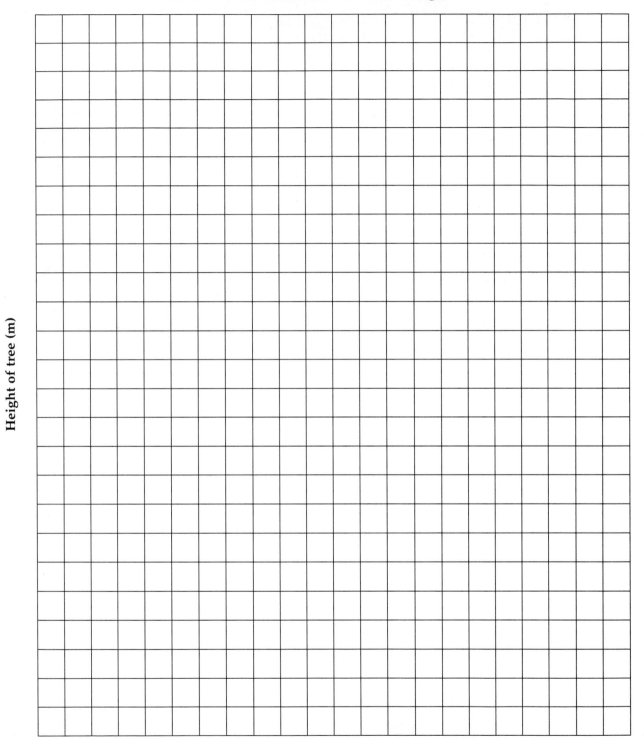

Height of tree (m)

Diameter of tree trunk (cm)

Questions (Method C)

1. From your graph, does the height of the tree show a correlation with the diameter of the stem? Is the relationship linear or curvilinear? Interpret this result.

_____

_____

_____

_____

2. In animals, the strength of a supporting bone is related to the cross-sectional area of the bone. This results in different proportions for legs of heavy animals, as seen in Fig. 2.6. Do you see a similar trend in larger trees, which bear a proportionately heavier weight?

_____

_____

_____

_____

3. If strength of a tree trunk is proportional to its cross-sectional area, and weight of a tree is related to its volume, then what powers of your measurement variables (diameter and height) would be expected to yield a straight line on the above graph?

_____

_____

_____

_____

4. Using the tangent to measure the height of a tree is a classic application of trigonometry in a field study. Find (or recall) a definition of the tangent. Why was this function used rather than a sine or cosine in our procedure?

_____

_____

_____

_____

5. On your campus, what factors other than the age of a tree could affect its height and diameter? Might one of these variables be affected more significantly than the other? How might these secondary factors affect your results?

_____

_____

_____

_____

Table of Tangents

ANGLE (DEGREES)	TANGENT	ANGLE (DEGREES)	TANGENT	ANGLE (DEGREES)	TANGENT
1°	.0175	16°	.2867	31°	.6009
2°	.0349	17°	.3057	32°	.6249
3°	.0524	18°	.3249	33°	.6494
4°	.0699	19°	.3443	34°	.6745
5°	.0875	20°	.3640	35°	.7002
6°	.1051	21°	.3839	36°	.7265
7°	.1228	22°	.4040	37°	.7536
8°	.1405	23°	.4245	38°	.7813
9°	.1584	24°	.4452	39°	.8098
10°	.1763	25°	.4663	40°	.8391
11°	.1944	26°	.4877	41°	.8693
12°	.2126	27°	.5095	42°	.9004
13°	.2309	28°	.5317	43°	.9325
14°	.2493	29°	.5543	44°	.9657
15°	.2679	30°	.5774	45°	1.0000

FOR FURTHER INVESTIGATION

1. Using the procedure of Method B, measure the weight of each beetle separately, and record its time. Use a spreadsheet program with statistical functions to determine the coefficient of correlation for beetle weight vs. warming time. (See Appendix 3—Correlation and Regression.) Explain your results.

2. For endotherms, larger animals have an advantage in colder climates. Bergmann's rule states that, within a species, the colder the climate, the larger the average size of the animals. Test this theory by looking at record body size for game animals such as deer or black bear by state or province in North America.

3. Use a spreadsheet program with statistical functions to determine the coefficient of correlation for tree diameter vs. tree height. (See Appendix 3—Correlation and Regression.) Since strength of the trunk is proportional to the square of its diameter, and since volume of the tree is proportional to the cube of its height, try plotting $(diameter)^2$ vs. $(height)^3$. Performing the same calculation on each number in a data set is called **transforming the data**. Repeat the correlation analysis to find out if the transformed data produce a higher coefficient of correlation than the original data set did.

4. Do all species of trees exhibit the same allometric relationship between height and trunk diameter? Do the proportions change more quickly in softwood species such as a pine than in hardwoods such as an oak? Gather data on more than one species for comparison.

FOR FURTHER READING

Gould, S. J. 1977. "Size and Shape." *Ever Since Darwin: Reflections in Natural History.* Norton, New York, 171–178.

Haldane, J. B. S. 1928. "On being the Right Size." Phys Link
http://www.physlink.com/Education/essay_haldane.cfm

Schmidt-Nielson, K. 1984. *Scaling: Why Is Animal Size So Important?* Cambridge University Press, New York, NY.

Schreer, J. F. and K. M. Kovacs. 1997. Allometry of diving capacity in air-breathing vertebrates. *Canadian Journal of Zoology* 75:339–358.

Chapter 3

Estimating Population Size

Figure 3.1   Banding Canada geese.

INTRODUCTION

One of the first questions an ecologist asks about a population is, "How many individuals are here?" This question is trickier than it appears. First, defining an individual is easier for some organisms than others. In the Canada geese shown in Figure 3.1, a "head count" of geese captured on the ground during their summer molt gives a clear indication of adult numbers, but should eggs be counted as members of the population or not? In plants, reproduction may occur sexually by seed, or asexually by offshoots that can remain connected to the parent plant. This reproductive strategy, called **clonal reproduction,** makes it difficult to say where one individual stops and the next one begins.

Consider the Kentucky bluegrass plant (Fig. 3.2), which is a common turf grass on college campuses. If it escapes mowing long enough, the plant can produce seed in the inflorescence depicted at the top of the drawing, but bluegrass also sends out vegetative shoots, called tillers, which run laterally just under the surface of the ground. Patches of bluegrass many feet in diameter may be interconnected parts of the same genetic individual. Botanists use the term **genet** to refer to the entire clonally produced patch of grass, and the term **ramet** for a standard unit of growth such as a bluegrass tiller.

Check your progress:

If you were asked to count the number of bluegrass plants in a quad area on your campus, which definition would you use? How would you go about it?

> Hint: The purpose of your count determines the best methodology. Are you more concerned with area covered by this species or by the number of genetic individuals in the population?

Once the individual is defined, ecologists working with stationary organisms such as trees or corals can use spatial samples, called **quadrats**, to estimate the number of individuals in a larger area. We will return to this method in Chapter 7.

Mobile animals are usually simpler to define as individuals, but harder to count, because they tend to move around, mix together, and hide from ecologists. Quadrats are not a good approach with mobile animals because **immigration** and **emigration** in and out of the study site make it hard to know what area the entire population occupies. For largemouth bass in a farm pond, you could easily draw a line around a map of the

population, but how would you define the edges of a population of house sparrows in your community? Although house sparrows tend to be more concentrated in towns and urban areas, they do not stop and turn back at the city limit sign. For zoologists, a fuzzy definition of the space occupied by the population often forces an arbitrary designation of the survey group, such as the "population" of robins nesting on your campus in the spring. Knowing the number of animals in a designated study area is interesting, but we must bear in mind that the **ecological population** is defined in terms of interactions among organisms of the same species, and not by the ecologist's convenience.

After defining the individual and establishing the limits of the population you wish to count, your next task is to choose a counting method. Arctic and prairie habitats such as the tundra in Figure 3.1 lend themselves to accurate survey by aerial reconnaissance. This approach works poorly in forests, at night, underwater, or in soil habitats. If animals can be collected or observed in a standard time or collecting effort, you can get an idea of relative abundance, but not absolute numbers. For example, the number of grasshoppers collected in 50 swings with an insect net through an old field community produces data that could be used to compare relative abundance in different fields, but would not tell you how many grasshoppers were in the population.

Figure 3.2   Kentucky bluegrass, *Poa pratensis*.

For estimates of absolute numbers, **mark-recapture methods** can be very effective. The first step is to capture and mark a sample of individuals. Marking methods depend on the species: birds can be banded with a small aluminum ankle bracelet, snails can be marked with waterproof paint on their shells, butterflies can have labels taped to their wings, large mammals can be fitted with collars, fish fins can be notched, and amphibians can have nontoxic dyes injected under the skin. Marked animals are immediately released as close as possible to the collection site. After giving the animals time to recover and to mix randomly with the whole population, the ecologist goes out on a second collecting trip and gathers a second sample of the organisms. The size of the population can then be estimated from the number of marked individuals recaptured on the second day.

The assumption behind mark-recapture methods is that the proportion of marked individuals recaptured in the second sample represents the proportion of marked individuals in the population as a whole. In algebraic terms,

$$\frac{R}{S} = \frac{M}{N}$$

  M = animals marked and released

  N = population size

  R = animals recaptured on a second day

  S = size of the sample on the second day

Let's consider an example. Suppose you want to know how many box turtles are in a wooded park. On the first day, you hunt through the woods and capture 24 turtles. You place a spot of paint on each turtle's shell and release all turtles back where you found them. A week later you return, and with an extraordinary effort, catch 60 turtles. Of these, 15 are marked and 45 are unmarked. Since you know how many turtles you marked, sampled, and recaptured, you can figure out the size of the whole population. By the definitions above, M = 24 marked and released, S = 60 in the second sample, and R = 15 recaptures. If the second sample is representative of the whole population, then:

$$\frac{15}{60} = \frac{24}{N}$$ **This can be rearranged to:** $N = \frac{(24)(60)}{15} = 96$ **turtles.**

This method is called the
Lincoln-Peterson Index of population size.
(See box at right.)

Lincoln-Peterson Index:

$$N = \frac{M \cdot S}{R}$$

N = population size estimate
M = marked individuals released
S = size of second sample
R = marked animals recaptured

In the rearranged version of the general formula, notice that *the smaller the number of recaptures, the larger the estimate of population size.* This makes good biological sense, because if the population is very large, the marked animals you release into the wild will be mixing with a greater number of unmarked animals, so you will recapture a lower percentage of them in your second sample.

Check your progress:

A biologist nets 45 largemouth bass from a farm pond, tags their fins, and releases them unharmed. A week later, she nets 58 bass from the pond, including 26 with tags. Based on the Lincoln-Peterson index, estimate the number of bass in the pond.

Answer: 100.4

Note that the estimate is carried out to one decimal place.

The Lincoln-Peterson method is fairly simple, and its calculations are straightforward, but it does depend on several assumptions. Violating the conditions of the Lincoln-Peterson model can seriously affect the accuracy of your estimate, so it is very important to bear these assumptions in mind as you interpret your results:

1. *Individuals with marks have the same probability of survival as other members of the population.* It is important to choose a marking method that does not harm your animal. If a predator used your paint marks to locate and capture marked turtles at a higher rate than other turtles, your number of recaptures would be lower, and the estimate would therefore be too high.
2. *Births and deaths do not occur in significant numbers between the time of release and the time of recapture.* If marked individuals die and are replaced with newborns, then you will recapture few or no marked individuals, and your estimate will be too high. This is not a large concern in studies of box turtles, but can significantly affect estimates for rapidly breeding organisms.
3. *Immigration and emigration do not occur in significant numbers between the time of release and the time of recapture.* If marked individuals leave the study area and new unmarked individuals come in to replace them, you will get fewer recaptures than the equilibrium population size would lead you to expect. To think about this another way, the real population covers a much larger area than the habitat you thought you were studying.
4. *Marked individuals mix randomly with the population at large.* If your marked turtles do not move among unmarked turtles, and you recapture them near the place you released them, then recaptured turtles may be overrepresented in your second sample, driving down your population estimate.
5. *Marked animals are neither easier, nor harder, to capture a second time.* If marking an animal frightens it so that it hides from you a second time, then recaptures will be underrepresented in a second sample. If animals become tame and are easier to recapture, then the opposite error is introduced.
6. *Marks do not come off of your marked organisms.* Invertebrates molt and shed marks, mammals can wriggle out of their collars, and many things can happen to obscure your marks. If this happens, recaptures will be undercounted, and your estimate will be too high.
7. *Recapture rates are high enough to support an accurate estimate.* The Lincoln-Peterson calculation tends to overestimate the population size, especially if the number of recaptures is small.

Assumption 7 is often violated, because it is difficult to generate sufficient recaptures in large populations.

To correct for this source of error, ecologists often use a slightly modified form of the Lincoln-Peterson index, called the Bailey correction (Begon, 1979).

There are other mark-recapture methods designed to account for violations of several other assumptions of the Lincoln-Peterson method, but most of these require repeated sampling and day-specific marks. To begin simply, we will use the Lincoln-Peterson method with the Bailey correction to estimate the size of a population.

Bailey Correction for the Lincoln-Peterson Index:

$$N = \frac{M(S+1)}{R+1}$$

N = population size estimate
M = marked individuals released
S = size of second sample
R = marked animals recaptured

Check your progress:

Recalculate the number of bass in the pond in the last progress check, using M = 45, S = 58, and R = 26, but with the Bailey correction formula. Did the correction significantly change your estimate?

Answer: N = 98.3

After you calculate a value for N, you should use the following formula to calculate the 95% confidence intervals on your estimate of population size. Recall that confidence intervals give you an idea of the accuracy of your estimate. By adding a standard error term for the upper limit, and subtracting the same amount for a lower limit, you know with 95% confidence that the true population size falls within this range. In common language, we could say the true population size is our estimated value of N, "give or take" the confidence interval. With a 95% confidence limit we would be wrong only 5% of the time. (See Chapter 1 for a discussion of confidence intervals.) The formula for 95% confidence intervals, using the Bailey method (Begon, p. 7) is shown in the box on this page.

Confidence Interval for population size estimate:

$$95\% \text{ Confidence Interval } = N \pm 1.96 \sqrt{\frac{M^2(S+1)(S-R)}{(R+1)^2(R+2)}}$$

N = population estimate
S = size of second sample
M = marked individuals
R = recaptured individuals

Check your progress:

Using the same data for bass mark-recaptures, estimate a 95% confidence limit for your population estimate. (Use the Bailey-corrected estimate of N.)

Answer: N = 98.3 ± 26.8

In summary, to estimate the size of a population: 1) define what you mean by an individual, 2) determine the area you will study, 3) decide on a counting method, and then 4) calculate your estimate and confidence intervals.

METHOD A: MARK-RECAPTURE SIMULATION

[Laboratory activity]

Research Question
How can mark-recapture methods be used to estimate the number of beans in a container?

Preparation
At a grocery or health food store, purchase dried white beans. Navy or Great Northern varieties in bags of approximately one pound are ideal.

Materials (per laboratory team)
1-pound bag of dried white beans

1-liter beaker, jar, or wide-mouthed flask

Felt-tip marker

Calculator

Procedure
1. Partially fill the container with beans, making sure there is room to shake and mix them. This jar of beans represents your population in the wild.
2. Remove a sample of 100 individuals from your population. Mark each bean on both sides with a felt-tip marker so that the marked beans are always easy to see. Return the beans to the jar. Shake well to simulate release and mixing of animals in the wild. Enter the number you marked (number = M) on the calculation page at the end of the chapter.
3. Take a second sample by shaking about 200 beans out of the jar onto a table or lab bench. After you have shaken your sample out, count the total number sampled (this number = S in the index formula) and the number of marked beans in your sample (this number representing recaptures = R in the formula). If you do not have at least 10 recaptures, shake out some more beans and recount. Enter the size of the second sample (S) and the number of recaptures (R) in the calculation page at the end of this chapter.
4. Calculate N and 95% confidence intervals on your estimate of population size, as explained in the Introduction.
5. Count all the beans in the jar, dividing up the task with your lab partners. How close was your estimate? Was the actual population size within the confidence intervals you calculated? Compare your results with others in the class to interpret your results.

Questions (Method A)

1. Why was it important to shake beans out onto the table in your second sample, rather than picking them out of the jar by hand? Can you imagine similar sampling issues arising with the box turtle example from the Introduction?

    _____

    _____

    _____

    _____

2. If you did not shake up the beans very well after returning them to the jar, how would that have affected your estimate of population size? Can you imagine a comparable problem arising in a field study?

    _____

    _____

    _____

    _____

3. Looking at the formula for confidence intervals, how would you expect the 95% interval to be affected by increasing the size of the second sample? How would it be affected by marking a larger number of individuals?

    _____

    _____

    _____

    _____

4. Suppose one of your classmates simulated a predator by removing some of the beans from the jar between the release of your marked individuals and your second sample. Would your estimate of population size be too high or too low? Explain why.

    _____

    _____

    _____

    _____

5. Suppose one of your lab partners simulated **emigration** and **immigration** between the release of your marked individuals and your second sample. Your partner removed a third of the population and replaced it with an equal number of new beans from the grocery store. Would your estimate of population size be too high or too low? How would your error compare with the previous example?

    _____

    _____

    _____

    _____

METHOD B: MARK-RECAPTURE ESTIMATE OF A STUDENT POPULATION

[Campus activity]

Research Question

How can mark-recapture methods be used to estimate the number of students on campus?

Preparation

This exercise uses human subjects, but in a nominal and nonintrusive way. However, if this method is adapted for use in a study leading to a student talk or publication, be sure to follow protocols for research on human subjects established for your institution. For best results, at least 5% of the student body should be "marked." If the number of participating students is much less than 5% of the population as a whole, estimates will vary substantially.

Materials (per student)

Colored yarn cut to 30-cm (12-inch) length

Procedure

1. During a lecture session before lab, ask all volunteers for this experiment to tie a piece of yarn on her/his left wrist, (not too tight) and to leave it on for 24 hours. The wristband can be removed at night, but should be displayed whenever students are out on campus. Get a count of the "marked" students in your class, and enter this number as (M) on the calculation page. If there are multiple sections of your class participating in this experiment, ask your professor to sum all the sections for a grand total of marked participants.
2. After marked students are "released" from class, allow them time to go their separate ways and mix with the population as a whole. Then take a second sample. Your second sample should be in a place where you can observe a large number of students without counting the same person twice. The library, a cafeteria, another class, or a sporting event might be reasonable choices. To take your sample, observe the left wrist of every student you encounter. Count both the total number of students in your sampling effort and the number who have a wristband. Enter the total number in your sample as (S) on the calculation page at the end of this chapter. Also record the number of students with wristbands as "recaptured" individuals (R).
3. Use the Lincoln-Peterson Index to calculate the population size of students on your campus. Also calculate confidence intervals on your estimate.

Questions (Method B)

1. When students left the classroom after being "marked," did they really mix randomly with the entire student body? How do the daily routines of ecology students differ from the average student on your campus? How does this difference in behavior potentially affect your estimate?

    _____

    _____

    _____

    _____

2. When you selected your sample site, was it a good place for observing a random sample of the student body, or were some categories of students overrepresented or underrepresented? Was the likelihood of a "recapture" higher or lower than the likelihood of encountering a nonparticipating student? How do these considerations compare with ecologists' concerns about biased sampling in a field study?

    _____

    _____

    _____

    _____

3. If several students in your class became irritated with the wristband and took them off too soon, how would your estimate of population size be affected? How does this concern compare with the field ecologist's concerns about marking methods for animals in the wild?

    _____

    _____

    _____

    _____

4. If some "marked" commuter students leave campus after class and other commuter students arrive before your sampling effort, how will this affect your estimate? How does this source of error compare with the field ecologist's concerns about emigration and immigration of animals?

    _____

    _____

    _____

    _____

5. This method has actually been used, with personal identification data rather than wristbands, to estimate numbers of transient and homeless persons in urban areas. (See Peterson, 1999, and Fisher et al., 1994.) Based on what you have learned in this study, do you think this method could help develop a more accurate census of the city's population? What sources of error would you anticipate?

    _____

    _____

    _____

    _____

METHOD C: MARK-RECAPTURE ESTIMATE OF MEALWORM POPULATION

[Laboratory activity]

Research Question

How can mark-recapture methods be used to estimate the number of insects in a mealworm culture?

Preparation

Mealworm cultures must be started well in advance, preferably several months before students do the exercise. Mealworms (actually larvae of the beetle *Tenebrio molitor* L.) are easily cultured in plastic shoe-boxes or sweater boxes half-filled with wheat bran. An apple or potato cut up and layered in the bran provides all the moisture needed by the mealworms. Larvae are most useful for the experiment. These can be purchased from biological supply companies, or reared from a starting population of adults if you allow sufficient time for egg laying and maturation.

Plan two short laboratory periods for this exercise: one to mark the larvae, and a second day for recapturing them and calculating results. This could easily be combined with Method B.

Materials (per laboratory team)

Mealworm culture, with large numbers of larvae

Screen sieve, #5, with 4-mm openings

Acrylic paint, in several colors

Small paintbrush

A sheet of newspaper

Procedure

1. Observe the life cycle of *Tenebrio molitor*, the mealworm or darkling beetle, in Figure 3.3. You may be able to find adults, pupae, and larvae of various sizes in a well-established mealworm culture. If you have adults, you probably also have eggs, but they are about the size of the period at the end of this sentence, and so are difficult to see. The purpose of this exercise is to count numbers of larvae, but you may also decide to include other life stages.
2. With your #5 sieve, scoop out a sample of the mealworm culture. Shake the sieve carefully over the box, allowing the bran meal to fall back into the culture. Remove bits of potato, bread, or apple by hand if they are included in your culture. Do you see any mealworms in the sieve?
3. Use water-based acrylic paint to mark the insects in your first sample. (Volatile solvents in oil-based paints are more likely to penetrate the cuticle and harm the mealworms.) Place the captured mealworms on a sheet of newspaper on the laboratory bench. Use a small paintbrush and acrylic paint to put a small mark on each insect. (See Figure 3.4.) Leave the insects on the newspaper while the paint dries. Make sure the paint is dry to the touch before releasing the mealworms back into their culture.

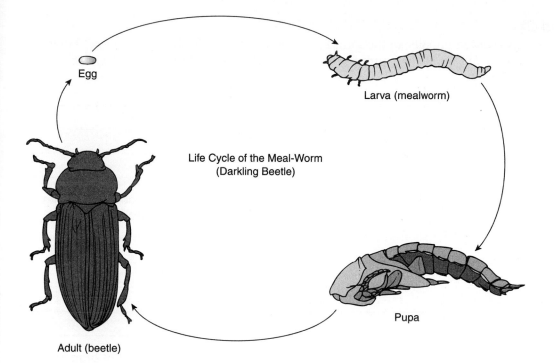

Figure 3.3   Mealworm life cycle, *Tenebrio molitor*.

4.   Record the number marked (M) in the calculation page, and let the cultures sit overnight.

5.   A day or two later, repeat the sampling procedure by sieving about half of the mealworm culture. As you collect mealworms, place them on the newspaper or in the shoebox lid. Count and record on the calculation page how many mealworms are included in your second sample (S), and how many of these are marked individuals that you recaptured (R).

6.   Use the Bailey correction of the Lincoln-Peterson Index to calculate the population size of mealworms in the culture. Also calculate confidence intervals on your estimate.

Figure 3.4   Marked mealworm larvae.

Questions (Method C)

1. Are all life stages represented equally in the sampling procedure used in this experiment? How might life history complicate mark-recapture estimates in field studies of birds, reptiles, fish, or mammals?

    _____

    _____

    _____

    _____

2. Like all animals with an exoskeleton, larval mealworms must shed their skin periodically as they grow in size. This is called molting. If some of the larval mealworms molted between your first and second sample, how would this affect your estimate?

    _____

    _____

    _____

    _____

3. You could have estimated the size of the mealworm culture by measuring the volume of bran in your first sample, and simply calculating numbers of mealworms per unit volume of bran. If you measured the volume of the culture box, you could then come up with an estimate by multiplication, without using mark-recapture methods. What assumptions would you have to make in a volumetric calculation? Which method do you think would be more accurate?

    _____

    _____

    _____

    _____

4. Were mealworms equally distributed throughout the medium, or did you notice that concentrations varied with depth or proximity to the edge of the culture box? How might a clumped distribution of insects affect your results?

    _____

    _____

    _____

    _____

5. Although larvae and adult mealworms are mobile, the pupae do not move through the culture. If you marked a large number of pupae, and sampled in the same part of the culture on your second sampling effort, how would your estimate of population size be affected? How could you avoid this source of error?

    _____

    _____

    _____

    _____

CALCULATIONS (METHODS A, B, OR C)

M = Total number marked and released =

S = Total number in second sample =

R = Number marked individuals recaptured =

Using the Bailey correction for the Lincoln-Peterson formula, estimate N.
(Show calculations.)

$$N = \frac{M(S+1)}{R+1}$$

N =

$$\text{95\% Confidence Interval} = N \pm 1.96 \sqrt{\frac{M^2(S+1)(S-R)}{(R+1)^2(R+2)}}$$

(Show calculations.)

= N ± 1.96

In other words, we can say with 95% confidence that the true population lies between:

High end of estimate   Low end of estimate.

and

If available, the **actual count** of the population size =

Does this fall within the 95% confidence interval?   _____

FOR FURTHER INVESTIGATION

1. Use "Mark-recapture" and "Lincoln-Peterson" as keywords in a literature search to find field studies that used this method to estimate population size. Did the authors use the Bailey correction, or not? Based on what you have learned in this exercise, did the authors mark enough individuals? Did the number of recaptures merit their conclusions? Try to calculate confidence intervals to compare with any published in the study.

2. From the college web site or catalog, find the total number of students attending your college or university. How does this compare with your estimate from Method B? Is the "student population" published by the institution really the same as the number of people on campus at any given time? How many definitions of the term "student population" could be applied to your institution?

3. In early autumn before the first frost, most campus environments are home to adult insects which could be sampled and counted in a mark-recapture experiment. Moths can be collected around security lights at night. Beetles can be collected by burying a sample jar up to its rim to make a pit trap. (You would need to find an undisturbed location, behind a fence or beneath shrubs, for this method.) Hopping insects can be collected with sweep nets if there is tall grass anywhere on your campus, and many small invertebrates can be collected beneath leaf litter. Try marking adult insects of the most abundant species with acrylic paint as you did in Method C, releasing them, and estimating population size by the Lincoln-Peterson method.

FOR FURTHER READING

Begon, M. 1979. *Investigating animal abundance: capture-recapture for biologists.* Edward Arnold, London.

Fisher, N., S. Turner, R. Pugh, and C. Taylor. 1994. Estimating the number of homeless and homeless mentally ill people in north east Westminster by using capture-recapture analysis. *British Medical Journal* 308: 27–30.

Litzgus, J. D. and T. A. Mousseau. 2004. Demography of a southern population of the spotted turtle (*Clemmys guttata*). *Southeastern Naturalist* 3(3): 391–400.

Peterson, I. 1999. Census sampling confusion. *Science News* 155(10): 152–154.

Chapter 4
Population Growth

Figure 4.1 Elephants reproduce slowly, but have the potential to generate large populations.

INTRODUCTION

The great 19th century biologist **Charles Darwin** developed an important ecological idea while reading a book on population growth by **Thomas Malthus** (*An Essay on the Principle of Population*, 1798). Although Malthus was an economist concerned about an expanding human population with a limited food supply, Darwin saw in Malthus' thesis a more general ecological principle. He realized that populations of all species have the potential to grow, and if that growth continues unchecked, competition for resources will eventually limit the size of every population in nature. Darwin documented the staggering growth rate of rapidly breeding organisms such as insects and weedy plants in his *Origin of Species* (1859), but he most effectively illustrated the universal impact of population growth with an example at the opposite end of the life history spectrum: the long-lived and slowly reproducing elephant. In Darwin's own words, "The elephant is reckoned to be the slowest breeder of all known animals, and I have taken some pains to estimate its probable minimum rate of natural increase: it will be under the mark to assume that it breeds when thirty years old, and goes on breeding till ninety years old, bringing forth three pair of young in this interval; if this be so, at the end of the fifth century there would be alive fifteen million elephants, descended from the first pair" (*Origin of Species*, Ch. 3).

How did Darwin arrive at this amazing result? Let's try to duplicate his calculations. If an elephant begins breeding at age 30, then let's assume its generation length is close to 40 years of age. If a pair of elephants has an average family size of "three pair of young," then the population triples every generation. In other words, Darwin's elephants would have the capacity to triple their numbers in approximately 40 years. To keep our model simple, we will ignore post-reproductive survival and mortality, simply counting each new generation as it replaces the old. With these simplifying assumptions, let's look at the projected growth for Charles Darwin's hypothetical population of elephants:

Year	Population Size
0	2
40	6
80	18
120	54
160	162
200	486
240	1458
280	4374
320	13,122
360	39,366
400	118,098
440	354,294
480	1,062,882
520	3,188,646
560	9,565,938

Indeed, before year 600, Darwin's elephants have passed the nine million mark! Darwin probably used a more complex population model, keeping track of surviving adult elephants and overlapping elephant generations, but even in our oversimplified calculation, the conclusion is clear. Any population can overrun its environment if parents continue to produce more offspring than are needed to replace themselves.

Exponential Growth

Let's look at our calculations of elephant growth in the form of a graph (Figure 4.2). The x-axis shows time in years, with year 0 marking the beginning of our calculations. The y-axis shows population numbers, in millions. Although the rate of growth is a constant, tripling every 40 years, the consequences begin to show dramatically after generation 12. This kind of increase is called **exponential population growth,** and a standard plot of its numbers always produces this kind of J-shaped curve. Although we would need different time scales to graph population growth for fruit flies or dandelions, their exponential growth curves would follow a similar pattern. Darwin's prolific pachyderms provide a good general model for population growth any time resources are in abundant supply and surviving offspring outnumber their parents.

Population Explosion of Darwin's Elephants

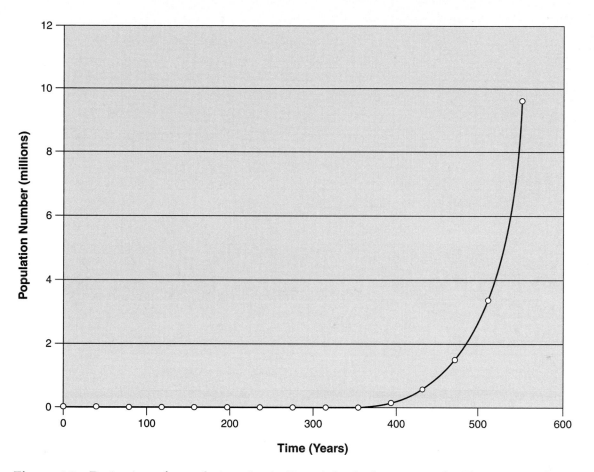

Figure 4.2 Projection of population size in Darwin's elephant example. The curve indicating population size bends upward in a J-shape, as is typical of exponential growth.

Why use the term *exponential growth?* Take a look at Figure 4.3. This graph shows exactly the same population growth numbers, and the x-axis is identical to Figure 4.2. You will notice, however, that the y-axis shows numbers in powers of ten (1, 10, 100, 1000, 10000, etc.). Another way of writing this series is to use *exponents* of ten. The y-axis would then show 10^0, 10^1, 10^2, 10^3, 10^4, and so on. These *exponents* increase at a constant rate on the population growth line in Figure 4.2, so we say this population demonstrates *exponential* growth. Another name for a variable exponent attached to a constant base (base 10 in this example) is a *logarithm*. For this reason, graphs in the form of Figure 4.3 are called ***logarithmic plots***. *Population ecologists find logarithmic graphs (or log plots for short) very handy, since they straighten out exponential curves, and display a broad range of data on a single axis.*

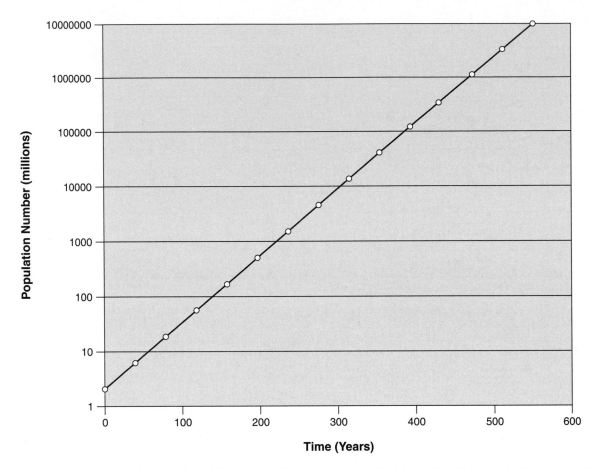

Figure 4.3 Logarithmic plot of Darwin's elephant example. Note that the y-axis shows numbers in a scale that increases by factors of ten. This changes the exponential growth curve to a straight line.

To describe the growth of an exponentially reproducing population accurately, we need a measure of its growth rate. The symbol used for this measure by ecologists is **r**, the **intrinsic rate of population increase**. One way to calculate r is to compare birth rates and death rates. The birth rate is simply the number of new individuals added to the population over a unit time period divided by the total population size. In the human population of a city, for example:

Birth rate = (newborns in the past year)/(total city population)

Similarly, the death rate is the number of deaths over a unit of time divided by the population size. For the city example:

Death rate = (deaths in the past year)/(total city population)

For humans, we chose a year as a convenient unit of time, but the appropriate time scale depends on the organism's life history. For fruit flies, we may want to calculate birth and death rate in days; for bacteria, the appropriate time scale may be minutes. Calculating r over a short time interval relative to the life of the organism is important, since r theoretically provides an instantaneous measure of the population's growth trajectory at a particular moment in its history.

Check your progress:

If the number of newborns in a city of 300,000 was 15,000 this year, and the number of deaths during the same period is 9000, what is the intrinsic rate of increase for the population?

Answer: r = 0.02

Since a population's growth depends on its birth rate in comparison to its death rate, we can define r for our city in these terms, assuming no net change due to people moving in or out of town:

r = population birth rate – population death rate

In exponential growth calculations, r is a constant, usually a small fraction near zero. The following table shows values of r for different biological situations:

r value	Population status
r < 0	Population is declining. Death rate exceeds birth rate.
r = 0	Population is stable. Birth rate equals death rate.
r > 0	Population is growing. Birth rate exceeds death rate.

Once you know the value of r, you can use a simple exponential equation to predict next year's population size:

To state this equation in words, we would say the population next year is equal to the population this year plus the new individuals added. The number of new individuals added equals the current population size times the intrinsic rate of growth. We will call this the "discrete form" of the equation, since it calculates growth in a series of discrete steps. To predict two years ahead, you must use the equation to generate N_1, and then plug that number in as the new value of N_t and repeat the calculation to get N_2. For a hundred years of growth, you would have to repeat the calculation 100 times!

Exponential Growth Equation (Discrete Form)

$$N_{t+1} = N_t + N_t(r)$$

N_{t+1} = population size at the next time interval
N_t = population size at the current time
r = intrinsic rate of population increase

Check your progress:

If r = 0.02 for a city, and the current population is 300,000, then what will the population be next year?

Answer: N_{t+1} = 306,000

The discrete form of the exponential growth equation can be repeated as many times as you wish to make long-range projections, but this is a cumbersome method, and it tends to magnify small errors as you repeat the calculations. A form of the exponential growth equation that provides continuous tracking of population growth over time is as follows:

To demonstrate the power of this formula, let's return to the city example. For a population of 300,000 with r = 0.02, what is the projected population size 10 years in the future?

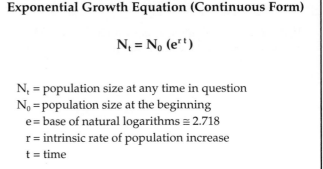

Exponential Growth Equation (Continuous Form)

$$N_t = N_0 \, (e^{rt})$$

N_t = population size at any time in question
N_0 = population size at the beginning
  e = base of natural logarithms $\cong 2.718$
  r = intrinsic rate of population increase
  t = time

$$N_{10} = N_0 \, (e^{rt}) = 300{,}000 \, (2.718)^{.02 \, (10)}$$
$$= 300{,}000 \, (2.718)^{.2}$$
$$= 300{,}000 \, (1.22)$$
$$= 366{,}000$$

Note that the number of individuals added in ten years (66,000) is more than ten times the number added in one year (6000). Like compound interest, the number of babies born each year increases as the number of parents increases, so additions to the population accelerate over time to produce a J-shaped curve.

There is one more form of the exponential growth equation you will find useful. By taking the natural logarithm of both sides of the equation, we can convert the J-shaped curve to a straight line, just as we saw in the log 10 plot of elephant numbers in Figure 4.3.

$$ln \; N_t = ln \; N_0 + r \, t$$

ln   means "natural log of"
N_t = population size at time t
N_0 = population size at the beginning
  r = intrinsic rate of population increase
  t = time

Note that $ln \; N_0$ is a constant, as is r. The equation is therefore in the form of a straight line, y = a + bx. Furthermore, r represents the slope of that line. To calculate r from population numbers over time, simply take the natural logarithm of all population size data, and plot time on the x-axis vs. *ln* (population size) on the y-axis. Then calculate the slope of that line, which you may recall is the change in y divided by the change in x. The slope of the resulting line is a good estimate of the intrinsic rate of increase, r. (See Figure 4.4.)

Check your progress:

Use the formula to predict the number of individuals in a city of 300,000 with r = 0.02 after 20 years of exponential growth.

Answer: $N_{20} = 447{,}529$

Calculating r from Darwin's Elephant Example

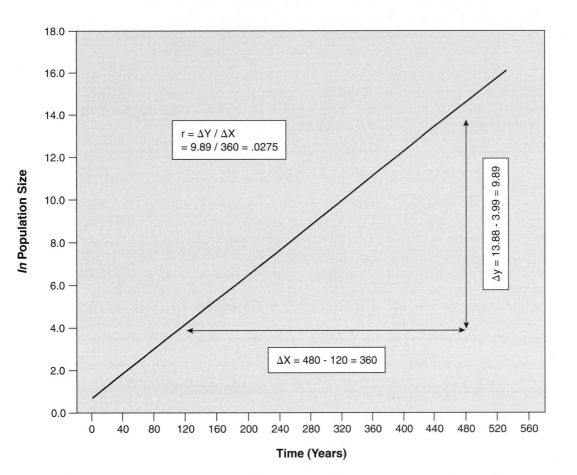

Figure 4.4 Calculating r from exponentially growing population numbers. If time in years is plotted against the natural log of population size, then r is the slope of the resulting straight line. To calculate the slope, pick two points on the line. Change in Y (denoted as Δ Y) is simply the difference between the *ln* population sizes of the two points. Change in X (denoted as Δ X) is the difference between the years designated by the two points. To calculate the slope, divide Δ Y/Δ X, as shown in the figure. For Darwin's elephants, r is approximately 0.0275.

You may choose any two points on the line to calculate the slope; let's choose two points near the beginning and end of the elephant growth calculations as follows:

	x value (year)	y value (*ln* population size)
Point 1	120	3.99
Point 2	480	13.88

To calculate the slope, $\quad r = \Delta y / \Delta x$

$= (y_2 - y_1)/(x_2 - x_1)$

$= (13.88 - 3.99)/(480 - 120) = 0.0275$

Check your progress:

Use the formula $ln\ N_t = ln\ N_0 + r\ t$ to check the numbers in the elephant growth table at the beginning of this chapter. Since $N_0 = 2$ in Darwin's example, use $ln\ N_0 = ln\ 2 = 0.693$ in the formula. Plug in the calculated value of $r = 0.0275$, and you can project the population for any year you wish.

Remember that your calculation will yield the ln of population size, so you will have to find the inverse ln (or anti-log base e) button on your calculator or computer spreadsheet to convert your answer back to the number of elephants in the year you selected. Expect minor discrepancies due to rounding errors, but your answer should be fairly close to the number presented in the table.

Logistic Growth

The exponential growth equation is a good model for rapidly expanding populations, but as Darwin pointed out in the *Origin of Species*, it is ridiculous to assume growth can continue indefinitely. Whether made up of fruit flies or elephants, sooner or later, an increasing population will outstrip the natural resource base it needs to keep growing. Inevitably, the number of individuals in the population reaches the maximum number that the environment can sustain. This population level is called the **carrying capacity** of the environment. It is important to note that carrying capacity depends on the quality of the habitat, but also on the body size and resource requirements of the organism. Carrying capacity for elk in a meadow may be only one or two individuals, while carrying capacity for field mice in the same location may number in the hundreds.

If the population grows past carrying capacity, resources are being used faster than they are being regenerated, and the habitat becomes unsuitable for further survival and reproduction. Overpopulation by elephants may mean the woody plants they feed on are stripped from the savannah. Overpopulation by rabbits may mean a shortage of hiding places, and easy hunting for their predators. Overpopulation by oak trees may mean too much shade for seedlings to survive.

Some species tend to overshoot carrying capacity, and then crash to local extinction when they degrade their own habitat. These organisms tend to exhibit drastic fluctuations in population size, with "boom and bust" cycles throughout their range. Other species tend to slow their reproductive rate as they approach carrying capacity. Reproductive success in these organisms is **density-dependent**, that is, sensitive to the numbers of individuals per unit area. Population growth in these species is called **logistic growth**, because their rate of increase is adjusted as the population grows. As an example, many songbirds defend a territory large enough to rear a nest of young. If the population size grows too large, many birds fail to reproduce because all the nesting sites are taken. As a result, per-capita production of nestlings stabilizes as the population size approaches carrying capacity.

Ecologists use the symbol **K** to represent carrying capacity. If we can imagine a population occupying resource space in its habitat, then K represents the number of individuals that would completely fill that space. Recall that N represents the size of the population. The number of "empty spaces" in the environment could therefore be expressed as $K - N$. If we wanted to measure the *proportion* of habitat space still available for newcomers, we could divide empty spaces by total capacity, which would be:

Proportion of habitat space available $= \dfrac{K-N}{K}$ 　　　K = carrying capacity
　　　　　　　　　　　　　　　　　　　　　　　N = population size

For example, if carrying capacity $K = 100$, and there are 80 animals in the population, then the proportion of the habitat still open for additions to the population would be $(K - N)/K = 0.20$, or 20%.

Check your progress:

If carrying capacity = 1500, and the population size is 600, what proportion of the resource base is still available?

Answer: 0.60 or 60%

To create an equation for population growth that takes available resources into account, we simply take the discrete form of the exponential growth equation, and multiply the growth term by our expression for available resources as follows:

The Logistic equation produces an S-shaped population growth curve, beginning with an accelerating increase in size, and ending with a gradual approach toward carrying capacity. To demonstrate with Darwin's elephants, if we presume the intrinsic rate of increase is still 0.0275, and that the beginning number of elephants is still 2, but the population occupies a grassland biome that can support a carrying capacity of

Logistic Growth Equation

$$N_{t+1} = N_t + N_t \cdot r \cdot \frac{K - N_t}{K}$$

N_t = population size at time t
N_0 = population size at the beginning
K = carrying capacity
r = intrinsic rate of population increase
t = time

no more than 500,000 individuals, then Figure 4.5 illustrates the resulting logistic population growth reaching a plateau at about 600 years. Compare Figures 4.2 and 4.5 to see the difference between exponential and logistic growth.

Check your progress:

If carrying capacity = 500,000, the intrinsic rate of increase is 0.0275, and the population size is 350,000 in time t, what does the logistic model predict for time t+1?

Answer: N_{t+1} = 352,887.5

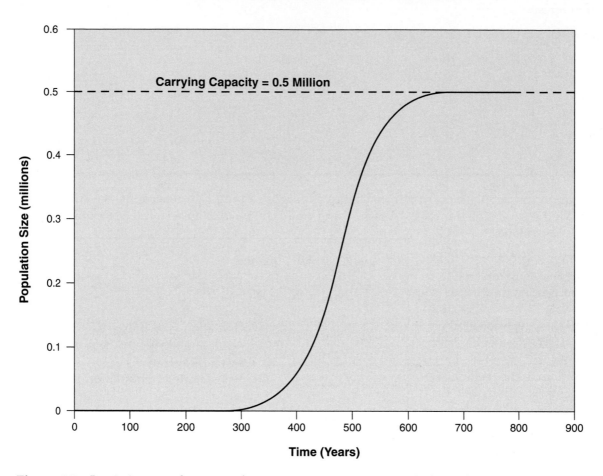

Figure 4.5   Logistic growth approaches carrying capacity as an S-shaped curve.

One final thought: in logistic growth models, K is treated as a constant. For general predictions of population behavior in relatively stable ecosystems, this is an appropriate assumption. We must realize, however, that carrying capacities in nature fluctuate with the weather, with changes in prey or predator numbers, and with many other kinds of environmental disturbance. Ecologists use growth models to understand basic patterns of population biology, but superimposed on these patterns we almost always see an overlay of random perturbation and continuous population adjustment. If a population fluctuates within definable limits, we say it is in **dynamic equilibrium**. The changing nature of carrying capacities does not negate the value of models such as the logistic growth equation, but variable upper limits to population size do need to be considered in applications to ecological problems such as wildlife management and biological pest control.

METHOD A: CALCULATING r FROM A PUBLISHED DATA SET

[Laboratory activity]

Research Question
How rapidly did populations of the Egyptian goose grow in the Netherlands?

Preparation
Students will need calculators with natural log functions or access to computers with spreadsheet software for this exercise. Extra graph paper may be useful for students attempting log plots.

Materials (per laboratory team)
Calculator or computer with spreadsheet software to calculate *ln* values

Ruler

Background
Figure 4.6 illustrates an Egyptian goose (*Alopochen aegyptiacus*), a species broadly distributed in Africa. Population biologist Rob Lensink documented exponential population growth of the Egyptian goose in the Netherlands in the years following its introduction to that country (Lensink, 1998). The following population numbers are cumulative census data from Lensink's observations along three rivers during the period 1985–1994. Each count included an entire winter's observations. For example, the population number for 1985 is the number observed during the winter of 1985–86.

Figure 4.6   Egyptian goose.

Year	Population Size
1985	259
1986	277
1987	501
1988	626
1989	897
1990	1324
1991	2475
1992	2955
1993	5849
1994	7259

Procedure

1. Enter the year and population size for each of the numbers above in the calculation page for *Method A*.
2. Calculate the log base e, also called *ln*, of each population number, and enter those data in the calculation page. This number is *ln* N.
3. Plot time in years (on the x-axis) vs. *ln* N (on the y-axis) on the graph following the calculation page.
4. Using a ruler, draw the best straight line you can through the points on the graph. If some points fall off the line, try to leave the same number of points above and below the line you draw.
5. Use the method explained in the Introduction to calculate the slope of the line. The slope is an estimate of r for Egyptian geese in the Netherlands during this period. Be sure to choose two points on your line for the calculations.
6. Use the population growth equation to project the numbers of geese in the winter of 2000, assuming continued exponential growth.

CALCULATIONS (METHOD A)

Year	Population Size	*ln* Population Size
1985	259	
1986	277	
1987	501	
1988	626	
1989	897	
1990	1324	
1991	2475	
1992	2955	
1993	5849	
1994	7259	

Data Analysis: Calculating r from the Slope of a Log Plot

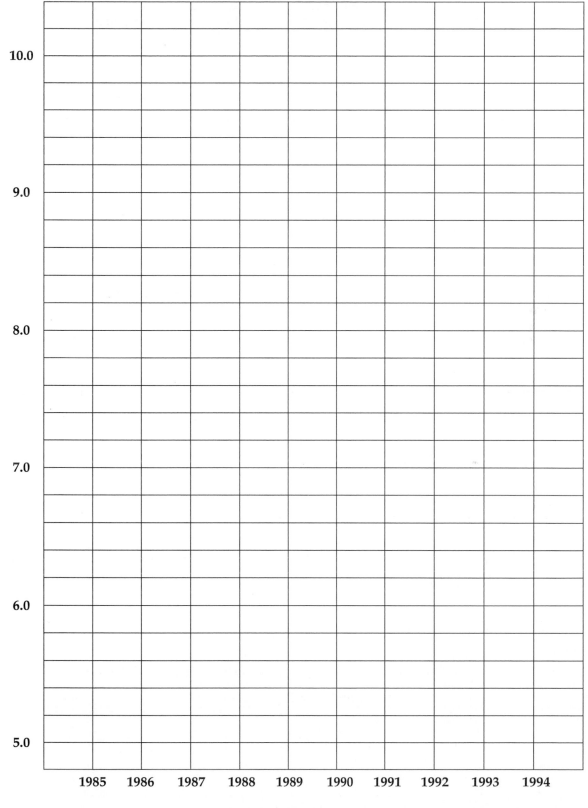

In POPULATION SIZE

10.0
9.0
8.0
7.0
6.0
5.0

1985 1986 1987 1988 1989 1990 1991 1992 1993 1994

TIME (YEARS)

Slope Calculation (From your graph of years vs. *In* population size)

	x value (year)	y value (*ln* population size)
Point 1		
Point 2		

To calculate the slope, $r = \Delta y / \Delta x$
                            $= (y_2 - y_1)/(x_2 - x_1)$

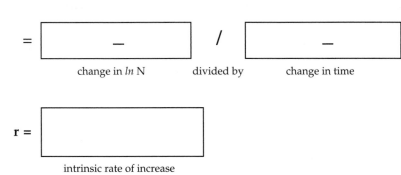

$$= \boxed{\qquad -- \qquad} \; / \; \boxed{\qquad -- \qquad}$$

change in *ln* N   divided by   change in time

$$r = \boxed{\qquad\qquad}$$

intrinsic rate of increase

Comment on the biological meaning of the r value you have calculated:

_____

_____

_____

_____

_____

_____

_____

_____

_____

_____

Questions (Method A)

1. In the exponential population growth equation, $N_t = N_0 (e^{rt})$, identify what each of the symbols stands for, and explain whether it is a variable or a constant for a given growing population.

___

___

___

___

2. In 1985, the number of Egyptian geese observed was 259. Starting with this as an initial population size, $N_0 = 259$, and using the value of r that you calculated in Method A, use the exponential growth equation to project numbers of geese in year 1994. Since that date is nine years later, use $t = 9$ in the equation. Does the calculated number approximate the number actually observed in 1994? Explain any discrepancies you encounter.

___

___

___

___

3. To calculate r, you plotted ln (population size) as a function of years, drew the best straight line through the points, and used two points on that line to determine a slope. Why is this method more reliable than simply choosing two points from the data table to determine $\Delta y / \Delta x$?

___

___

___

___

4. Why do you think Egyptian goose populations are increasing exponentially in the Netherlands, but not in Africa where they originated?

___

___

___

___

5. When species are introduced to a new continent, they often grow so quickly that they out-compete native species. It may be too soon to tell if this is the case for the Egyptian goose, but there are North American examples of introduced (exotic) species that have become an ecological problem. Can you name an example, and explain why this species is an ecological threat?

___

___

___

___

METHOD B: COMPARING EXPONENTIAL AND LOGISTIC GROWTH

[Computer activity]

Research Question

How do projections of the exponential and logistic growth models compare?

Preparation

Students will need a basic understanding of spreadsheet software such as Microsoft's Excel©. Corel's Quattro Pro© is also suitable, but the cell formulas in Quattro will require an opening parenthesis (rather than an equals sign =) as a first character. Graphing population data works best with the scatter-plot option in Excel.

Materials (per laboratory team)

One computer work station, equipped with spreadsheet software

Background

Wildlife biologists Anna Whitehouse and Anthony Hall-Martin studied growth of the elephant population of Addo National Elephant Park in South Africa over a 23-year time span. Hunting prior to 1950 had reduced elephants to low numbers, but the population recovered after Addo Park was established for their protection. The study generated the following population data (Whitehouse and Hall-Martin, 2000).

Year	Population Size		Year	Population size
1976	94		1988	160
1977	96		1989	170
1978	96		1990	181
1979	98		1991	189
1980	103		1992	199
1981	111		1993	205
1982	113		1994	220
1983	120		1995	232
1984	128		1996	249
1985	138		1997	261
1986	142		1998	284
1987	151			

Procedure

1. Using a spreadsheet program, set up a spreadsheet, labeled as shown in the following table.
2. Use the first row as column labels, "Year" in A1, "Observed Number" in B1, "Exponential Model" in C1, and "Logistic Model" in D1 as shown.
3. In column A, enter years 1976 through 1998 in cells A2 through A24. Make sure you enter all of the 23 years of data in your spreadsheet.
4. In column C, enter the beginning population size of 94 in cell C2. This number is N_t in the exponential population growth equation. Likewise, enter the beginning population size of 94 in cell D2 to supply a starting population number for the Logistic growth model.

	A	B	C	D
1	Year	Observed Population Size	Exponential Model r = 0.0525	Logistic Model r = 0.0525 K = 500
2	1976	94	94	94
3	1977	96	= C2 + (C2*0.0525)	= D2 + (D2*0.0525*(500–D2)/500)
4	1978	96	copy previous cell	copy previous cell
5	1979	98	copy previous cell	copy previous cell
6	1980	103	copy previous cell	copy previous cell
7	1981	111	copy previous cell	copy previous cell
8	etc.	etc.	etc.	etc.

5. Type the equation **= C2 + (C2*0.0525)** in cell C3. This is the spreadsheet version of the discrete form of the exponential growth equation $N_{t+1} = N_t + N_t (r)$ discussed in the Introduction. Compare the equation with the spreadsheet formula. We have already said that C2 is the number we have entered to represent N_t. The number **0.0525** is an estimate of this elephant population's value for **r**, calculated from these data estimated from the slope of a log plot, as demonstrated in Method A. Note that we are using the population size in cell C2, which stands for N_t, and placing the result in cell C3, which holds the value for N_{t+1}.

6. Now COPY the spreadsheet formula in cell C3, and PASTE it in every cell from C4 to C24. By dragging to highlight the entire area before you select PASTE, you can do this in one step. You have now commanded the spreadsheet to make a series of annual calculations, computing the population size for each year from the previous year's population size, located just above it in column C.

7. Type the equation **= D2 + (D2*0.0525*(500 – D2)/500)** in cell D3. This is the spreadsheet version of the discrete form of the logistic growth equation $N_{t+1} = N_t + [N_t \cdot r \cdot (K - N)/K]$ which was discussed in the Introduction. Note that the cell D2 holds the initial population size represented in the equation by N_t. We have used 0.0525 as our value for the intrinsic rate of increase **r**, and 500 as our value for the carrying capacity **K**. Note that we are basing our calculation on the population number in cell D2, which holds the value for N_t, and placing the result in cell D3, which holds the value for N_{t+1}.

8. Now COPY the spreadsheet formula in cell D3, and PASTE it in every cell from D4 to D24. By dragging to highlight the entire area before you select PASTE, you can do this in one step. You have now commanded the spreadsheet to make a series of annual calculations, computing the population size for each year from the previous year's population size, located just above it in column D.

9. Use your spreadsheet software to produce a graph. The x-axis should show time in years, the y-axis population numbers. To simplify the y-axis labels, you can display population numbers in millions, or billions if you choose. You should be able to produce three lines on the graph: the actual population growth over time, an exponential curve, and a logistic curve. Which model best projects the actual growth of this population?

10. Copy and paste the formulas in columns C and D through C49 and D49 to predict the fate of the Addo elephant population 25 years into the future. Repeat the graphing procedure in step 9 to show the fate of the elephants in Addo Park, as described by these two models of population growth. Remember that in column D you presumed the carrying capacity of the park was 500. This carrying capacity figure is NOT based on data from Whitehouse and Hall-Martin, so you may try substituting other values for K, and recopying the formulas in column D to produce alternative logistic projections if you wish.

11. Print out your graph and spreadsheet to attach to the Questions for Method B.

Questions (Method B)

1. Compare the tables for Addo Elephants in this exercise with the table at the beginning of the Introduction for Darwin's hypothetical elephants. Was Darwin as conservative in his calculations for elephant reproductive rates as he claimed to be?

_____

_____

_____

_____

2. If the Addo elephants do eventually reach carrying capacity, what factors might limit the further growth of the population? As a park manager trying to maintain biodiversity and ecosystem health, would you prefer to limit population growth at some point below carrying capacity? Explore some of the practical and ethical issues involved.

_____

_____

_____

_____

3. Since we really do not know what K is for Addo Elephant park, we presumed a carrying capacity of 500 for the purpose of this exercise. How might you determine a more realistic value for K in a field study of these animals?

_____

_____

_____

_____

4. Thomas Malthus was concerned that the global human population was growing exponentially, doubling every 50 years or less, and that we would eventually outgrow our food supply. What scientific advances since 1798 have altered **r** and increased **K** for human beings? Have these technologies actually solved the problem, or just postponed the famine Malthus was concerned about?

_____

_____

_____

_____

5. The global human population is over 6 billion and still climbing. What do you think the global carrying capacity might be for humans? Consider agricultural production as a possible limiting resource, but also fresh water, regeneration of oxygen by plants, energy, minerals, and cultural factors. How might our lifestyle choices and diet affect the maximum number of people the earth can sustain?

_____

_____

_____

_____

METHOD C: POPULATION GROWTH OF YEAST
[Laboratory activity]

Research Question
What kind of population growth do yeasts exhibit in laboratory cultures?

Preparation
Use potato dextrose broth in large (e.g., 20 mm × 150 mm) culture tubes to grow yeast cultures in advance. Make sure the tubes are sterilized and capped for aerobic culture of a yeast. From a stock culture of the common yeast *Saccharomyces cerevisiae*, transfer a small number of yeast cells to a sterilized culture tube using a flame-sterilized inoculation loop. For several days before this lab begins, set up a schedule to inoculate two yeast cultures a day, each day, until the lab begins. A five-day series would produce 10 cultures, which is certainly enough to count in one laboratory session. Be sure to label the hour of the day as well as the date on each yeast culture you inoculate. Cultures can be incubated at room temperature. In an incubator, a temperature of 21° C allows for population growth over a period of days. A higher setting of 28° C is recommended for rapid growth, but incubation at higher temperatures will require more frequent inoculation and observation to ensure that yeasts do not overgrow the cultures before the laboratory time.

Materials (per laboratory team)
Compound microscope with maximum magnification 400× or higher

Access to a series of yeast cultures of different ages

Pasteur pipettes for each culture in the series

Standard glass microscope slides and cover slips

Wax pencil or fine-point marker

Vortex test-tube mixer to suspend yeast cells (or lots of patience)

Background
Yeasts rarely if ever exhibit sexual reproduction. Instead, yeast cells produce asexual offspring by budding (Figure 4.7). A bud begins as a weak point in the cell wall. The cytoplasm then bulges out through the hole, forms a new cell, and builds its own cell wall as it grows. DNA is replicated and passed into the new cell before it separates from its parent, so that each cell retains a copy of the yeast genome. This process occurs frequently in the life of the cell, so yeasts have a prodigious reproductive rate. In some cases, the bud has another bud before it separates, so yeast cells often appear as short chains of cells producing smaller cells.

In this laboratory, you will work with a series of cultures of the common nonpathogenic yeast *Saccharomyces cerevisiae*. The cultures are of varying ages, so you can observe populations at different stages of growth. These yeasts do not normally cause disease, but you should always handle microbial cultures carefully. Keep test tubes upright, and keep pipettes oriented tip-down to avoid contamination. Wash your hands and clean up lab benches at the end of the laboratory.

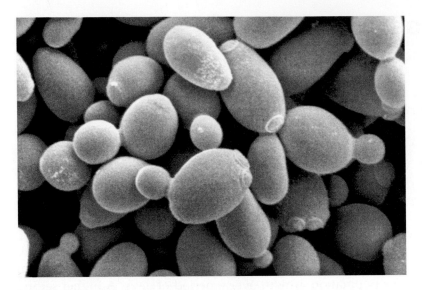

Figure 4.7   Yeasts reproduce by budding.

Procedure

1. Label a glass slide with a wax pencil or fine-point marker for each culture you will examine.
2. Obtain a culture tube, taking note of the time it was inoculated. Record the date and time this culture was started on the calculation page for *Method C*.
3. Using a vortex test-tube mixer, thoroughly suspend the contents of the yeast culture. Be careful not to set the mixer's speed too high; it is easy to spill your culture. Thorough mixing is important before obtaining your sample, since cells tend to settle in the bottom of the tube. If you do not have a vortex tube mixer, you can hold the tube near the top with one hand, and tap the contents repeatedly with the fingers of the other hand to suspend the cells. Try a drumming motion, as if you were drumming your fingers on the table top.
4. When cells are thoroughly mixed, have a lab partner hold the tube and remove the cap. Quickly insert a Pasteur pipette (with bulb attached) into the middle of the tube, about halfway down into the culture medium, and withdraw a small sample.
5. Without giving your cells too much time to settle, place the pipette over a clean slide and squeeze out exactly one drop onto the slide. A drop is 1/20 ml, so this method ensures you are counting cells in the same volume of culture from each sample.
6. Place the edge of a cover slip against the glass slide at the edge of your culture drop, and lower it gently over the drop, trying not to trap any air bubbles.
7. Under a compound microscope, find the cells under low power, then move to high power (400×). Compare the cells you see with Figure 4.7.
8. Sample 10 fields of view as follows: move the slide slightly, look in the eyepiece, and count the number of cells you see in this field of view. Move the slide again, and recount. After 10 counts, compute an average number of cells per field and record your result on the calculation page.
9. Repeat steps 2 through 8 for each yeast culture in the series. You may want to share the counting tasks with others in the lab to make sure all the counts are completed. If so, be sure everyone agrees on a convention for counting buds: do they count as a cell, and if so, how big should a bud be before it is counted?

10. Calculate the age of each yeast culture in hours. Add 24 hours for each day since the culture was inoculated, and then add hours if the culture was started earlier, or subtract hours if it was started later in the day than your lab time. Record age in hours for each culture on the calculation page.
11. Plot time (in hours) vs. cell density (mean number per field).
12. Convert your mean cell counts to a log scale by taking the natural log (base e) of each mean number of cells per field. Record these transformed cell density numbers in the calculation page.
13. Plot time in hours vs. *ln* (cell density).
14. If growth is exponential, calculate r from the slope, as shown on the calculation page.
15. Interpret your graphs, and answer Questions for Method C.

CALCULATIONS (METHOD C)

Culture Number	Date of culture inoculation (Date and time)	Age of culture (hours)	Mean number of yeast cells per field	*ln* (number of cells per field)
1				
2				
3				
4				
5				
6				
7				
8				
9				
10				

Results: Yeast Cell Density vs. Time

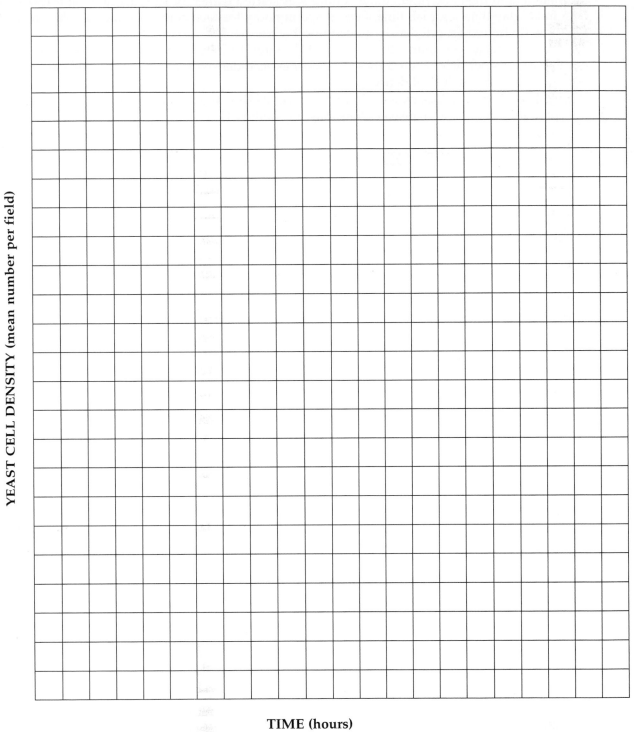

YEAST CELL DENSITY (mean number per field)

TIME (hours)

Results: *In* Yeast Cell Density vs. Time

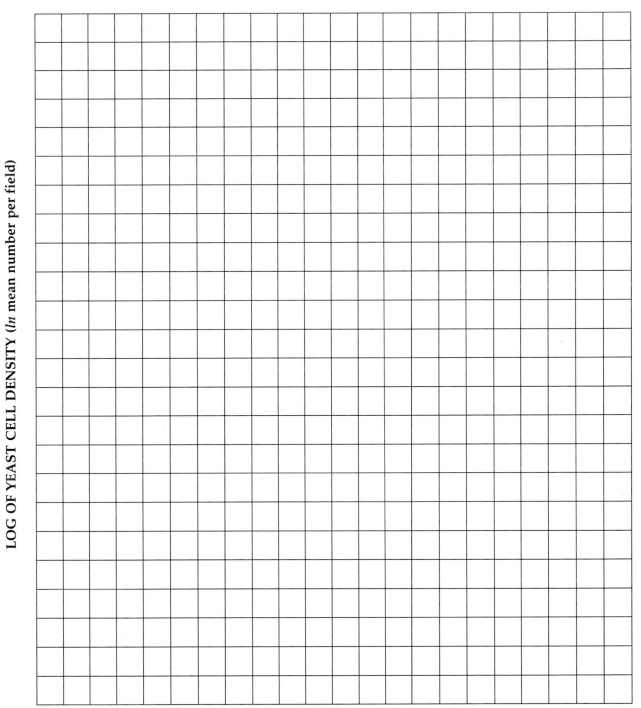

LOG OF YEAST CELL DENSITY (*ln* mean number per field)

TIME (hours)

Slope Calculation (From your graph of hours vs. *ln* population size)

	x value (year)	y value (*ln* population size)
Point 1		
Point 2		

To calculate the slope, $r = \Delta y / \Delta x$
$$= (y_2 - y_1)/(x_2 - x_1)$$

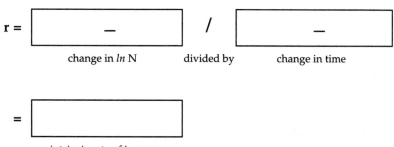

$r = $ [   ] — / [   ] —

change in *ln* N   divided by   change in time

$= $ [   ]

intrinsic rate of increase

Questions (Method C)

1. Many rapidly growing organisms, including yeasts and bacteria, exploit temporary habitats such as a rotting apple. Compare advantages and disadvantages of asexual reproduction for organisms exploiting these kinds of resources.

    _____

    _____

    _____

    _____

2. Was the logistic or the exponential growth model best for describing your yeast population growth? Could the exponential model work for early stages of population growth, even if cell growth slows and stops as the population approaches carrying capacity?

    _____

    _____

    _____

    _____

3. In this experiment, you did not actually estimate population size, but only the number of cells per microscope field. Under carefully replicated conditions, are these two variables related? What errors in procedure could compromise the value of cell density per field as an index of population numbers?

    _____

    _____

    _____

    _____

4. Most microorganisms exhibit faster growth at higher temperatures, up to the point that their enzyme functions are compromised. If you incubated cultures at a higher temperature, how might the population growth curve be affected in your experiment?

    _____

    _____

    _____

    _____

5. Suppose an r value is calculated for yeast cells by plotting *ln* population number vs. time in hours, and the same kind of calculation is performed for elephants, plotting *ln* population number vs. time in years. Can the two r values be compared? Explain your answer.

    _____

    _____

    _____

    _____

FOR FURTHER INVESTIGATION

1. Try using exponential and logistic equations to model growth of the human population over the past 55 years, using the following data from the United Nations Population Division, World Population Prospects 2002 revision (http://esa.un.org/unpp/p2k0data.asp).

    After you have set up your spreadsheet, try projecting the human population 100 years into the future—a time span that will include lives of today's young adult generation and their children. How do your two models compare? What criteria did you choose to determine the global carrying capacity for human beings?

Year	Global Human Population
1950	2 518 629 000
1955	2 755 823 000
1960	3 021 475 000
1965	3 334 874 000
1970	3 692 492 000
1975	4 068 109 000
1980	4 434 682 000
1985	4 830 979 000
1990	5 263 593 000
1995	5 674 380 000
2000	6 070 581 000
2005	6 498 225 000

2. When yeast populations fill their test-tube habitat, their growth is eventually slowed by depletion of food such as sugars in the broth medium, but also by metabolic wastes (alcohols) that accumulate in the culture. To show logistic growth over a longer period of time, start your own series of yeast cultures, and sample them periodically over an extended time. Repeat your census of cells per field, calculating standard errors on your population estimates (see Chapter 1). Then try constructing a logistic growth model, with **r** derived from the "log phase" of fastest population growth as we did in Method C, but also with **K** calculated from the maximum number of cells per field you see at carrying capacity. If followed long enough, do yeast cultures exhibit logistic growth?

FOR FURTHER READING

Darwin, Charles R. 1859. The Origin of Species. In: *The Harvard Classics, Vol. XI.* New York, P. F. Collier & Son. On Line Edition: Bartleby.com, 2001. www.bartleby.com.

Lensink, Rob. 1998. Temporal and spatial expansion of the Egyptian goose *Alopochen aegyptiacus* in The Netherlands, 1967–94. *Journal of Biogeography* 25(2): 251–263.

Malthus, Thomas Robert. 1798. *An Essay on the Principle of Population.* J. Johnson. Library of Economics and Liberty. 31 December 2004. On Line Edition: www.econlib.org/library/Malthus/malPop1.

United Nations Population Division, *World Population Prospects, 2002 Revision.* http://esa.un.org/unpp/p2k0data.asp.

Whitehouse, Anna M. and Anthony J. Hall-Martin. 2000. Elephants in Addo Elephant National Park, South Africa: reconstruction of the population's history. *Oryx* 34(1): 46–55.

Chapter **5**

Demography

Figure 5.1 Great Blue Heron, with nestlings.

INTRODUCTION

Simple models for calculating population growth assume that all individuals in a group contribute equally to survival and reproduction of the population. Is this a reasonable assumption? Let's consider the exponential growth equation, first introduced in Chapter 4, and apply it to a rookery (or colonial nesting area) of Great Blue Herons (Figure 5.1).

The exponential growth equation states that the number of herons next year (N_{t+1}) will equal the current population size (N_t), plus the new recruits to the population, calculated as $N_t(r)$. This last term contains a questionable assumption, because new recruits are calculated by multiplying the **intrinsic rate of increase (r)** by every individual in the population, regardless of age, sex, or breeding condition. If heron populations always had the same composition, this would be a reasonable approach, because **r** is calibrated to reflect the reproductive potential of the average individual. However, it is easy to imagine that the composition of a heron population could change from year to year or

---

Exponential Growth Equation (Discrete Form)

$$N_{t+1} = N_t + N_t(r)$$

N_{t+1} = population size at the next time interval
N_t = population size at the current time
r = intrinsic rate of population increase

from place to place. Would a 400-bird population composed of 55% females build the same number of nests as a 400-bird population composed of 45% females? Would a population of young adult birds nesting for the first time enjoy the same breeding success as a population of more experienced parents? Ecologists attempting to make more accurate assessments of growth and survival of populations, particularly for species with longer lives and more complex life histories, have developed more accurate ways to track population characteristics such as age composition and sex ratio. Many, in fact, were adapted from mathematical techniques previously developed by insurance and pension analysts. This approach to the study of populations is called **demography**.

Check your progress:

If you were asked to project the birth rate for a human population of 1500 people on a small Pacific island having an average family size of 2.4 children, what additional information would be most helpful in making an accurate estimate?

Answer: Sex ratio and age structure are among the most important factors.

Survivorship

If we census the number of herons in the rookery this year, and ask how many will return to breed next year, we need to know probabilities of survival for each bird in the population. Since a bird's age is an important consideration in this calculation, we can simplify the task by grouping the birds according to age. As a first step, we can ask, what is the age-dependent probability of survival, or **survivorship schedule** for these birds? A survivorship schedule calculates year-to-year probabilities of survival throughout the life cycle, from birth or hatching, to fledging, maturation, and senescence.

Suppose we could place leg bands on a sample of 1000 nestling herons, and follow up with careful observation throughout their lives. We could keep records on the number that died and the number that survived each year. This sample of 1000 herons is called a **cohort**, defined as a subset of the population all of the same age. Cohort data can be used to calculate mortality and survival rates as shown in the following table, which is based on heron life history information from R. W. Butler (1992).

AGE IN YEARS X	NUMBER OF HERONS ALIVE AT BEGINNING OF YEAR n_x	DEATHS DURING THIS YEAR d_x	SURVIVAL RATE = (NUMBER ALIVE AT BEGINNING OF YEAR) / 1000 L_x
0	1000	690	1.000
1	310	112	0.310
2	198	44	0.198
3	155	34	0.155
4	121	27	0.121
5	94	21	0.094
6	73	16	0.073
7	57	13	0.057
8	45	10	0.045
9	35	8	0.035
10	27	6	0.027
11	21	5	0.021
12	17	4	0.017
13	13	3	0.013
14	10	2	0.010
15	8	2	0.008
16	6	1	0.006
17	5	5	0.005
18	0		0.000

In this table, each row represents a year in the life of our cohort of herons. We use x to represent the herons' age. Since n is the symbol used for a census count, we can designate n_x as the size of our cohort population at age x. Each year, the number dying during that year is subtracted from the population size to calculate the number alive in the following year. In mathematical terms, $n_x - d_x = n_{x+1}$. The proportion of the population still living at age x, designated as L_x, is calculated by dividing the number still alive by the original cohort of 1000. In mathematical terms, $L_x = n_x/1000$. In survivorship tables, L_x values always begin with 1.000, representing 100% alive at the cohort's beginning, and end with a value of 0 when the last member of the cohort dies.

Check your progress:

If a cohort that began with 1000 newborns had 650 living members at age 5, and declined to 530 at age 6, what are d_x and L_x for age 5?

Answer: $d_x = 120$, and $L_x = 0.650$

What life history information can we derive from the heron survivorship table? The riskiest time of life for a Great Blue Heron is its first year, with 69% of the hatchlings failing to survive to age 1. The death rate falls to 36% in the second year, and then settles down to a constant 22% per year after that. A constant death rate is fairly typical of adult birds, since most mortality is due to accidents independent of the bird's age, such as getting caught in a storm or encountering a predator. To illustrate the pattern of survivorship, we can plot age in years on the x-axis, and numbers left alive per 1000 on the y-axis (Figure 5.2).

Great Blue Heron Survivorship

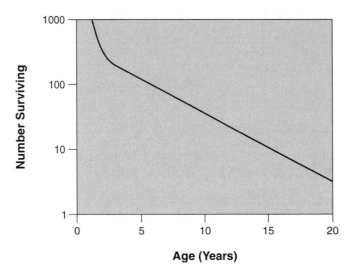

Figure 5.2   Survivorship curve for Great Blue Heron (based on data from Butler, 1992).

Following a system proposed by E. S. Deevey (1947), population ecologists classify organisms according to their patterns of survivorship (Figure 5.3). **Type I** organisms, such as humans, exhibit comparatively low juvenile and young adult mortality, with most deaths occurring at the end of the life span. This produces a convex survivorship curve, bowing up and to the right on the log plot. **Type II** organisms, including most birds, experience a constant risk of mortality throughout their lives. On a log scale, the constant risk of death generates a straight line for Type II species, descending from birth to the end of the life span. **Type III** organisms, such as trees, experience very high mortality of juveniles (seeds and seedlings), but an exceptionally low death rate once established as adults. Their survivorship curve is said to be concave, since it bows downward toward the origin of the graph.

Check your progress:

Which type of survivorship curve in the Deevey classification best describes the life history of the Great Blue Heron? If Deevey's model is not a perfect fit for heron survivorship, comment on reasons for any differences that you see.

Answer: Type II, except for juveniles

Types of Survivorship Curves

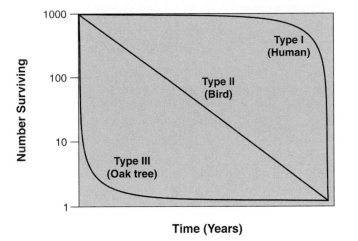

Figure 5.3   Three types of survivorship curves. Type I species, such as humans, have low mortality until the end of the life span. Type II species have constant mortality rates throughout the life span. Type III species have very high juvenile mortality, but high survival after establishment as adults. (Terminology from Deevey, 1947.)

Fecundity

To forecast the future of our heron rookery, we need to know more than the survivorship schedule. Reproductive rates, also influenced by age, are equally important. In Great Blue Herons, we know that breeding begins at age two, and continues throughout the birds' lives. Adults returning to the rookery choose a new mate each year, and breeding pairs cooperate in the feeding of young. Nests are built of sticks and vegetation high in the tops of trees to avoid mammalian predators. The number of nestlings produced by each breeding pair varies from three to seven, depending on available resources, with four being typical during favorable conditions (Butler, 1992).

In demography studies, the number of offspring produced by a parent, called **fecundity**, is measured as a function of age. We will let b_x stand for the mean number of offspring produced by a parent of age x. In making this calculation, it is important to remember that two parents are involved in each nest, so numbers of fledglings in a nest reared by a *pair* of herons would have to be divided by 2 to calculate the number of offspring per parent.

For our hypothetical heron population, we will let $b_x = 0$ for age 0 and age 1, since herons do not begin nesting until they are two years of age. We will further assume young breeding adults are somewhat less successful than older birds, so $b_x = 1.5$ for ages 2 and 3. This would reflect a pair of young herons successfully rearing three nestlings in those years. After age 3, we will assume more experienced birds rear an average of four fledglings per nest, so $b_x = 2.0$ for herons age 4 and above. We will terminate the table at age 17, which is the maximum life span of the bird. From these assumptions, we construct a fecundity schedule as follows:

AGE IN YEARS	NUMBER OF OFFSPRING PER PARENT OF AGE X
x	b_X
0	0
1	0
2	1.5
3	1.5
4	2.0
5	2.0
6	2.0
7	2.0
8	2.0
9	2.0
10	2.0
11	2.0
12	2.0
13	2.0
14	2.0
15	2.0
16	2.0
17	2.0

Like the survivorship schedule, age-specific fecundity rates for our heron population can be represented on a graph (Figure 5.4).

Check your progress:

In a human population, if the birth rate for women between ages of 25–30 is 0.16 per year, what is b_x for this age group?

Answer: $b_x = 0.08$, assuming females make up half of the population

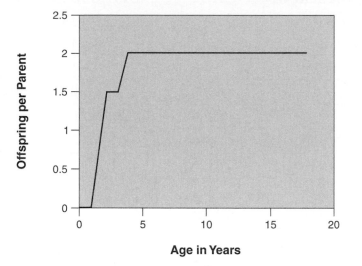

Figure 5.4   Fecundity schedule for Great Blue Heron population, showing numbers of offspring per parent as a function of parental ages.

Life Table

A life table combines survivorship and fecundity information to generate a comprehensive demographic model. In addition to survivorship and fecundity columns from earlier tables in this chapter, notice that the final three columns are made by multiplying values from the preceding columns. Column sums are shown at the bottom.

After the familiar survivorship and fecundity columns, find the column labeled "expected offspring." Each figure in this column represents the number of offspring that a newly hatched heron can be expected to produce at the indicated age. To calculate this number, two questions must be asked: 1) what is the likelihood that the heron will survive to this age? and 2) if it attains this age, how many offspring is it expected to produce? Multiplying the L_x and b_x numbers together yields the joint probability of living x years and then producing the number of offspring typical of x-aged herons. The sum of the $L_x b_x$ column represents the expected number of young a typical heron is expected to produce, summed over its entire lifetime. Demographers call this total the **net reproductive rate**, represented by the symbol $\mathbf{R_o}$.

CALCULATING NET REPRODUCTIVE RATE:

Net reproductive rate, $\mathbf{R_o}$, is defined as the lifetime production of offspring per parent. It is calculated by summing the $L_x b_x$ column in a life table.

$$\mathbf{R_o} = \Sigma \, [\, L_x \, b_x \,]$$

The symbol
$\mathbf{R_o}$ = net reproductive rate
Σ means the bracketed numbers are summed over all ages.
$L_x b_x$ = survivorship at age x times fecundity at age x

AGE IN YEARS X	SURVIVORSHIP L_X	FECUNDITY b_X	EXPECTED OFFSPRING $(L_X)(b_X)$	WEIGHTED AGE $(X)(L_X)(b_X)$	STABLE AGE DISTRIBUTION $L_X / \Sigma L_X$
0	1.000	0	0.000	0.000	0.454
1	0.310	0	0.000	0.000	0.141
2	0.198	1.5	0.298	0.595	0.090
3	0.155	1.5	0.232	0.696	0.070
4	0.121	2.0	0.241	0.966	0.055
5	0.094	2.0	0.188	0.942	0.043
6	0.073	2.0	0.147	0.881	0.033
7	0.057	2.0	0.115	0.802	0.026
8	0.045	2.0	0.089	0.715	0.020
9	0.035	2.0	0.070	0.627	0.016
10	0.027	2.0	0.054	0.544	0.012
11	0.021	2.0	0.042	0.466	0.010
12	0.017	2.0	0.033	0.397	0.008
13	0.013	2.0	0.026	0.335	0.006
14	0.010	2.0	0.020	0.282	0.005
15	0.008	2.0	0.016	0.235	0.004
16	0.006	2.0	0.012	0.196	0.003
17	0.005	2.0	0.010	0.162	0.002
18	0.000	0	0.000	0.000	0.000
TOTALS:	2.204		1.611	9.177	1.000

In our example, R_0 is equal to 1.611, which means that the average heron produces 1.611 offspring over its lifetime. Remember that a pair of herons would produce double this number of offspring on average, which would be 3.222 baby herons.

Check your progress:

Which makes the most significant contribution to family size in Great Blue Herons, offspring produced by herons younger than eight years of age, or offspring produced by herons older than eight? Why is reproduction in the first half of the life span more important than in the second half?

Hint: Compare values in the $L_X b_X$ column of the life table.

Next, we need to calculate a **weighted age** column by multiplying x, L_x, and b_x values together for each year in the life table. By itself, the weighted age column does not have much intuitive meaning, but its sum can be used to calculate an important population parameter: the **generation length**. Generation length, symbolized as **T**, is the mean age at which parents produce offspring. For a human population, you could get a good estimate of **T** from a hospital's birth records if you wrote down the mother's age for every birth in the past 10 years, and calculated an average age for all the mothers who delivered a child. To calculate generation length, divide the weighted age column total by the expected offspring column total as follows:

CALCULATING GENERATION LENGTH:

Generation length, **T**, is defined as the mean age at which parents produce offspring. It is calculated by dividing the $L_x b_x$ column in a life table by the sum of the weighted age column in the life table.

$$T = \Sigma\,[X\,L_x\,b_x]\,/\,\Sigma\,[L_x\,b_x]$$

T = generation length
$\Sigma\,[X\,L_x\,b_x]$ = sum of the weighted age column
$\Sigma\,[L_x\,b_x]$ = sum of the expected offspring column

In our example, the weighted age total for Great Blue Herons is 9.177, and the expected offspring total is 1.611, so the generation length is 9.177/1.611 = 5.7 years.

Once we have calculated the net reproductive rate R_o and generation length **T**, we can calculate a fair approximation of the **intrinsic rate of population increase**, represented by **r** in the population growth equations. To find **r**, first realize that after a generation of population growth, the number of new individuals is equal to the population size in the previous generation, multiplied by the net reproductive rate. If we let N_o designate the original population size, and N_T be the population size after one generation,

$$N_T = N_o\,R_o$$

Now from the continuous equation for population growth (see Chapter 4), we know that:

$$N_T = N_o\,e^{r\,T}$$

By substitution, we can equate the two expressions for N_T as follows:

$$N_o\,R_o = N_o\,e^{r\,T}$$

Check your progress:

If you cancel units of measurement in the numerator and denominator of the equation for **T**, what units are you left with? Does this make sense as a measure of generation length?

Answer: Units for L_x and b_x cancel, leaving x (in years) for generation length.

ESTIMATING THE INTRINSIC RATE OF INCREASE (r) FROM LIFE TABLE DATA:

$$r \cong (ln\ R_o) / T$$

r = intrinsic rate of population increase
R_o = net reproductive rate
T = generation length

Dividing both sides of the equation by N_o we get:

$$R_o = e^{r\,T}$$

Then we can take the natural log of both sides:

$$ln\ R_o = r\ T$$

And finally rearranging for **r**,

$$r = (ln\ R_o) / T$$

This method is simple to calculate, and is very accurate in short-lived organisms that reproduce only once, such as annual plants and many insects. If generations overlap as we see in the Great Blue Heron, this method gives only an approximation of **r** because the new generation does not completely replace the old at time **T**. A more accurate calculation of **r** can be derived from a more complex equality, called the **Euler equation**. An explanation of this method can be found in population biology texts (e.g., Ebert, 1999). For this laboratory exercise, we will use the simpler approximation.

For our example, family size in Great Blue Herons is 1.611, and the generation length is 5.7 years. Therefore, a good estimate of **r** for these birds is:

$$r \cong ln(1.611) / 5.7 = .084$$

Check your progress:

Organizations concerned with global human overpopulation often discuss ways to reduce family size. From the equation for **r** above, can you see a second way to reduce the intrinsic rate of increase? What social factors are involved in this approach to the problem?

Hint: Since **T** is in the denominator, increasing **T** reduces **r**.

Finally, we can use the column at the extreme right of the life table to project the **stable age distribution** of the population. This column forecasts what portion of our herons will be newborns, one-year-olds, two-year-olds, three-year-olds, etc., after the population reaches equilibrium. This column is calculated by dividing this year's survivorship by the total for all years:

$$L_X / \Sigma\ L_X$$

We can think of this fraction as the portion of the total life span that a typical heron lives in year x. If we extrapolate from our cohort of 1000 herons to the whole population, this same fraction approximates the proportion of the population expected to be x years old. Of course, a population might start out with a different age distribution, but if the survivorship and fecundity schedules remain constant as written in the life table, the population will achieve a stable age distribution within a few generations. Interestingly, a stable age distribution is achieved whether the population is growing or not; stable population size and stable age distribution are distinct properties of the population.

To demonstrate from our example, the $L_x/\Sigma\, L_x$ value for three-year-old herons is 0.070, which means that 7% of the heron population will be members of the three-year-old age class. We can use all the entries in this column to generate proportional representation for the whole array of age classes. The $L_x/\Sigma\, L_x$ column always sums to 1.000, since its column total represents 100% of the age classes.

Demographic Patterns

A population's history of survival and reproduction is reflected in its age structure. Trees provide good examples for demographic analysis, since individuals can be aged by the rings of xylem laid down in the wood with each year's growth (Figure 5.5). It is not necessary to cut down the tree to see the rings. A core sample taken from the trunk can be used to determine age without harming the tree (Figure 5.6). If you used core samples to age trees in a stand of pines, and found all the trees were 32 years old, you could guess the area had been logged 32 years ago. Recruitment of new seedlings after logging resulted in a population dominated by a single cohort. From the logger's perspective, this age structure has advantages. All the trees in this forest will mature at the same time, so the trees can be clear-cut again at some point in the future. From the resource management perspective, this is not generally accepted as good forestry practice, since even-aged stands of trees provide a less diverse habitat for forest species, and clear-cutting leaves the soil exposed to erosion and nutrient loss. A more balanced use of woodlands for wildlife management, soil conservation, watershed protection, and forestry can be achieved by selectively cutting mature trees from mixed-aged stands. Even this practice removes the hyper-mature trees needed by some woodland species. For example, the red-cockaded woodpecker nests only in the hollow cavities of old pines, and became endangered as mature trees were cut out of forests throughout the Southeastern U.S. A tree population protected from logging typically develops a broad range of age classes, reflecting constant recruitment of seedlings over time, and providing habitat for a diverse forest community (Figure 5.7).

Check your progress:

After the population reaches a stable age distribution, what percentage of this heron rookery will be composed of one-year-old birds?

Answer: 14.1%

Figure 5.6 Ecologist removing a core sample to analyze tree ring data.

Figure 5.5 Growth rings in a tree trunk's cross section indicate age of the tree.

In many plant species, the size of the organism is a better predictor of survival and fecundity than the individual's age. Demographers working with these species use size classes or life stages rather than age in their models. Stage-specific survivorship and fecundity can be analyzed to make population projections, just as we have done with ages in this chapter.

Demographic analysis has proven quite useful in understanding human populations. Figure 5.8 illustrates demographic patterns for the United States population, based on year 2000 census data. In this kind of graph, each horizontal bar represents an age class, with males shown on the left side and females on the right. Age classes are arrayed from babies (age 0–4) at the bottom to the oldest members of the population (age 85+) at the top. Notice that the U.S. population graph is roughly rectangular, with some tapering at the top due to age-related mortality. The asymmetry at the top of the graph is caused by a longer life expectancy for females than for males. You will notice a bulge in the middle of the graph, caused by the "baby boom" post–WWII generation, who were primarily between age 35 and 49 during the 2000 census. The demographic "echo" of the baby boomers' children, ages 5–19, is not quite as large as the baby boom cohort, but still creates a significant "bulge" in the age structure. Although it has some cohorts larger than others, the general shape of the U.S. age distribution is called "rectangular." This shape is characteristic of stable populations maintaining a fairly constant population size from one generation to the next.

Figure 5.7 A maple forest exhibits a broad range of tree ages.

Now examine the age distribution for India (Figure 5.9). For many years, family size in India has been large, so each generation has more people than the last. This creates a pyramid-shaped age distribution dominated by a very large base of young people at the bottom. Note that Indian society has taken some steps to curb populationgrowth prior to the 2000 census, so the rate of expansion in the 0–4 age class is beginning to show signs of slower growth. Because of the demographic structure, however, the population of India is projected to continue growing as the large number of children in that country mature into their childbearing years.

Compare the structure of India with the age structure of Italy, which has a contracting population. The bottom of the age structure is smaller than the middle, because family size averages fewer than two children per couple. If this trend continues, Italian society will be dominated by older people, with fewer demands on schools but greater demand for health care and retirement benefits as the population ages. Clearly, a view of the age structure of a population, whether herons, trees, or humans, tells us a lot about a population's past, and helps us predict its future as well.

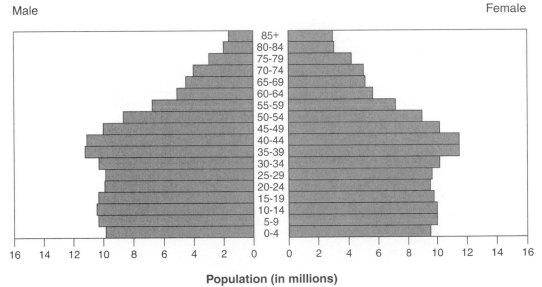

Figure 5.8 Demographic profile of U.S. population, from the 2000 census.

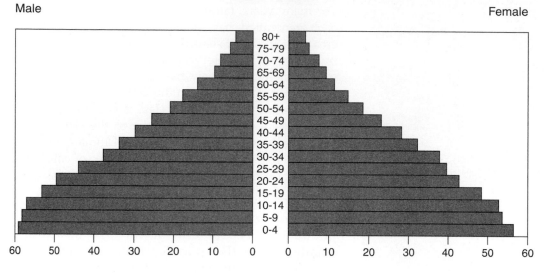

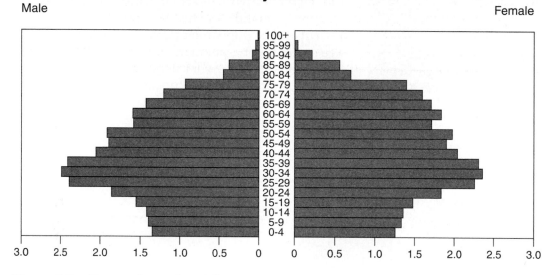

Figure 5.9 Comparison of rapidly growing population (India) with contracting population (Italy).

METHOD A: DEMOGRAPHY SIMULATION

[Computer activity]

Research Question

How do populations achieve a stable age distribution?

Preparation

Because demographic concepts are built on ideas about population growth, it is highly recommended that students complete readings and exercises from Chapter 4 before attempting exercises in this chapter. Instructions for this simulation are appropriate for Microsoft Excel software. If you use other spreadsheet software, instructions for writing cell formulas given in this procedure may require modification.

Materials (per laboratory team)

A computer station, with spreadsheet software

Background

This exercise simulates age-specific survivorship and reproduction for an imaginary population of medium-sized burrowing mammals; let's call them marmunks. Female marmunks begin breeding at age 1, and produce one viable offspring in their first year. In their second year, they produce three viable offspring. In their third and subsequent years, they produce five offspring per year. Newborn marmunks experience 65% mortality before age 1, but annual mortality rates are only 35% per year after that. The maximum life span is six years. Our task is to develop a model that shows age distributions for a marmunk population year by year as the population grows.

Procedure

1. Based on the background information, and following the heron illustration in this chapter, complete the following life table for marmunks. Don't forget that the L_X column begins with 1.000 alive at age 0, and that the b_X numbers equal half the family sizes for each age group.

AGE IN YEARS X	SURVIVORSHIP L_X	FECUNDITY b_X	EXPECTED OFFSPRING $(L_X)(b_X)$	WEIGHTED AGE $(X)(L_X)(b_X)$	STABLE AGE DISTRIBUTION $L_X / \Sigma L_X$
0					
1					
2					
3					
4					
5					
6					

If you have completed the life table correctly, it will predict some of the results of the simulation to follow.

2. Set up a computer spreadsheet with the following column headings. Numbers in the first column indicate the year in our simulation, so each row will show the state of the population in another year as we go down the page. Numbers in columns 2 through 8 represent the number of marmunks in the population that are newborns, one-year-old, two-year-old, etc. The column on the right is a sum of all the marmunks in the population.

	A	B	C	D	E	F	G	H	I
1	YEAR	AGE 0	AGE 1	AGE 2	AGE 3	AGE 4	AGE 5	AGE 6	TOTAL POPULATION
2									

3. The first entry in cell A2, under "year" should be 0, the starting point of our simulation. Go down this column, numbering years 1, 2, 3, 4, etc. If you do not want to type all those numbers in, just enter 0 in cell A2, and then in cell A3 type the formula **= A2+1**. If you copy and paste this formula down the column to the thirty-second row, you will get a series of numbered years from 0 through 30. Format all the cells below the first row to display numerical values, with no decimal places. The top of your spreadsheet should now look like this:

	A	B	C	D	E	F	G	H	I
1	YEAR	AGE 0	AGE 1	AGE 2	AGE 3	AGE 4	AGE 5	AGE 6	TOTAL POPULATION
2	0								
3	1								
4	2								
5	3								
6	4								

4. To begin our population, let's release 100 juvenile marmunks, just weaned and ready to start life on their own. To do this, enter the number 100 in cell B2, and zeroes in cells C2 through H2. In cell G2, enter the formula **= SUM(B2:H2)**. Copy and paste this formula down all the cells in column G, to the 30th row. This will give you the population size each year of your simulation. Your spreadsheet should now look like this:

	A	B	C	D	E	F	G	H	I
1	YEAR	AGE 0	AGE 1	AGE 2	AGE 3	AGE 4	AGE 5	AGE 6	TOTAL POPULATION
2	0	100	0	0	0	0	0	0	100
3	1								0
4	2								0
5	3								0
6	4								0

5. Next we need to model survival of newborns through their first year. The 100 newborns in year 0 will become one-year-olds in year 1. Thus, the survivors from our 100 released marmunks will be in cell C3 as one-year-olds. To calculate how many marmunks survive, recall that the death rate for newborns is 65%. This means that 0.35 of the newborns will be one-year-olds the following year. We can calculate this number by entering the formula = **B2 * 0.35** in cell C3. Copy and paste this formula all the way down column C so that your spreadsheet will calculate the number of one-year-olds each year. Your spreadsheet should now look like this:

	A	B	C	D	E	F	G	H	I
	YEAR	AGE 0	AGE 1	AGE 2	AGE 3	AGE 4	AGE 5	AGE 6	TOTAL POPULATION
1
2	0	100	0	0	0	0	0	0	100
3	1		35						35
4	2		0						0
5	3		0						0
6	4		0						0

6. To calculate survival of all the other age classes, recall that survivorship is 65% for older marmunks. In cell D3, enter the formula = **C2*0.65**. Copy and paste the formula into all the cells in columns D, E, F, G, and H from rows 3 through 32. The top of your spreadsheet should now look like this:

	A	B	C	D	E	F	G	H	I
	YEAR	AGE 0	AGE 1	AGE 2	AGE 3	AGE 4	AGE 5	AGE 6	TOTAL POPULATION
1
2	0	100	0	0	0	0	0	0	100
3	1		35	0	0	0	0	0	35
4	2		0	23	0	0	0	0	23
5	3		0	0	15	0	0	0	15
6	4		0	0	0	10	0	0	10

As you follow the diagonal from your 100 young marmunks down and to the right, you are following the fate of the cohort as it ages through years 1, 2, 3, 4, 5, and 6.

5. To complete the life history, we need to model reproduction. In cell B3, enter the following formula:

$$=C3*0.5 + D3*1.5 + SUM(E3:H3)*2.5$$

This equation states that the number of newborn marmunks this year equals ½ times the number of one-year-old parents, plus 1½ times the number of two-year-old parents, plus 2½ times the number of parents over two years of age. The offspring produced *per parent* correspond to family sizes of one, three, or five per pair, depending on parental ages, as described in the background section . Copy and paste this formula all the way down column B. Your spreadsheet now reflects both survival and reproduction, and should look like this:

	A	B	C	D	E	F	G	H	I
1	YEAR	AGE 0	AGE 1	AGE 2	AGE 3	AGE 4	AGE 5	AGE 6	TOTAL POPULATION
2	0	100	0	0	0	0	0	0	100
3	1	18	35	0	0	0	0	0	53
4	2	37	6	23	0	0	0	0	66
5	3	49	13	4	15	0	0	0	81
6	4	52	17	8	3	10	0	0	90

6. Finally, create a row at the very bottom of the spreadsheet to calculate the age distribution at year 30. In cell A34, put the label "Age Distribution." Change the format in row 34 to show numbers in three decimal places. Then, in cell B34, enter the formula: **=B32/I32**.
   Copy and past this formula into cells C34 through H34. This formula takes the numbers of marmunks of each age in the 32nd row (30th year) and converts numbers to proportions. In theory, this row will give you the stable age distribution predicted by the $L_x/\Sigma L_x$ column in your marmunk life table.

Questions (Method A)

1. From your life table, calculate the net reproductive rate $R_o = \Sigma[\,L_x\,b_x\,]$. Does your answer make sense, given what you know about family sizes and survivorship in marmunks? Does this net reproductive rate allow each marmunk to replace itself to ensure continuity of the population?

2. From your life table, calculate the generation length $T = \Sigma[x\,L_x\,b_x\,]\,/\,\Sigma[L_x\,b_x\,]$. Comment on the meaning of this answer, noting that marmunks begin to reproduce at age 1 and continue bearing offspring until age 6.

3. Use the values of R_o and T calculated above to compute the intrinsic rate of increase for marmunks, using $r \cong (ln\,R_o)\,/\,T$.

4. How accurate is your estimate of **r**? To find out, use the exponential growth equation, solved for r. To ensure that the population has reached a stable age structure and is growing in a smooth exponential curve, let's look at the final 10 years of population growth. Using N_{20} as a starting point, and measuring population growth over the 10 years to N_{30}, we can use:

$$N_{30} = N_{20}\, e^{r\,(10)}$$

Taking the $\log_e$ of both sides and rearranging, we get:

$$r = (ln\ N_{30} - ln\ N_{20})\ /\ 10$$

Plug in values from your simulation to calculate the rate of increase, using natural logs of the population sizes at N_{20} and N_{30} to calculate marmunk population growth over the final decade of your simulation. Now you have calculated **r** in two ways. The first is a theoretical value estimated from the life table, and the second is an empirical determination from the simulation results. How do the two **r** values compare? Comment on reasons for any discrepancies in your two calculations of **r**.

_____

_____

_____

_____

_____

_____

_____

5. Compare the final age distribution of marmunks in year 30 with the stable age distribution projected from the $L_X/\Sigma\ L_X$ column in your marmunk life table. How do the two compare? Did it take 30 years for the population to reach a stable age distribution? Check frequencies across the row in earlier years to find out.

_____

_____

_____

_____

_____

_____

_____

METHOD B: DEVELOPING A LIFE TABLE FROM THE ALUMNI NEWSLETTER

[Laboratory activity]

Research Question

What are the demographic characteristics of the alumni population of your school?

Preparation

Because alumni are not a population in the biological sense, this exercise should be viewed as a demonstration of demographic principles. Alumni newsletters vary significantly in the amount and kinds of information they present. The best source for this study would list obituaries and birth announcements for a representative sample of alums—not just the more successful or famous ones. This exercise also works best if your institution is over 70 years old, to ensure that alumni from all portions of the life span are represented. Obituaries that list age at death and all surviving children are best, but ages can be estimated from the graduating class year if necessary. This entails several assumptions which introduce error, such as age at matriculation during the historical period included in your study. Since many alumni magazines are available in an on-line format, students may find it easier to search back issues posted on web sites rather than paper copies. If your alumni magazine does not carry birth and death announcements, a city newspaper might provide an alternative source of data.

Materials (per laboratory team)

Access to back issues of the alumni newsletter, with at least 50 birth announcements and 50 obituaries available to each laboratory team

Background

This exercise demonstrates how to build a life table from information supplied by your alumni newsletter. Because access to this information may be limited to alums and students at your school, be sure to respect the confidentiality of the families involved, and do not identify individual names in any reports you produce.

Be aware that an alumni newsletter reports data from a biased sample of the human population as a whole. Since one must be a college student before becoming an alum, deaths occurring before the age of 18 will not be represented in your data set. If your university or college has many nontraditional students, a significant number of children born before their parents matriculate will be excluded from the sample. Finally, college-educated persons typically exhibit different demographic characteristics from the population at large.

The purpose of the exercise is not to estimate demographic parameters for the population as a whole, but to demonstrate in general how birth and death records can be used to construct a life table for a particular group of interest.

Procedure

1. *From obituary notices*, complete the longevity and fecundity tables on the calculation pages for Method B. Assign each obituary that you read a unique record number to preserve the anonymity of the people included in the study. Longevity can be estimated in any of three ways: first, the year of birth may be listed in the notice. If so, subtract the year of birth from the year of death to determine longevity. Second, the class year may be listed. If so, you can subtract 22 from the graduation class year as an estimate of the date of birth, assuming a typical college student is 22 years old at the time of graduation. Then subtract estimated year of birth from the year of death. Third, if the age at death is reported, skip the year of birth and year of death columns, and enter age at death under the "Longevity" heading.

Enter the gender of the deceased under the column titled "sex." Finally, under the heading "No. Offspring," enter the number of children listed as survivors. This number may not include all children, but provides a reasonable estimate of family size in most cases. Sample records are illustrated in the following abbreviated table:

Record Number	Year of Birth	Year of Death	Longevity	Sex	Number of Offspring
100	1943	2004	61	F	3
101	1930	2005	75	M	2

2. For the next step, go down the table you have just created line by line, reading only the records for female alums. In the table titled "Grouped Longevity Data for Females," place tally marks in the appropriate boxes to count the number of females who died in each of the 10-year age classes: 0–9, 10–19, 20–29, etc. Sample entries are shown in the abbreviated table below. If you like, you can pool data from the entire class at this step for a larger sample size and greater accuracy. After you have gone through all the obituary records, add up the tally marks and enter a total number of females who died in each age class as shown. Then sum the column of numbers to find the grand total of females of all ages in your obituary sample.

AGE CLASS	FEMALE LONGEVITY Tally marks for females dying within this age class.	Total	PROPORTION DYING IN THIS AGE CLASS d_x	AGE-SPECIFIC SURVIVOR-SHIP L_x
0–9		0	0	1.000
10–19	/	1	.020	1.000
20–29	///	3	.060	.980
30–39	////	4	.080	.920

Next, compute d_x values by dividing the number who died in each age class by the total number of females in the sample. (Note that we are expressing d_x as a proportion and not as a whole number here.) For example, if our sample had a total of 50 females, the three women who died between the ages of 20 and 29 would comprise 3/50 = 0.060 of the total population. Finally, calculate L_x values by subtracting d_x from the L_x value in one line to get the L_{x+1} value for the next line, as shown in the table. For example, in the line representing age class 20–29, the subtraction gives us 0.980 – 0.060 = 0.920, which is the L_x value for the next age class. Complete these calculations all the way through the table. If calculated correctly, entries in the L_x column should decline toward 0 in the 100+ age class.

3. Repeat step 2 for the records on males in your obituary data. Complete the table titled "Grouped Longevity Data for Males" in the calculation pages.

4. Calculate net reproductive rate R_0 by summing all the numbers of surviving children in the "Number of Offspring" column in the obituary data, and dividing by 2 times the total number of people in your obituary sample. The 2 in the denominator accounts for two parents in each family, so the result represents the number of offspring per parent. Enter these calculations in the small table entitled "Estimating Net Reproductive Rate."

5. Draw survivorship curves on the graph following the grouped longevity data tables. Plot the survivorship L_x as a function of age class separately for males and females. Connect male points with a dotted line, and female points with a solid line. Compare your results.

6. *From birth notices*, determine the mother's birth year, and whether the baby is male or female. You will need to estimate the mother's birth year by subtracting 22 from the graduation year. Enter the year of birth of the child. Subtract the mother's birth year from the baby's birth year to determine the age of the mother. Finally, record the sex of the child as female (F), male (M), or not stated (NS). If it is reasonable to use first names to infer the sex of the child, do so, with an awareness that some error will be introduced by wrong guesses. In cases of twins, fill out a separate line for each of the two children. A sample record is included below for illustration.

Record Number	Mother's graduation year	Mother's birth year	Baby's birth year	Mother's age at time of birth	Sex of child (F) (M) (NS)
1	2000	1978	2005	27	M

7. In the table titled "Grouped Fecundity Data," place tally marks in the appropriate boxes to count the number of females who gave birth in each of the 10-year age classes: 0–9, 10–19, 20–29, etc. Sample entries are shown in the abbreviated table below.

AGE CLASS	FECUNDITY Tally marks for females giving birth within this age class.	Total	PROPORTION OF BIRTHS OCCURRING IN THIS AGE CLASS	EXPECTED OFFSPRING $L_x b_x$
0–9		0	0	0
10–19	//	2	.040	0.072
20–29	///// ///// ///// ///// ///// /	26	.520	0.936
30–39	///// ///// ///// ///	18	.360	0.648

Divide each of the age-specific birth counts by the total births in the entire sample to calculate the proportion of births occurring in each age class. For example, if the whole sample included 50 births, the proportion born to mothers in the 20–29 age class would be $26/50 = 0.520$, as shown in the sample table. After you have calculated all the proportions, multiply each of these fractions by R_o to convert expected proportions to expected offspring. For example, if R_o were 1.8, then the number of expected offspring per 20–29-year-old in the population would be $(0.520)(1.8) = 0.936$ children per parent.

8. In the table titled "Calculation of Sex Ratios," enter numbers of records you found for males and females in the obituary data in the top row, and from birth announcements in the second row. Divide numbers of each sex by total sample size to calculate a sex ratio for deceased alumni and a sex ratio for alumni children as indicated in the table.

9. Use your data to complete a "Life Table for Alumni" at the end of the calculation pages for Method B. You have L_x values from the obituary data, and $L_x b_x$ values from the birth announcements. If you like, you can divide each $L_x b_x$ value by L_x to get a b_x value for that age class. At the bottom of the table are questions about the generation length, net reproductive rate, intrinsic rate of reproduction, and stable age distribution for the alumni population. Refer to the Introduction if you have forgotten how to make these calculations from a life table. Answer the questions that conclude this demography exercise.

Data From Obituary Notices (Page 1)

Record Number	Year of Birth	Year of Death	Longevity	Sex	No. Offspring
1					
2					
3					
4					
5					
6					
7					
8					
9					
10					
11					
12					
13					
14					
15					
16					
17					
18					
19					
20					
21					
22					
23					
24					
25					

Data From Obituary Notices (Page 2)

Record Number	Year of Birth	Year of Death	Longevity	Sex	No. Offspring
26					
27					
28					
29					
30					
31					
32					
33					
34					
35					
36					
37					
38					
39					
40					
41					
42					
43					
44					
45					
46					
47					
48					
49					
50					

Grouped Longevity Data For Females

AGE CLASS	FEMALE LONGEVITY — Tally marks for females dying within this age class.	Total	PROPORTION DYING IN THIS AGE CLASS d_x	AGE-SPECIFIC SURVIVOR-SHIP L_x
0–9				1.000
10–19				
20–29				
30–39				
40–49				
50–59				
60–69				
70–79				
80–89				
90–99				
100+				

Total Females =

Grouped Longevity Data For Males

AGE CLASS	MALE LONGEVITY — Tally marks for males dying within this age class.	Total	PROPORTION DYING IN THIS AGE CLASS d_x	AGE-SPECIFIC SURVIVOR-SHIP L_x
0–9				1.000
10–19				
20–29				
30–39				
40–49				
50–59				
60–69				
70–79				
80–89				
90–99				
100+				

Total Males =

Estimating Net Reproductive Rate

TOTAL NUMBER OF RECORDS FROM OBITUARY DATA = n	TOTAL NUMBER OF SURVIVORS LISTED = s	$R_o = s / (2n)$

Survivorship Curves for Alumni
- - - - - - - - Males ——————Females

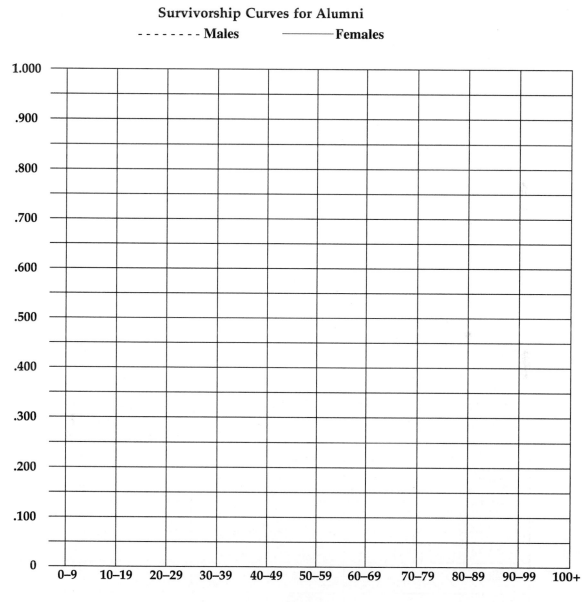

AGE CLASS (YEARS)

Data From Birth Announcements (Page 1)

Record Number	Mother's graduation year	Mother's birth year	Baby's birth year	Mother's age at time of birth	Sex of child (F) (M) (NS)
1					
2					
3					
4					
5					
6					
7					
8					
9					
10					
11					
12					
13					
14					
15					
16					
17					
18					
19					
20					
21					
22					
23					
24					
25					

Data From Birth Announcements (Page 2)

Record Number	Mother's graduation year	Mother's birth year	Baby's birth year	Mother's age at time of birth	Sex of child (F) (M) (NS)
26					
27					
28					
29					
30					
31					
32					
33					
34					
35					
36					
37					
38					
39					
40					
41					
42					
43					
44					
45					
46					
47					
48					
49					
50					

Grouped Fecundity Data

AGE CLASS	FECUNDITY Tally marks for females giving birth within this age class.	Total	PROPORTION OF BIRTHS OCCURRING IN THIS AGE CLASS	EXPECTED OFFSPRING $L_x b_x$
0–9				
10–19				
20–29				
30–39				
40–49				
50–59				
60–69				
70–79				
80–89				
90–99				
100+				

Total Births =

Calculation of Sex Ratios

DATA SOURCE	TOTAL SAMPLE SIZE	MALES NUMBER	%	FEMALES NUMBER	%
OBITUARY DATA (SEX OF DECEASED)					
BIRTH ANNOUNCEMENT DATA (SEX OF CHILDREN)					

Life Table for Alumni

AGE CLASS X	SURVIVORSHIP L_X	FECUNDITY b_X	EXPECTED OFFSPRING $(L_X)(b_X)$	WEIGHTED AGE $(X)(L_X)(b_X)$	STABLE AGE DISTRIBUTION $L_X / \Sigma L_X$
0–9					
10–19					
20–29					
30–39					
40–49					
50–59					
60–69					
70–79					
80–89					
90–99					
100+					
TOTALS:					

1. Net Reproductive Rate =

2. Generation Length =

3. Estimate of intrinsic rate of population growth =

4. Stable age distribution =

Questions (Method B)

1. Compare the survivorship curves for male and female alums. If they differ, comment on biological and sociological explanations. Do you expect survivorship curves for your own cohort to resemble these, or do you expect changes due to diet, lifestyle, medical advances, or environmental factors? Explain your reasoning.

   _____

   _____

   _____

   _____

2. How did your method of data collection potentially affect the shape of the survivorship curves? How might missing data be collected?

   _____

   _____

   _____

   _____

3. Comment on the sex ratio for alums vs. the sex ratio for infants born to alums. Are they the same? What social factors or reporting bias might be responsible for these differences?

   _____

   _____

   _____

   _____

4. To develop an accurate life table from past birth and death records, rather than from cohort data, one must assume the population is in a stable age distribution, and not changing rapidly in overall size. Is this true of the alumni population of your institution? If not, what kinds of error did you introduce in your life table? For example, how would your age-specific fecundity data be affected if the number of female graduates has been increasing every year over the past two decades?

   _____

   _____

   _____

   _____

5. How might demographic analysis of a human population help state and national leaders develop plans for education, health care, economic development, or retirement benefits? Give one or two specific examples, based on information gathered in this exercise.

   _____

   _____

   _____

   _____

METHOD C: DEMOGRAPHY OF CAMPUS TREES

[Outdoor activity, any season]

Research Question

What size-class structure do shade trees exhibit on campus?

Preparation

This outdoor exercise treats shade trees on your campus as a population of carefully managed organisms of varying ages. Size classes are used rather than ages, so you will not have to take core samples of your trees. To get sufficient data, there should be at least 30 individuals of the same tree species for a lab group to measure. Ideally, trees of different ages should be included in your sample, with some of the largest ones nearing the maximum life span. If more than one species is abundant on your campus, assign different species to different laboratory groups.

Materials (per laboratory team)

Metric measuring tape, 3 meters or longer

Background

Consider the life history of a shade tree on your campus. New trees typically enter the population as transplanted saplings. If recruits survive transplanting, they have a high survival rate in comparison to forest trees, because they are spaced far enough apart to avoid serious competition. However, at the end of the life span, campus trees probably die more quickly than they would in a forest, because hyper-mature specimens with dead limbs or hollow trunks are generally removed before they can fall on someone.

Although campus trees do not include the very beginning or the very end of a wild tree's life history, we can examine their demographic characteristics between those extremes. Because trees add xylem to their trunks at a fairly constant rate in annual growth rings, the diameter of a tree trunk is positively correlated with the tree's age. We will use trunk diameter to place trees in size classes, and develop a size-class structure diagram, similar to the age-structure diagrams shown for human populations in the Introduction. Both the history and the future of this tree population can be examined through your demographic analysis.

Procedure

1. Select a tree species abundant on your campus. You will need 30–50 individuals to generate a size-class structure diagram. If you cannot include all of these trees in your study, develop a sampling plan to generate data representative of the whole population, without bias toward larger or smaller individuals.
2. For each tree in your sample, use a metric tape to measure the circumference of the trunk at chest height to the nearest cm. In the column labeled "Observations," note any dead limbs, peeling bark, or other indications of the tree's health. If you find a plaque or marker indicating the year the tree was planted, be sure to record that information as well.
3. Return to the lab or classroom to analyze your data. Because the circumference of a circle is π times its diameter, you can divide the circumference by 3.14 to calculate a tree's diameter, as shown in the data table. Round off your calculation to the nearest cm.

4. To group your tree diameter records into size classes, place a tally mark in the appropriate size-class box for each tree in your sample as shown in the following table fragment. Count the tally marks and record the number in the "Totals" column. Divide this number by the sample size (50 in this example) to get a frequency. Multiply by 100% to convert the frequency to a percentage. The percentage expresses what portion of the whole population resides in this size class. Complete the table with data for all size classes you observed on the campus.

| SIZE CLASS (cm) | NUMBER OF TREES | | FREQUENCY (Percentage) |
	Tally marks for trees with diameters falling within this size class.	Totals	$\dfrac{\text{(number)}}{\text{(sample size)}} \times 100\%$
0–9	///// ///// ///// //	17	34%
10–19	///// ///// ///	13	26%
20–29	///// ///	8	16%

5. Finally, color in squares in the graph labeled "Size Class Diagram for Campus Trees" to show the results of this demography study. If you have fewer than 20% of the whole sample in any given size class, let each square in the graph stand for 1%, and color in squares from left to right to show proportions in each size class. If you have more than 20% in a single class as shown in the illustration above, let each square represent 2%. Your graph will not be symmetrical like the human "population pyramids" illustrated in the Introduction, because we are not keeping separate records for male and female trees. (Some plant species, including Ginkgo and holly trees, do have separate sexes, but many do not.) A sample result for the bottom three rows of the graph is shown here. Interpret your graph and answer the questions at the end of the chapter.

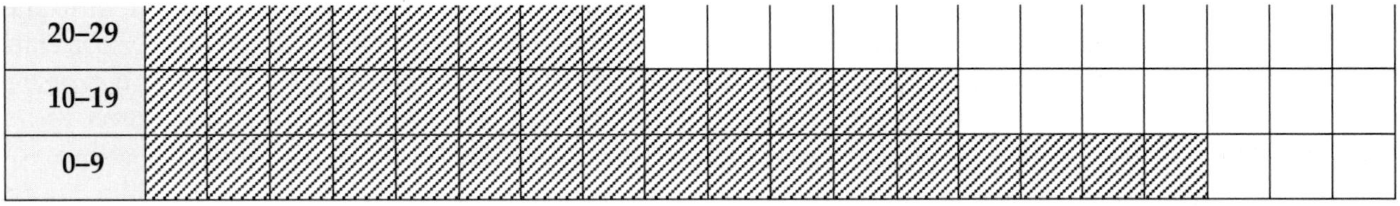

Data from Tree Demography Study (Page 1)

Record Number	Circumference at chest height (cm)		Diameter (cm)	Observations
1		÷3.14 =		
2		÷3.14 =		
3		÷3.14 =		
4		÷3.14 =		
5		÷3.14 =		
6		÷3.14 =		
7		÷3.14 =		
8		÷3.14 =		
9		÷3.14 =		
10		÷3.14 =		
11		÷3.14 =		
12		÷3.14 =		
13		÷3.14 =		
14		÷3.14 =		
15		÷3.14 =		
16		÷3.14 =		
17		÷3.14 =		
18		÷3.14 =		
19		÷3.14 =		
20		÷3.14 =		
21		÷3.14 =		
22		÷3.14 =		
23		÷3.14 =		
24		÷3.14 =		
25		÷3.14 =		

Data from Tree Demography Study (Page 2)

Record Number	Circumference at chest height (cm)		Diameter (cm)	Observations
26		÷3.14 =		
27		÷3.14 =		
28		÷3.14 =		
29		÷3.14 =		
30		÷3.14 =		
31		÷3.14 =		
32		÷3.14 =		
33		÷3.14 =		
34		÷3.14 =		
35		÷3.14 =		
36		÷3.14 =		
37		÷3.14 =		
38		÷3.14 =		
39		÷3.14 =		
40		÷3.14 =		
41		÷3.14 =		
42		÷3.14 =		
43		÷3.14 =		
44		÷3.14 =		
45		÷3.14 =		
46		÷3.14 =		
47		÷3.14 =		
48		÷3.14 =		
49		÷3.14 =		
50		÷3.14 =		

Grouped Tree Demography Data

Tree species sampled:_____

SIZE CLASS (cm)	NUMBER OF TREES Tally marks for trees with diameters falling within this size class.	Totals	FREQUENCY (number) / (sample size)
0–9			
10–19			
20–29			
30–39			
40–49			
50–59			
60–69			
70–79			
80–89			
90–99			
100+			

Total Sample Size = _____

SIZE CLASS (cm)	SIZE CLASS DIAGRAM FOR CAMPUS TREES SPECIES_____ DATE_____
100+	
90–99	
80–89	
70–79	
60–69	
50–59	
40–49	
30–39	
20–29	
10–19	
0–9	

Questions (Method C)

1. Do you think you have measured any trees near the end of their life span in this study? Do the largest trees show any signs of decay, peeling bark, or dead limbs?

   _____

   _____

   _____

   _____

2. From the shape of the size-class distribution, what can you guess about the history of tree planting on your campus? Does the size-class distribution show that trees have been continuously planted, or are there large cohorts indicating periods of intense tree planting, followed by years of no new additions to the population?

   _____

   _____

   _____

   _____

3. From the shape of the size-class distribution, what can you guess about the future of the shade tree population on your campus? When existing large trees finally die, are there younger trees growing into the larger size classes so that the campus will keep some shade?

   _____

   _____

   _____

   _____

4. In our study of size classes, we assume all trees grow at a similar rate so that a larger diameter indicates an older tree. However, growth rates in trees depend on soil quality, access to water, shading, severe pruning, and many other factors. Do you think differences in growth rate could compromise the accuracy of your study? If so, how could you document the actual ages of these trees?

   _____

   _____

   _____

   _____

5. Since forest trees compete for light, moisture, and nutrients, juvenile mortality is quite high in a natural environment. How would you expect a size-class distribution for a forest population to differ from the size-class distribution you documented on campus?

   _____

   _____

   _____

   _____

FOR FURTHER INVESTIGATION

1. Rerun the simulation in Method A by typing different initial numbers into the "Year 0" row at the top of your spreadsheet. Does the population get a faster start if you introduce adults as well as newborn marmunks to begin the simulation? Does the initial age structure have any effect on the ultimate stable age distribution?
2. Try different combinations of survivorship and age-specific fecundity schedules to achieve a stable population size in the marmunk simulation (Method A). As you change the formulae for survival or for calculating births, remember to copy and paste them in the appropriate areas of the spreadsheet. Compare the relative importance of survivorship and fecundity in determining the value of **r**.
3. Look up the Euler method for calculating the intrinsic rate of increase from life table data. (See Ebert, T. A. 1999, p. 15.) Use this approach to develop a more accurate calculation of **r**, and check it against your simulation results. Does this method yield a significantly improved estimate?
4. On the web site for the U.S. Census Bureau (http://www.census.gov), look up demographic parameters such as intrinsic rate of increase, longevity, and family size for the population at large. Develop a life table for the U.S. population, and compare it with the life table you developed in this exercise. In what ways are alumni atypical of the population at large?
5. If you have records that go back far enough, try compiling a life table based on only one graduating class, using Method B calculations. You should choose a graduating class from about 60 years ago to ensure that you have data through the entire life span. This amounts to a cohort study, and provides more accurate demographic data than assessments based on a mixture of cohorts. Do you see any significant changes when you compare newer or older cohorts?
6. Compare demographic information found in alumni newsletters published on-line from other universities across the country and around the world. Does a demographic analysis yield meaningful information about the history and mission of a university?
7. Look at memorial plaques, old photographs, or university records to find out when some of the trees in your sample were planted. Then you can divide the diameter of the tree by its age in years to determine the growth rate, expressed in cm per year. If you can age more than one tree of the same species, calculate an average growth rate for the species in this environment. Then divide each tree diameter by the mean growth rate to convert all sizes in your data set to equivalent ages. Use the converted numbers to construct an age-structure diagram for the population.
8. Consult with your physical facilities director, landscaping committees, and ecology or botany faculty to find out if there is a long-range plan for planting trees on your campus. If not, contribute ideas for tree planting and get involved in making landscaping improvements as a class project. A collection of tree species, called an **arboretum**, is a significant asset for a campus. If possible, think about adding species native to your region. They tend to require less maintenance, and provide a valuable and convenient resource for botany instruction.

FOR FURTHER READING

Butler, R. W. 1992. Great Blue Heron (*Ardea herodias*). In: *The birds of North America*, No. 25 (A. Poole, P. Stettenheim, and F. Gill, eds.) Philadelphia: The Academy of Natural Sciences, Washington, D.C.: The American Ornithologists' Union.

Deevey, E. S., 1947. Life tables for natural populations of animals. *Quarterly Review of Biology* 22: 283–314.

Ebert, T. A. 1999. *Plant and animal populations: methods in demography*. Academic Press, San Diego.

Preston, Richard J. 1989. *North American Trees*. Iowa State University Press, Ames.

United States Census Bureau. http://www.census.gov

Chapter **6**

Population Genetics

Figure 6.1 Genetic variation in the coquina clams, *Donax variabilis*.

INTRODUCTION

Coquina clams (Figure 6.1) live in dense populations on sandy beaches of the Atlantic and Gulf coasts of the United States. They occupy the active surf zone, constantly digging in against the action of the waves, filter feeding on **plankton**, and providing food for a variety of shore birds. As their scientific name *Donax variabilis* implies, these little bivalves exhibit noticeable variations in color and pattern among members of the same species found on the same beach. Ecologists call this a **polymorphic** population, because genetic differences among individuals create more than one common local phenotype. Although polymorphism often involves biochemical traits less easily observed than the colors of coquina shells, genetic variation among individuals is a widespread and ecologically relevant feature of natural populations of plants and animals.

Population genetics pioneer Ernst Mayr pointed to polymorphism as an essential, but frequently unappreciated source of difference between the life sciences and physical sciences (Lewin, 1982). The study of life, Mayr said, requires "**population** thinking," because life scientists always have to consider genetic variations among subjects of their research. While the chemist may assume one oxygen atom reacts in pretty much the same way as any other, biologists cannot assume all black bears exhibit similar feeding behavior, or that all Douglas fir trees respond identically to climate change. Genetic diversity is a driving force in populations, affecting all sorts of ecological processes from competition for resources to the formation of new species. Emerging theory in conservation ecology relies heavily on population genetics to develop effective strategies for protecting vanishing organisms (e.g., Loew, 2002). Ecologists are therefore wise to investigate how much genetic variation exists in the species they study, and how this variation is distributed within and among populations.

Genetics: A Brief Review

We will begin with a brief review of the basic principles of genetics. If you are confident that you understand and remember the concepts covered in the next two "Check your Progress" boxes, you may wish to skip to the next section on genetics and probability.

Every organism in every population carries a set of genetic instructions called its **genome**. In multicellular species, the genome is replicated in every cell, encoded by the DNA in each cell's nucleus. **Diploid** organisms have two sets of chromosomes, one set inherited from each parent. As a result, each gene is represented twice in the genome. We call the two chromosomes carrying copies of the same genes a **homologous pair**. Prior to reproduction, a specialized kind of cell division called meiosis divides the homologous chromosomes, sending them into separate **haploid** cells called **gametes** (Figure 6.2). Gametes are the sex cells, either sperm or eggs. Any given chromosome carries a predictable sequence of genetic instructions from one end of its DNA code to the other. For example, a particular gene for flower color would be found in a predictable place on a particular chromosome of a given plant species. That place is called the gene's **locus**. If you examined the corresponding locus on the other homologous chromosome, you would find the other copy of the flower color gene.

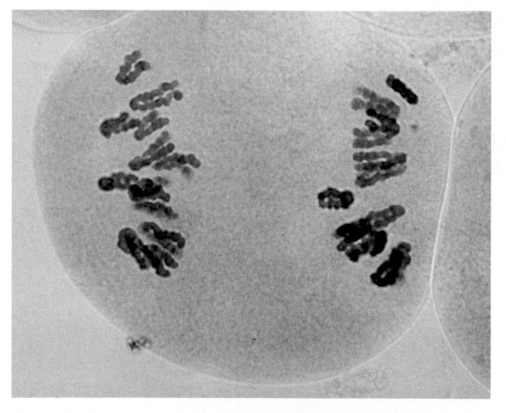

Figure 6.2 Homologous chromosomes separating during gamete formation in lily cells.

If a gene comes in more than one form, we say there are two **alleles** at this locus. For example, a gene for flower color in snapdragons can carry instructions to make a red pigment or a white pigment (Figure 6.3). Let's use the symbol C^R to represent the red allele and the symbol C^W to represent the white allele. The genes a plant carries for this trait are called its **genotype**; the actual flower color we see when we look at the plant is its **phenotype**. Snapdragon plants with two copies of the same allele for red pigment, designated as genotype $C^R C^R$, have red flowers as their phenotype. Snapdragon plants with two copies of the allele for white pigment, designated as genotype $C^W C^W$, have white flowers as their phenotype. Whether red or white, any snapdragon having two copies of the same allele is said to be **homozygous**. If the snapdragon inherits one allele from its male parent and a different allele from its female parent, the plant is said to be **heterozygous**. These plants have the genotype $C^R C^W$ and their phenotype is pink—intermediate between red and white.

The phenotype of a heterozygote is not always intermediate between the homozygous types. For many genetic traits, the instructions on a **dominant allele** override the instructions on a **recessive allele** so that the heterozygous phenotype looks just like the homozygous dominant. The recessive allele is still present in the genome, and is inherited by half the offspring, but its presence is "masked" or hidden in heterozygous individuals. Note that dominant genes are *not* necessarily better or more prevalent than recessive genes. Dominance refers *only* to the way one allele overrides the other in determining the phenotype in the heterozygous condition.

Figure 6.3   Snapdragons.

Within a population, there can be more than two alleles for the same gene. For example, for the **ABO blood groups** in the human genome, there are three alleles: I^A, I^B, and i. All three of these alleles can be found in a typical human population, but any individual in the population can have a maximum of two—one on each homologous chromosome. Individuals having the homozygous $I^A I^A$ genotype produce a molecule on the surface of the red blood cells called the A **antigen**. Their phenotype is "type A blood." (See Table 6.1.) Individuals with the $I^B I^B$ genotype produce a type B antigen on their blood cells, so their phenotype is "type B blood." Homozygous i i individuals have neither type A antigens nor type B antigens, so their phenotype is called "type O blood." The i allele is recessive to both of the others, so the I^A i genotype produces type A antigens, resulting in type A blood. The recessive i allele is still present on the chromosome, and can be inherited by offspring, but is masked by its dominant counterpart. Similarly, the I^B i genotype results in type B blood. The I^A and I^B alleles express no dominance over one another, so persons with the genotype $I^A I^B$ produce both kinds of antigens, and their phenotype is "type AB."

Table 6.1

GENOTYPES	PHENOTYPES
$I^A I^A$ or I^A i	Type A blood
$I^B I^B$ or I^B i	Type B blood
$I^A I^B$	Type AB blood
i i	Type O blood

Check your progress:

Make sure you can explain the difference between the following pairs of terms:

1. Genome and chromosome
2. Haploid and diploid
3. Locus and allele
4. Genotype and phenotype
5. Homozygous and heterozygous
6. Dominant and recessive
7. ABO blood groups and Rh factor

Hint: Consult the glossary if terms are not clear.

You may have heard the words "positive" and "negative" along with A, B, and O in reference to blood types. These descriptors refer to a totally different trait that was first found in rhesus monkeys, appropriately named "**Rh factor**." Since the human genome and the genomes of other primates contain many DNA sequences in common, it is not surprising that the Rh gene was later discovered in humans as well. The Rh locus has two alleles, (+) and (−). This is where the "positive" and "negative" terms come from. Rh factor is important for matching blood types, but its inheritance is independent of the ABO system.

The preceding generalizations about chromosomal pairs are true of 22 of the 23 pairs of human chromosomes, but there is one exception. Human **sex chromosomes** come in two types: X and Y. Females have two X chromosomes, while males have an X and a Y. The **X chromosome** is quite large, and contains lots of information, some of which is essential for normal growth and development. The **Y chromosome** is much smaller, and carries few genetic instructions. Even though it is small, the Y chromosome is important for male development, because one of its genes triggers hormonal changes resulting in a male child. Thus, XY babies develop as males; XX babies develop as females.

Because the X chromosome contains many genes, we need a system for tracking the inheritance of these characters, which are called **sex-linked traits**. We can indicate a gene on the X chromosome with a superscript. The most common form of **color blindness**, for example, is caused by a recessive gene that resides on the X chromosome. X^B symbolizes an X chromosome carrying the dominant allele for normal color vision, and X^b represents an X chromosome carrying the recessive allele for color blindness. Since females have two X chromosomes, a woman can be homozygous for either trait, or heterozygous (having genotype $X^B X^b$). Since the allele for color blindness is recessive, heterozygous females have full color vision. Males, however, have only one X chromosome, so they are either color-blind (genotype $X^b Y$), or not color-blind (genotype $X^B Y$). Note that no superscript is attached to the Y chromosome, since it carries no genetic information about color vision.

Check your progress:

If a man is color-blind, which of his parents passed the allele for color blindness to him?

Hint: Men inherit their Y chromosome from their fathers.

The familiar fruit fly, ***Drosophila melanogaster***, is a useful model for the study of sex-linked traits, because it also has the XX chromosomal arrangement in females and an XY arrangement in males. Although the developmental mechanism for sex determination in fruit flies is not the same as for humans, inheritance of sex chromosomes and sex-linked traits works in a similar fashion. In fruit flies, a gene for eye color resides on the X chromosome. Most fruit flies have the dominant allele for red eyes, but the recessive allele at this locus produces white eyes. The following table shows all possible genotypes and phenotypes with regard to the white or red eye color gene.

Table 6.2

DROSOPHILA GENOTYPES	*DROSOPHILA* PHENOTYPES
$X^B X^B$ or $X^B X^b$	Red-eyed Female
$X^b X^b$	White-eyed Female
$X^B Y$	Red-eyed Male
$X^b Y$	White-eyed Male

Genetics and Probability

Genetic inheritance is a **stochastic** process. This means that some of the mechanisms controlling the inheritance of genes follow predictable laws, but others operate in a random fashion. **Random events** cannot be predicted in individual cases, but if we look at a large group of random events, we can make reliable observations about outcomes for the group as a whole. For example, the gender of a baby is unpredictable before conception. If a family has one baby girl, and the mother becomes pregnant with a second child, no one can say in advance whether to expect a baby sister or a baby brother. Determination of sex in each child is an independent random event with the probability of having a girl roughly equal to $\frac{1}{2}$, and the probability of a boy $\frac{1}{2}$. (The ratio is actually more like 1.06 boys for every girl at birth, but we will adopt $\frac{1}{2}$ as a reasonable approximation for this example.)

Even though gender outcome is unpredictable for individual pregnancies, the process becomes predictable at the population level. If you checked records for the past 1000 births at your local hospital, you could predict with some confidence 50% girls and 50% boys, or 500 of each sex in the sample of 1000. An outcome of 495 to 505 would not be surprising, but the prediction holds true within limits determined by sample size. If you expanded your study to include the past 100,000 babies born in your state or province, the predicted 50:50 ratio would be even more accurate. This is an important scientific principle misunderstood by most people: *random events are subject to natural laws, and can be predicted in the aggregate.* To the scientist, the word "random" does *not* mean "without cause." It refers instead to events that are more easily predicted at the level of the population than at the level of the individual.

Check your progress:

When 1 kg of 50° water is mixed with 1 kg of 40° water, we can accurately predict the resulting mixture will have a temperature of 45°. Water temperature is defined as the average kinetic energy of the H_2O molecules. Since individual water molecules move at random, we cannot say what the kinetic energy of any individual molecule will be at a given time. How can we know the outcome of a series of random events with such certainty?

Hint: Consider the sample size.

Population ecologists have found a few simple probability laws quite powerful in explaining inheritance at the population level. To introduce these useful principles, let's begin with some symbols and terms.

Let (A) represent an event that may or may not happen. The symbol f(A) stands for the frequency of A's occurrence from past experience. We can also use f(A) to represent the probability that A will occur under the same set of conditions in the future. We can use a number between 0 and 1 to represent the range of all possible frequencies of occurrence, from f(A) = 0 if the event is impossible, to f(A) = 1 if the event is absolutely certain. Between these end points, we can designate probabilities much more accurately with fractions between 0 and 1 than we could using English phrases such as "possibly," "maybe," or "in most cases."

Check your progress:

Match each numerical probability with a verbal description by writing in a phrase from the following list: "Always," "Half the time," "Never," "Rarely," "Pretty often," "Usually."

f(A) = 0	_____	f(A) = 0.5	_____
f(A) = 0.01	_____	f(A) = 0.9	_____
f(A) = 0.3	_____	f(A) = 1.0	_____

Hint: A higher number means greater likelihood.

In our example of the sex of a baby, two outcomes are possible. To examine this in probability terms, f(female) = frequency of female births, and f(male) = frequency of male births. Since we define no third outcome, the frequencies must sum to one. Assuming equal likelihoods for girls and boys,

f(male) = 0.5,   **f(female) = 0.5,**   **and**   **f(male) + f(female) = 1.0**

What if we wanted to describe more than one characteristic of these children? Handedness, here defined as the hand we choose to write with, is influenced by culture and training, but has innate biological causes as well. Right-handedness occurs in about 90% of the North American population. We may therefore designate f(right-handed) = 0.9 and f(left-handed) = 0.1, assuming for the purpose of this exercise that ambidextrous people will favor one hand or the other for writing. To calculate the likelihood that a randomly selected child will be female *and* left-handed, we can simply multiply the probability fractions together:

f(female) = 0.5

f(left-handed) = 0.1

Thus,   **f(female *and* left-handed) = (0.5) (0.1) = 0.05**

This means that we could expect 5% of the population to be made up of left-handed females. It is important to note that this demonstration assumes left-handedness is equally common among females and males. If this is true, we could say determination of gender and of handedness are **independent events**. If this were not the case, we would have to change the handedness probability to the frequency of left-handedness *given that the child is female*.

The product rule makes intuitive sense if you remember that frequencies (or probabilities) are expressed as fractions. The probability of two uncertain events happening at the same time has to be smaller than the probability of either event by itself. Since the product of fractions is always a smaller fraction, multiplying probabilities gives us the smaller probability of a joint occurrence.

PRODUCT RULE: (Calculating joint occurrence of two independent events)

$$f(A \text{ and } B) = f(A) * f(B)$$

f(A and B) = frequency of both events happening together
f(A) = frequency of event A
f(B) = frequency of event B

Check your progress:

Apply the product rule to predict the frequency of right-handed male babies.

Answer: f(male and right-handed) = 0.45

Often in biology, outcomes are grouped so that we need to know how often either of two events occurs. For our example, we could ask how often would a randomly selected child be *either* female *or* right-handed? Since this pooled frequency will be larger than the frequency of either females or right-handed children considered separately, it makes sense to add the frequencies together. However, simple addition would overestimate the frequency of children either female or right-handed. Adding f(female) + f(right-handed) would yield a frequency of 0.5 + 0.9 = 1.4, which would amount to 140% of the entire population!

Figure 6.4 shows how we obtained a fraction over 100%. The rectangle shaded with vertical stripes represents the frequency of right-handed children in the population, and the rectangle shaded with horizontal stripes represents the frequency of girls. By adding the frequency of all girls to the frequency of all right-handed children, we counted right-handed girls twice. (The overcounted individuals are represented by the cross-hatched area in the figure.) To calculate the area covered by the two rectangles without double-counting right-handed girls, we could add the areas of the two rectangles, and then subtract the area of overlap. By analogy, the frequency of children either female or right-handed would be calculated as follows:

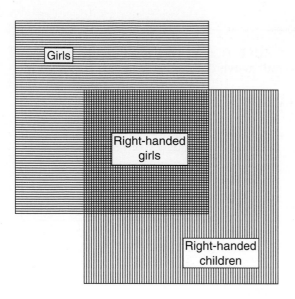

Figure 6.4   Adding frequencies of all right-handed children and of all girls in a population results in double-counting all the right-handed girls, who fit into both groups.

f(female *or* right-handed) = f(female) + f(right-handed) − f(female *and* right-handed)

Since we have already calculated f(female and right-handed) to be (0.5 ∗ 0.9), the equation can be solved as follows:

f(female *or* right-handed) = 0.5 + 0.9 − (0.5 ∗ 0.9) = 0.95 of all children.

This method for finding either/or probabilities is called the sum rule. Stated generally:

SUM RULE: (Calculating the frequency of either event A or event B)

f(A *or* B) = f(A) + f(B) − f(A *and* B)

f(A *or* B) = frequency of either one event or the other
f(A) = frequency of event A
f(B) = frequency of event B
f(A *and* B) = frequency of both events happening together
This last term can be dropped if the two events never occur together.

Check your progress:

Apply the sum rule to find the expected frequency of children who are either male or left-handed.

Answer: f(male *or* left-handed) = 0.55

It is worth noting that if the two events never occur at the same time, the third term in the sum rule equation is equal to zero and can be dropped. For example, f(male *or* female) = f(male) + f(female) = 0.5 + 0.5 = 1. Since children are not both male and female, there is no "overlap" to subtract in this case. In this situation, we say the two events are **mutually exclusive**, because the occurrence of one excludes the possibility of the other.

Always remember to use the product rule to find f(A and B), and use the sum rule to find f(A or B).

Population Genetics

Predicting distributions of genes in populations can be accomplished with a straightforward application of the two probability rules. You have already done this kind of calculation in introductory biology if you have used a Punnett square to generate frequencies in a genetics cross. For a cross of two pea plants heterozygous for plant height, where T is the dominant allele for tall plant size, and t the recessive allele for dwarf size:

	T f(T eggs) = 0.5	**t** f(t eggs) = 0.5
T f(T pollen) = 0.5	**TT** f(TT offspring) = (0.5)(0.5) = 0.25	**Tt** f(Tt offspring) = (0.5)(0.5) = 0.25
t f(t pollen) = 0.5	**Tt** f(Tt offspring) = (0.5)(0.5) = 0.25	**tt** f(tt offspring) = (0.5)(0.5) = 0.25

Because a heterozygous plant produces dominant (T) and recessive (t) pollen in equal proportions in its stamens, f(T pollen) = 0.5 and f(t pollen) = 0.5, as seen on the left side of the Punnett square. Similarly, since the other parent produces eggs in equal proportions in its ovaries, f(T eggs) = 0.5 and f(t eggs) = 0.5, as seen across the top of the square. Inside the Punnett square are expected offspring frequencies. To calculate the probability of a dwarf plant, for example, we first acknowledge that an offspring must inherit the t allele from both its male parent and its female parent. Since two events must co-occur to make a dwarf offspring, we apply the product rule as follows:

Check your progress:

Consider a genetic cross between a man with type B blood (genotype $I^B i$) and a woman with type A blood (genotype $I^A i$). Use the probability rules to calculate the expected frequency of female offspring with type O blood from this type of cross.

Answer: f(female *and* ii) = 1/8

f(tt offspring) = f(t pollen) ∗ f(t egg) = (0.5)(0.5) = 0.25

The equation predicts that 1/4 of the offspring are expected to exhibit the recessive phenotype in a cross between two heterozygous parents. The frequency of **TT** homozygous tall offspring can be calculated in the same way:

f(TT offspring) = f(T pollen) ∗ f(T egg) = (0.5)(0.5) = 0.25

To calculate the frequency of heterozygotes, we can use the sum rule:

f(Tt offspring) = f(T pollen meets t egg) + f(t pollen meets T egg) = (0.25) + (0.25) = 0.5

Note that the two ways to make a heterozygous offspring are mutually exclusive, so we do not have to worry about subtracting a third term in this case.

In summary, we use the probability rules every time we calculate frequencies for a genetic cross. Think of a Punnett square as a graphic representation of the probability rules, applied to inheritance in one "nuclear family" of two parents and their immediate offspring. Having mastered that analysis, you are ready to calculate expected genotypic frequencies for the whole population.

The Hardy-Weinberg Rule

To illustrate the Hardy-Weinberg calculation in a natural population, let's return to our polymorphic population of coquina clams (Figure 6.1). Coquinas reproduce by releasing eggs and sperm into the sea. Fertilization occurs when sperm meet eggs, and the resulting larvae drift with the currents before settling down on a beach to lead their adult lives. Because mating in coquinas is a random event, probability rules can be realistically applied to predict the genetic structure of its populations.

Assume for the sake of illustration that a dominant allele R produces purple rays (radiating stripes) on the coquina's shell. We will assume that rr genotypes are solid white clams with no purple rays. If the **gene pool**, defined as all the genes in all the members of a population, contains 70% R alleles and 30% r alleles, we would say that the **gene frequencies** are f(R) = 0.7 and f(r) = 0.3 in this population. Given this information, we could use the probability laws to predict the proportions of RR, Rr, and rr genotypes. These proportions, called **genotypic frequencies**, determine the ratio of shell phenotypes we could actually observe if we dug up a bucket of coquinas from the sand.

Imagine our population of coquinas releasing their eggs and sperm into the ocean, where they combine at random to produce the next generation. Frequencies of offspring genotypes can be predicted using the product rule. The following table illustrates frequencies of eggs and sperm cells as column and row headings, and probabilities for offspring genotypes in the cells. The resulting diagram resembles a Punnett square, but notice that the frequencies of R and r gametes are unequal in this case. To generalize this model, let p = f(R) and q = f(r). As long as there are only two alleles, p + q = 1.0 because the gene frequencies must add up to 100%.

	f(R eggs) **p = 0.7**	**f(r eggs)** **q = 0.3**
f(R sperm) **p = 0.7**	**f(RR offspring)** $p^2 = (0.7)(0.7) = 0.49$	**f(Rr offspring)** $pq = (0.7)(0.3)$ $= 0.21$
f(r sperm) **q = 0.3**	**f(Rr offspring)** $pq = (0.7)(0.3) = 0.21$	**f(rr offspring)** $q^2 = (0.3)(0.3)$ $= 0.09$

Following the rules of probability,

$$f(RR\ offspring) = f(R\ sperm\ meets\ R\ egg) = p^2 = (0.7)(0.7) = 0.49$$
$$f(Rr\ offspring) = f(R\ sperm\ meets\ r\ egg) + f(r\ sperm\ meets\ R\ egg) = 2pq = 2(0.7)(0.3) = 0.42$$
$$f(rr\ offspring) = f(r\ sperm\ meets\ R\ egg) = q^2 = (0.3)(0.3) = \underline{0.09}$$
$$Total = p^2 + 2pq + q^2 = 1.00$$

THE HARDY-WEINBERG CALCULATION

To determine genotypic frequencies from frequencies of alleles R and r:

$$f(RR) = p^2 \qquad f(Rr) = 2pq \qquad f(rr) = q^2$$

To sum genotypic frequencies in the population:

$$p^2 + 2pq + q^2 = 1.0 \qquad where:\ \ p = f(R\ allele)$$
$$q = f(r\ allele)$$

This set of equations is known as the Hardy-Weinberg law, named after a mathematician and a biologist who both came up with the same idea in 1908. Apparently, great minds do sometimes think alike!

Check your progress:

If p = f(R) = 0.6, and q = f(r) = 0.4, what proportion of the population would you expect to be heterozygous under Hardy-Weinberg rules?

Answer: 2 pq = 0.48

Genetic Equilibrium

For our coquina clam population having 70% R alleles and 30% r alleles, the Hardy-Weinberg equations predict that a collection of 1000 coquinas from the surf will be made up of 490 coquinas homozygous for purple rays, 420 heterozygous coquinas, and 90 plain white coquinas with the rr genotype.

Since we have generated these genotypic frequencies from the gene pool of gametes in the ocean, have the gene frequencies changed as a result of reproduction? No, they have not. When coquinas reproduce again, for every 1000 individuals, all of the gametes coming from the 490 RR clams and half of the gametes coming from the 420 heterozygous Rr clams will carry the R allele. Similarly, all of the gametes from the 90 rr clams and half of the gametes from 420 Rr clams will carry the r allele. If we divide observed numbers of clams by 1000 to convert these expected numbers of clams to genotypic frequencies f(RR) = 0.49, f(Rr) = 0.42, and f(rr) = 0.09, we can generate the next gamete pool as follows.

To calculate genotypic frequencies from gene frequencies

p = f(RR) + 1/2 f(Rr) =   0.49 + 0.5 (0.42) = 0.7

Similarly,

q = f(rr) + 1/2 f(Rr) =   0.09 + 0.5(0.42) = 0.3

Note that the gene frequencies remain *exactly where they were at the beginning of our calculation*: at 70% R and 30% r allelic frequencies. This is why we say the genes are in equilibrium. In each generation, the two alleles are separated by meiosis, and recombine according to the rules of probability, but neither (R) nor (r) changes its proportional representation in the gene pool.

Check your progress:

In a population of 1000 coquinas, if 360 are RR, 480 are Rr, and 160 are rr, what are the gene frequencies of R and r?

Answer: f(R) = 0.6 and f(r) = 0.4

Violating Hardy-Weinberg Equilibrium

The Hardy-Weinberg formula is what ecologists call a **neutral model**. It shows what to expect if no biological forces are changing gene frequencies in the population. In nature, gene frequencies are not always held in equilibrium. **Evolution**, defined as changes in a population's genetic makeup over time, happens constantly and has been demonstrated across a full range of biological species from bacteria to humans. To understand the origins of evolutionary change, we need to consider the following biological factors that can push gene frequencies out of Hardy-Weinberg equilibrium.

1. **Small population size:** Just as the sex ratio of newborn babies can depart from an expected 50:50 ratio in a small group of children, genotypic frequencies can depart from the expected proportions of p^2, $2pq$, and q^2 in a small population. Remember that genetic inheritance is a random process that becomes less predictable as the number of events decreases. A population reduced to a "bottleneck" of a few survivors can experience significant genetic change due to the small sample size represented in its gamete pool. The longer the population stays very small, the more significant are potential shifts in gene frequencies. Change of this type is called **genetic drift** because frequencies go up or down at random. If one allele or the other ever drifts to a frequency of 100%, we say the allele is **fixed** in the population. The other allele cannot drift back up from a frequency of 0, after it has been pushed to extinction. This is why very small populations tend to lose genetic variation over time.

2. **Non-random mating:** Coquina gametes combine at random in the sea, but many organisms are more selective in their choice of mates. For example, chemical reactions in the female tissues of many flowers block the passage of pollen too similar in genetic composition to the female parent. This promotes outcrossing to unrelated pollen donors, and also results in more heterozygosity than would be expected under the Hardy-Weinberg rule. On the other hand, some bird species tend to select mates that look like themselves. This kind of "**assortative mating**" reduces heterozygosity below Hardy-Weinberg expectations, because (AA) and (aa) genotypes do not mate with one another to produce (Aa) offspring as often as predicted.

3. **Immigration or emigration:** Imagine a coquina population that is primarily of the (rr) genotype. If (RR) larvae are brought in by ocean currents from another population down the coast, then the frequency of the R allele will increase, and more purple-rayed shells will appear than we would have predicted based on the genotypes of adult clams on the beach. Emigration (or outward migration) can also affect some animal species if one genotype tends to migrate out of the population with greater frequency than another. Movement of genes via gametes, juveniles, or breeding adults is called **gene flow**, and these movements have a large potential influence on genotypic frequencies.

4. **Mutation:** Although mutation is a rare and usually deleterious event, it is important to realize that all new alleles arise by mutation from a pre-existing gene. Mutation makes a small initial impact on gene frequencies, but in conjunction with other factors, mutation can be the beginning of significant genetic changes in the population.

5. **Natural Selection:** Many shore birds feed on coquinas. What would happen if the purple-rayed coquina shells were a little harder for predatory birds to see in the surf than the pure white shells? In this case, we would say the **fitness** of the white coquinas is smaller than the fitness of the purple-rayed shells. If white shells are taken by birds at a disproportionately high rate, then we would expect a shift in gene frequencies away from Hardy-Weinberg equilibrium as clams producing (r) gametes declined, and the proportional representation of (R) gametes increased. Depending on the environment and the adaptive advantage of one genotype over another, natural selection can exert measurable effects on the gene pool. Whether selection occurs among sperm competing to reach the eggs, among juveniles competing to survive, or among adults competing to produce more offspring, any advantage of one allelic type over another will alter the genetic balance that Hardy and Weinberg described.

In summary, a population is expected to remain in Hardy-Weinberg equilibrium only if there is no natural selection, no mutation, no immigration or emigration, and only as long as mating is random and population size is very large.

Check your progress:

If RR and rr genotypes are more common in a large coquina population, and Rr genotypes are less common than we would expect under Hardy-Weinberg equilibrium, what biological causes could explain this observation?

Hint: You can rule out mutation,
non-random mating, and selection.

METHOD A: POPULATION GENETICS SIMULATION
[Laboratory activity]

Research Question
How do population size and natural selection influence gene frequencies?

Preparation
In a craft store, find beads of two colors having holes large enough to slip over a toothpick. For convenience, styrofoam blocks can serve as toothpick holders.

Materials (per laboratory team)
Calculator

1 500-ml beaker

100 light-colored beads

100 dark-colored beads

10 toothpicks

Procedure 1: Modeling Genetic Drift
1. Select 10 toothpicks to represent a population of 10 animals.
2. Place 50 dark-colored beads and 50 light-colored beads into a beaker. These represent genes in a gamete pool with allelic frequencies 0.5 A and 0.5 a. Stir up the beads to ensure random genetic recombination.
3. Calculate expected frequencies of AA, Aa, and aa individuals, based on Hardy-Weinberg expectations. Multiply these expected frequencies by 10 and round off to the nearest whole numbers to calculate how many AA, Aa, and aa individuals to expect in your population.
4. Without looking at the beaker, draw out two beads and slip them onto a toothpick. If you draw two dark beads, this animal has genotype AA. Two light beads means aa. One of each means Aa. Repeat this process, placing two beads on each of the 10 toothpicks. How do your actual genotypic frequencies compare to the expected frequencies calculated in step 3? Can you explain the reason for any discrepancies?

5. Calculate gene frequencies in your population by counting "genes" in your population of 10 animals: $p = f(A) = $ (# dark beads)/20, and $q = f(a) = $ (# light beads)/20. Record your result for p and q on the calculation page.

6. Take all of the beads off your 10 toothpicks, and generate a new gamete pool by changing the ratio of light and dark beads in the beaker to match the p and q values you calculated in step 5. Keep the size of the gamete pool constant at 100 beads, but take out or put in beads to create the correct proportions. For example, if your population has values $p = 0.55$ and $q = 0.45$, you would fill the beaker with 55 dark-colored beads and 45 light-colored beads. Since you started with 50 of each, return all beads to the beaker, then take out five light-colored beads and replace them with 10 dark-colored ones to make a new $p = 55/100$ and $q = 45/100$.

7. Repeat steps 4 through 6 a total of 10 times to simulate 10 generations of genetic drift. Record p and q values on the calculation page each time you recalculate them. Since departures from original gene frequencies are random, you cannot predict which way genetic drift will take your small population. If p becomes 1.00, we say that the A gene has gone to fixation and the a gene to extinction. Fixation of the a gene could also happen.

8. Plot changes in p on the graph at the end of this section. The first point, showing $p = 0.5$, is already entered on the graph. Draw similar points for p in each generation, and connect the points with a line to show how p changes over time due to genetic drift. Then compare your results with those of other lab groups. Do you find different results in different groups?

Table of Genetic Drift Results (Method A, Procedure 1)

Generation number	Observed f(RR)	Observed f(Rr)	Observed f(rr)	p value $= \dfrac{\text{(\# dark beads)}}{\text{(total \# beads)}}$	q value $= 1 - p$
				0.50	0.50
1					
2					
3					
4					
5					
6					
7					
8					
9					
10					

Graph of Genetic Drift Results

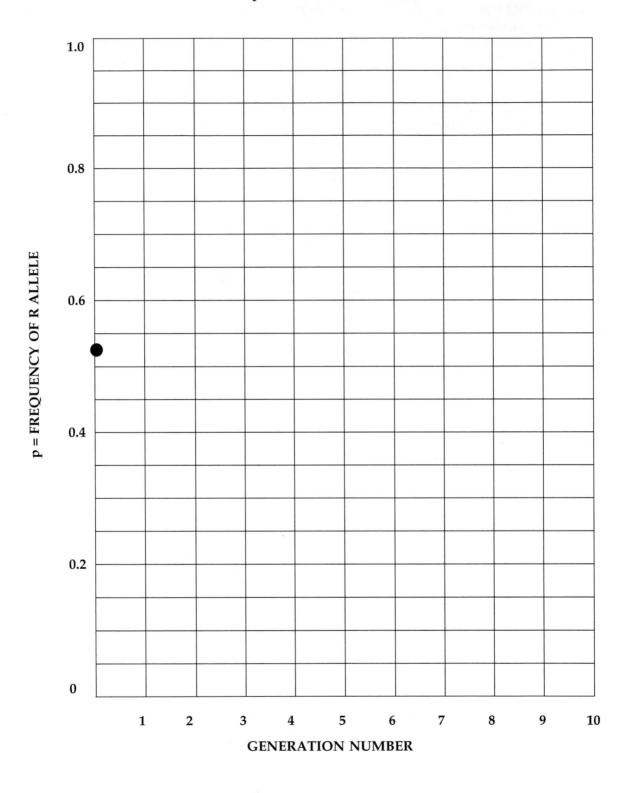

Questions (Method A, Procedure 1)

1. In horses, the genes for white coat color and red coat color are codominant. Heterozygotes have a light red coloration, called roan. If you located a population of wild mustangs in a valley that had 476 red horses, 323 roan horses, and 51 white horses, could you say the population is in Hardy-Weinberg equilibrium? First calculate p and q, then use the Hardy-Weinberg formula to calculate expected genotypic frequencies.

2. If we look at Rh factor genes in the United States, the dominant Rh^+ allele makes up about 60% of the gene pool and the recessive Rh^- allele makes up 40%. If the U.S. population were in Hardy-Weinberg equilibrium, what portion of people in the country would you expect to have Rh positive blood? What factors might result in departures from expected frequencies in this example?

3. Based on this exercise, can you explain why zoos go to so much trouble to exchange rare animals such as tigers or gorillas rather than maintaining small separate breeding populations?

_____

_____

_____

_____

_____

_____

_____

4. In a classic study of blood types in Italy, Dr. Luigi Cavalli-Sforza (1969) found that small, isolated towns in mountain regions had reduced genetic diversity within populations, but showed significant genetic differences from one town to the next. In valley regions with more movement of people from place to place, he found more diversity within the population of any given town, and fewer differences between towns. If you consider your genetic simulation to represent one mountain town, and other lab groups to represent different mountain towns, can you explain Cavalli-Sforza's results? How could you alter this lab procedure to simulate what happens in the valley towns?

    _____

    _____

    _____

    _____

    _____

    _____

    _____

5. Gene frequencies for the ABO alleles vary geographically. For example the I^A allele makes up nearly 50% of the genes for blood type among Australian Aborigines, about 25% in Central Asian populations, and 0% in peoples native to South America. No blood type has sufficient selective advantage over another to explain these regional differences. From these facts, what can you infer about the history of our species? Did humanity develop as a large and continuous population spreading out across the planet, or did we disperse while still living in small scattered groups?

    _____

    _____

    _____

    _____

    _____

    _____

    _____

Procedure 2: Modeling Selection

In this simulation, we will assume that a is a recessive lethal gene. It causes no harm in heterozygotes, but is fatal to aa individuals. Rather than a random change in gene frequency, here we will see how natural selection against one allele can make a directional change in gene frequencies.

1. Select 10 toothpicks to represent a population of 10 animals as you did in the first simulation.
2. Begin the simulation, as you did in Procedure 1, with 50 dark-colored beads and 50 light-colored beads in the beaker. These represent genes in a gamete pool with allelic frequencies 0.5 A and 0.5 a. Stir up the beads to ensure random genetic recombination.
3. Draw two beads to place on each toothpick as you did before. Look at the resulting genotypes. In this simulation, assume that aa genotypes produce a fatal genetic illness. Since A is dominant, Aa individuals do not have this disease. Record frequencies for all three genotypes in the table, then remove the aa toothpicks to represent selection against the aa genotype.
4. Calculate p and q for the surviving population. This time,
   p = f(A) = (# dark beads)/(all beads in surviving animals), and
   q = f(a) = (# light beads)/(all beads in surviving animals).
   Check your accuracy by confirming that p + q = 1. Record your values of p and q in the data table for natural selection results.
5. Remove beads from toothpicks and refill the beaker with dark and light beads to generate a gamete pool of 100 total beads, in the proportion of p and q that you just calculated. Mix the beads thoroughly.
6. Repeat steps 3 through 5 for 10 generations, recording the p and q values each time.
7. Complete graph 2 for the results of selection against a recessive lethal gene. The first point, showing p = 0.5, is already entered on the graph. Draw similar points for p in each generation, and draw a line connecting the dots to show the changes in p over time.

Table of Natural Selection Results (Method A, Procedure 2)

Generation number	Observed f(RR)	Observed f(Rr)	Observed f(rr)	p value $= \dfrac{\text{(\# dark beads)}}{\text{(total \# beads)}}$	q value $= 1 - p$
				0.50	0.50
1					
2					
3					
4					
5					
6					
7					
8					
9					
10					

Graph of Natural Selection Results

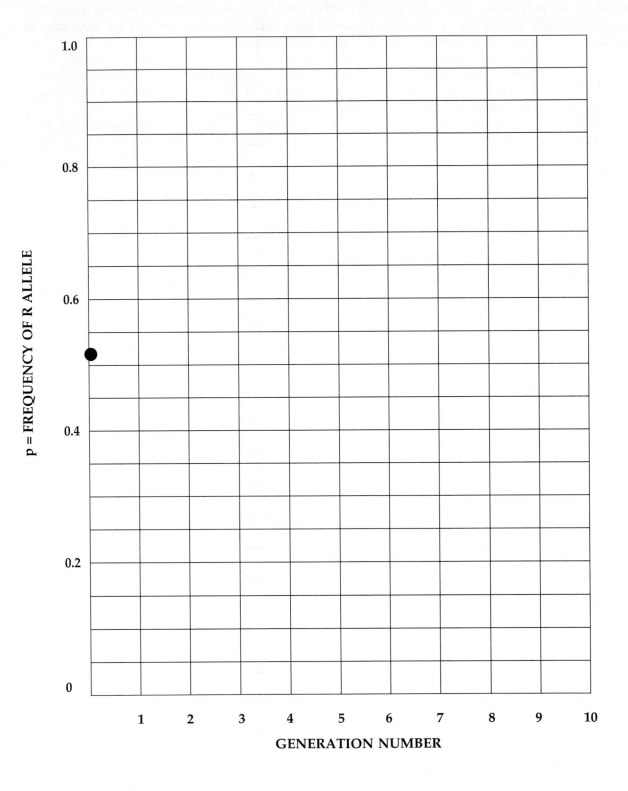

Questions (Method A, Procedure 2)

1. Most serious genetic diseases are caused by recessive alleles rather than dominant alleles. Based on this exercise, can you explain why a recessive lethal gene could persist in a population, while a dominant lethal gene could not?

    _____

    _____

    _____

    _____

    _____

    _____

2. Suppose your model of selection represents the change in frequency of plain white coquina clams because predatory birds see and remove them more quickly from a beach with dark-colored sand. If larval offspring from this population drift to a different beach made of light-colored sand, could selection go the other way? What does this say about fitness of a particular gene?

    _____

    _____

    _____

    _____

    _____

    _____

    _____

3. In this simulation, you established the fitness of the aa allele as 0, removing all aa individuals from the breeding population. This may be realistic in the case of a serious genetic disease, but selection is not always so drastic. In the case of coquinas on the beach, white shells may suffer a measurable disadvantage due to higher bird predation, but their fitness is obviously not zero. Some white clams do survive to reproductive age. Suppose predation rates on white clams were not total, but did occur at twice the rate of predation on purple-rayed clams. How would you alter this procedure to simulate a 50% survival rate of the aa genotype? How would you expect this change in the simulation rules to affect the outcome?

    _____

    _____

    _____

    _____

    _____

    _____

    _____

4. Many people incorrectly think of evolution as a completely random process. Based on the changes you observed in frequencies of alleles in this simulation, would you say natural selection is totally predictable, totally unpredictable, or something in between? Explain.

    _____

    _____

    _____

    _____

    _____

    _____

5. In this simulation, you sampled the gene pool without replacing beads in the beaker after you drew each one. Thus, f(A) and f(a) in the gene pool changed slightly after each bead was drawn. For example, if you begin with 50 light and 50 dark beads, the probability of drawing a dark bead the first time is $50/100 = 0.500$. The beaker would then contain 49 dark beads and 50 light beads, so the probability of drawing a second dark bead becomes $49/99 = 0.495$. Does this make your simulation slightly less realistic? In small natural populations, does one mating change the gene pool available for the next mating, or not? What biological factors must be considered in answering this question?

    _____

    _____

    _____

    _____

    _____

    _____

    _____

METHOD B: POPULATION GENETICS IN A FRUIT FLY CULTURE

[Laboratory activity]

Research Question

Do white and red eye color alleles exhibit Hardy-Weinberg equilibrium in a fruit fly population?

Preparation

Obtain a culture of wild-type (red-eyed) *Drosophila melanogaster* and a culture of the sex-linked white-eyed genotype. At least a month before this laboratory exercise, fill a large culture bottle to a depth of 4–5 cm with *Drosophila* medium. Add a few grains of dried baker's yeast if your growth medium does not already include it. You will need one culture bottle for each laboratory group. To each culture, introduce the same number of white-eyed and wild-type flies, making sure both males and females of both genotypes are included. You now have a polymorphic population with equal representation of the wild type and white alleles. Keep the cultures at room temperature (about $21°$ C) through two generations (about four weeks). If necessary, replace the culture medium by making up new bottles and moving all the adult flies from the old culture to the new.

Materials (per laboratory team)

A large fruit fly culture of mixed white-eyed and red-eyed genotypes

Dissecting microscope, preferably with 15× magnification

Small paintbrush for manipulating flies

5 × 7 index card for holding flies

Anesthetic for flies

Small fly vial for anesthesia

Calculator

Background

In this exercise, we will investigate frequencies of a sex-linked gene for eye color called white. The dominant allele, X^+, codes for a tomato-red pigmentation in the prominent compound eyes of the fruit fly. The recessive allele, X^w, produces a white eye. Since males have only one X chromosome, males are either X^+Y with red eyes or X^wY with white eyes. Assuming that males inherit X chromosomes in proportion to their frequency in the gene pool, the fraction of male flies exhibiting the white-eyed phenotype should indicate $f(X^w)$ in the population as a whole. Similarly, the proportion of males exhibiting the red-eyed phenotype should be a good indicator of $f(X^+)$.

Females, on the other hand, have two X chromosomes so they have three possible genotypes. Females with the X^+X^+ genotype are homozygous red-eyed flies. Females with the X^+X^w genotype are carriers of the white allele, but have red eyes, and X^wX^w females have white eyes. If you use the frequencies of X^+Y males vs. $X^w Y$ males to determine $p = f(X^+)$ and $q = f(X^w)$, you should be able to use the Hardy-Weinberg equations to calculate the frequencies of homozygous red, heterozygous, and homozygous white genotypes among the females.

Procedure

1. Get as many flies as you can out of the large culture and into an empty *Drosophila* vial. This is a critical step, because you do not want to introduce bias in your sample. You can place the vial over the mouth of the larger culture, tip both containers, and tap flies into the vial. Be careful, however, not to dislodge the *Drosophila* medium from the bottom of the culture. Alternatively, you can take advantage of fruit flies' tendency to walk upward and toward a light source. Cover the large container with paper or foil, place an empty vial over its mouth, and place the combined vessels, empty vial on top, under a laboratory lamp. With patience, you should be able to collect a large number of flies from the culture.

2. Use commercial fly anesthesia to put the flies to sleep. Although timing is critical for anesthesia if you want to revive the flies later for genetics crosses, it is not necessary for the purpose of this experiment to keep the flies alive. Make sure they are sufficiently anesthetized so that they do not awaken and fly away before you have time to count them.

3. When all the flies are well anesthetized, tap them out onto the 5 × 7 index card and place the card on the stage of the dissecting scope. Using your paintbrush, separate the flies by sex. Refer to Figure 6.5 to see differences between male and female *Drosophila*. The dark patches of bristles, called sex combs, on the front feet of males provide a good way to separate younger adult flies. Older males have sex combs too, but they also have a darker tip of the abdomen which makes them easier to distinguish from females.

4. When you have separated flies by sex, count the numbers of white-eyed vs. red-eyed males. Enter those numbers in the data table. Then count the numbers of white-eyed vs. red-eyed females. Enter those numbers as well.

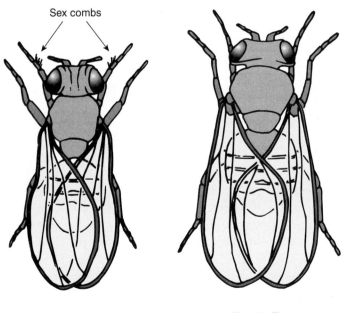

Sex combs

Male Fly   Female Fly

Figure 6.5   Fruit fly male (left) and female (right). Note differences in overall body size, banding pattern on the abdomen, and bundles of black bristles, called sex combs, on the front legs of the male.

5. Complete the data table for Method B, using frequencies of males to generate p and q. Then use the Hardy-Weinberg formula to calculate expected frequencies of red-eyed and white-eyed females.
6. Compare your observed frequencies of red-eyed and white-eyed females to the expected frequencies, based on values of p and q you calculated from the male data.

Table of Fruit Fly Population Genetics Results (Method B)

MALES	FEMALES
Number of red-eyed males collected from the population:	Number of red-eyed females collected from the population:
Number of white-eyed males collected from the population:	Number of white-eyed females collected from the population:
$p = f(X^+)$ = red-eyed males/total males	Observed frequency of red eyes in females = (red-eyed females/total females)
$q = f(X^w)$ = white-eyed males/total males	Observed frequency of white eyes in females = (white-eyed females/total females)
Use p and q from male data to calculate expected frequency of red-eyed females. $\Rightarrow$	Expected frequency of red eyes in females = $f(X^+X^+) + f(X^+X^w) = p^2 + 2pq$
Use q from male data to calculate expected frequency of white-eyed females. $\Rightarrow$	Expected frequency of white eyes in females = $f(X^wX^w) = q^2$

Questions (Method B)

1. How did your expected frequencies of red-eyed vs. white-eyed females compare with observed frequencies? Do your observations support the assumption that the female flies exhibit Hardy-Weinberg equilibrium? Explain.

    _____

    _____

    _____

    _____

2. Of the female red-eyed flies, how many would you expect to be homozygous X^+X^+ genotypes, and how many heterozygous X^+X^w, based on Hardy-Weinberg expectations? Show your calculations.

3. Consider each of the biological factors that can force a departure from Hardy-Weinberg equilibrium. Based on what you know about the fly culture, the behavior of fruit flies, and your observed phenotypic frequencies, which factors could have major effects on population genetics in this system? Which factors are less likely to exert a significant influence?

    _____

    _____

    _____

    _____

4. The Hardy-Weinberg model assumes no selection for or against the X^w allele. This may or may not be the case. Did you observe a 50:50 sex ratio in all flies? Was the number of white-eyed males equal to the number of red-eyed males? Based on initial frequencies of p and q = 0.5, these statements would be true in the absence of selection or drift. If these statements are not consistent with your observations, develop a hypothesis to explain what you see.

    _____

    _____

    _____

    _____

5. Based on the frequencies you calculated for heterozygous and homozygous females, what is the probability that a red-eyed female randomly chosen from this population and bred to a white-eyed male will produce a white-eyed offspring from the very first egg she lays? (Hint: list all statements that must be true for this to happen, along with the separate probability of each. Then using the sum rule, multiply the probabilities together to calculate the joint probability.)

FOR FURTHER INVESTIGATION

1. Repeat the genetic drift simulation in Method A with a population of 20 animals, represented by 20 toothpicks. Bear in mind that you will now need to divide the number of dark beads by 40 when calculating p. What difference do you see in the rate of genetic drift when the population size is doubled?

2. Sickle cell anemia, represented by the symbol (s), is a genetically determined change in the hemoglobin molecule, which carries oxygen in the blood. The sickle-cell trait is a good example of an allele whose selective value depends on the environment. Before modern medical therapies, the gene was often fatal in the homozygous ss condition, but Ss heterozygotes were protected from malaria. In many tropical regions, malaria is historically so serious that most SS genotypes would be too sick to survive and reproduce. In extreme cases, heterozygous Ss people would be the only survivors. Heterozygote advantage, also called **heterosis**, results in stable **polymorphism**. You can model this kind of natural selection using your toothpick animal system. Run the selection model as you did in Method A, Procedure 2, but remove both the homozygous recessive and the homozygous dominant individuals. You will need only a couple of generations to see what happens to gene frequencies. Do you think this kind of selection for heterozygotes could explain why so much polymorphism exists in natural populations?

3. As an independent study project, anesthetize flies carefully, and perform the above analysis as described. After sexing and counting your flies, select a random sample of 20 of the red-eyed female flies for further testing. You will need 20 culture vials, each with *Drosophila* medium prepared according to supplier's directions. Place each red-eyed female in a separate fly vial along with a white-eyed male. Allow several days for the flies to mate and for the females to lay eggs. As soon as you see larvae in the vials, remove the adults. The offspring will indicate the genotype of the female parents: X^+X^+ females will produce all red-eyed offspring, while X^+X^w females will produce half red-eyed and half white-eyed offspring. Draw out Punnett squares to demonstrate the two outcomes. After you have progeny in all 20 vials, use the data to calculate the proportion of X^+X^+ to X^+X^w females in the original population. Do these proportions conform to Hardy-Weinberg expectations?

4. For a long-term experiment, keep the polymorphic fly culture alive for several more weeks. Collect samples of flies periodically to see whether p and q shift due to drift or selection over time. Consider maintaining more than one culture for comparison so that you can determine whether genetic change is due to selection (where all bottles would be expected to change in the same direction) or drift (where the direction of change should vary at random).

FOR FURTHER READING

Cavalli-Sforza, L. L. 1969. Genetic drift in an Italian population. *Scientific American* 221(2):30–38.

Edgren, Richard A. 1959. Coquinas (*Donax variabilis*) on a Florida beach. *Ecology* 40(3): 498–502.

Lewin, Roger. 1982. Biology is not postage stamp collecting. *Science* 216:718–720.

Loew, S. 2002. Role of genetics in conservation biology. *Quantitative Methods for Conservation Biology* (Ferson, S. and M. A. Burgman eds.) Pp. 226–258. Springer-Verlag, N.Y.

Chapter 7
Spatial Pattern

Figure 7.1   Monarch butterflies congregate during their fall migration.

INTRODUCTION

Members of a population constantly interact with physical features of their environment, one another, and other species in the community. Distinctive **spatial patterns**, describing the distribution of individuals within their habitat, result from these interactions. Movements, family groupings, and differential survival create spatial patterns that vary from one population to another. A population can also change the way it is scattered through space as seasons or conditions change. As an example, monarch butterflies spread out to feed and reproduce during the summer, but congregate in dense assemblies during fall migration and winter dormancy (Figure 7.1). The physical arrangement of organisms is of interest to ecologists because it provides evidence of interactions that have occurred in the past, and because it can significantly affect the population's fate in the future. Analyzing spatial distributions can reveal a lot more about the organism's natural history than we could ever know from estimates of population size alone.

Since it is often impossible to map the location of every individual, ecologists measure features of spatial pattern that are of particular biological interest. One such feature is the dispersion of the population. **Dispersion** refers to the evenness of the population's distribution through space. (Dispersion should not be confused with **dispersal**, which describes movement rather than pattern.) A completely uniform distribution has maximal dispersion, a randomly scattered population has intermediate dispersion, and an aggregated population with clumps of individuals surrounded by empty space has minimal dispersion (Figure 7.2).

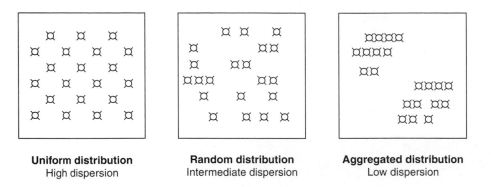

Uniform distribution
High dispersion

Random distribution
Intermediate dispersion

Aggregated distribution
Low dispersion

Figure 7.2   Three types of spatial distribution. Individuals spread evenly through the environment are highly dispersed, individuals clumped together exhibit low dispersion.

How can we measure dispersion in populations? A typical approach involves **quadrat** sampling. Quadrats are small plots, of uniform shape and size, placed in randomly selected sites for sampling purposes. By counting the number of individuals within each sampling plot, we can see how the density of individuals changes from one part of the habitat to another. The word "quadrat" implies a rectangular shape, like a "quad" bounded by four campus buildings. Any shape will work, however, as long as quadrats are all alike and sized appropriately for the species under investigation. For creatures as small as barnacles, an ecologist may construct a sampling frame a few centimeters across, and simply drop it repeatedly along the rocky shore, counting numbers of individuals within the quadrat frame each time. For larger organisms such as trees, global positioning equipment and survey stakes may be needed to create quadrats of appropriate scale.

The number of individuals counted within each quadrat is recorded and averaged. The mean of all those quadrat counts (symbolized as $\bar{x}$) yields the **population density**, expressed in numbers of individuals per quadrat area (barnacles per square meter, for example, or pine trees per hectare). An alternative approach is to measure **ecological density**, expressed in numbers of individuals per resource unit (numbers of deer ticks per host, for example, or numbers of apple maggots per fruit). To get a measure of dispersion in our population, we need to know how much variation exists among the samples. In other words, how much do the numbers of individuals per sampling unit vary from one sample to the next? The **sample variance** (symbolized as s^2) gives us a good measure of the evenness of our distribution. (For an introduction to the variance, see Appendix 1—Variance.) Consider our three hypothetical populations, now sampled with randomly placed quadrats (Figure 7.3).

Notice that the more aggregated the distribution, the greater the variance among quadrat counts. To standardize our measurements for different populations, we can divide the variance by the mean number of individuals per quadrat. This gives us a reliable way to measure aggregation. Statisticians have demonstrated that the **variance/mean ratio,** symbolized as $s^2/\bar{x}$ yields a value close to 1 in a randomly dispersed population, because in samples from a random distribution the variance is equal to the mean. Any ratio significantly greater than 1 indicates aggregation, and a ratio less than 1 indicates a trend toward uniformity. We could therefore call the variance/mean ratio an index of aggregation, because it is positively related to the "clumping" of individuals in the population. The variance/mean ratio is also called an **index of dispersion**, even though dispersion is inversely related to $s^2/\bar{x}$. It is good to remember: a high value of $s^2/\bar{x}$ means high aggregation, but low dispersion.

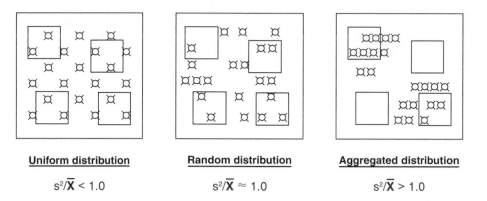

Uniform distribution   **Random distribution**   **Aggregated distribution**

$$s^2/\overline{X} < 1.0 \qquad s^2/\overline{X} \approx 1.0 \qquad s^2/\overline{X} > 1.0$$

Figure 7.3   Quadrat sampling allows measurement of dispersion by counting numbers of individuals within each sampling frame, and then comparing the variance to the mean. Note that the mean number per frame is the same for all three patterns, but variance increases with aggregation.

Bear in mind that the size of the sampling frame can significantly influence the results of this kind of analysis. A population may be clumped at one scale of measurement, but uniform at another. For example, ant colonies represent dense aggregations of insects, but the colonies themselves can be uniformly distributed in space. Whether we consider the distribution of ants to be patchy or uniform depends on the scale of our investigation. Figure 7.4 illustrates a population that would be considered uniformly distributed if sampled with large quadrats, but aggregated if sampled with smaller quadrats. For organisms distributed in clusters, the $s^2/\overline{x}$ ratio will be maximized when the size of the sampling frame is equal to the size of the clusters.

Check your progress:

If densities are equal, which would yield a higher variance in numbers of individuals per quadrat: a highly aggregated population or a highly dispersed population?

Answer: High variance in relation to the mean results from highly aggregated spatial pattern.

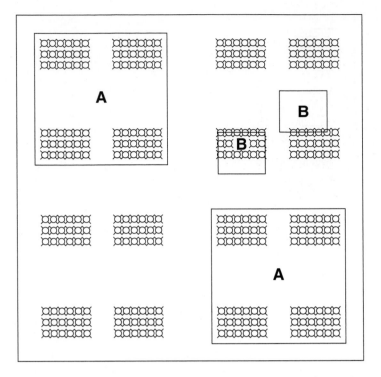

Figure 7.4   The calculated index of dispersion depends on the size of the quadrats used to sample the population. This hypothetical spatial pattern would exhibit a uniform dispersion index if sampled with large quadrats (A), but a clumped dispersion index if sampled with small quadrats (B).

The significance of aggregation or dispersion of populations has been demonstrated in many kinds of animal and plant populations. **Intraspecific competition**, for example, tends to separate individuals and create higher dispersion. Territorial animals, such as male robins on campus lawns in the spring, provide an excellent example. As each male defends a plot of lawn large enough to secure food for his nestlings, spaces between competitors increase, and the population becomes less aggregated. Competition can also create uniform plant distributions. In arid habitats, trees and shrubs become uniformly distributed if competition for soil moisture eliminates plants growing too close together (Figure 7.5).

Figure 7.5 Saguaro and cholla cactus exhibit highly dispersed spatial patterns because of intense competition for water in a desert ecosystem.

If organisms are attracted to one another, their population shows increased aggregation. Schooling fish may limit the chance that any individual within the group is attacked by a predator. Bats in temperate climates conserve energy by roosting in tightly packed groups (Figure 7.6). Cloning plants and animals with large litter sizes create aggregation as they reproduce clusters of offspring. For example, the Eastern wildflower called mayapple generates large clusters of shoots topped by characteristic umbrella-like leaves as it spreads vegetatively across the forest floor (Figure 7.7). By setting up quadrats, and calculating the variance/mean ratio of the quadrat counts, you can gain significant insights about the biology of your organism.

Check your progress:

Give an animal example of high dispersion; of low dispersion.
Give a plant example of high aggregation; of low aggregation.

Hint: Dispersion is the opposite of aggregation.

Figure 7.6 Lesser horseshoe bats roost in clusters to conserve body heat.

Figure 7.7 Mayapple (*Podophyllum peltatum*) forms highly aggregated stands through asexual reproduction.

METHOD A: DISPERSION OF PLANTS IN A LAWN COMMUNITY

[Outdoors, spring, summer, or fall]

Research Question

What can we infer about the natural history of lawn species from their spatial distribution?

Preparation

Before laboratory, carefully examine lawns on campus. Regardless of maintenance efforts, few lawns are actually monocultures. Almost all lawn communities include some broad-leaved plants such as dandelions, plantain, or clover growing among the turf grasses (Figure 7.8).

Check with your facilities management department to ensure that pesticides have not been applied to campus lawns within a few days prior to this laboratory.

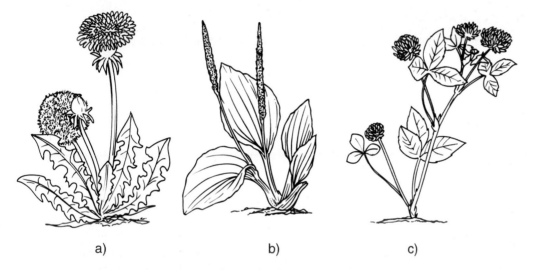

a)   b)   c)

Figure 7.8   Three common broad-leaved lawn weeds: (A) dandelion (*Taraxicum*), (B) Plantain (*Plantago*), and (C) clover (*Trifolium*).

Materials (per laboratory team)

1 large nail (#16 galvanized or aluminum gutter nail)

1 meter stick

1 piece of nylon string, about 1-1/2 m long

Procedure

1. Make quadrat sampler by tying one end of the string around the nail, tightly enough to stay on, but loosely enough to swivel around the head (Figure 7.9). Then using the meter stick, mark a point on the string 56 cm from the nail by tying an overhand knot at that position. Repeat the procedure to make a second knot 80 cm from the nail, then a third knot 98 cm from the nail, and a fourth knot 113 cm from the nail. The distance from the nail to each knot will become the radius of a circle in your sampling plots. Distance to the first knot represents the radius of a circle of area 1 m^2. (Try verifying this calculation, using the formula Area = π r^2 for a radius of 0.56 m.) The knots farther along the string will be used to sample circles of areas 2 m^2, 3 m^2, and 4 m^2, respectively. Take your sampler with you to a lawn area.

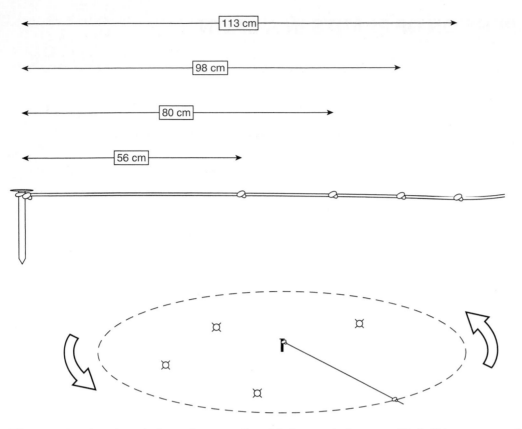

Figure 7.9 A string tied to a large nail, with knots tied at specified distances, can be used to sample a fixed area of lawn. Put the nail in the ground and pull the string taut. Moving the knot around the nail, count how many of your organisms fall within the circle.

2. Choose one lawn species exhibiting an interesting spatial pattern and common enough to find some specimens growing less than a meter apart. Decide what vegetative unit of this plant you will designate as an individual for the purpose of counting plots. For non-cloning plants such as dandelions, one rosette of leaves constitutes one individual. For cloning plants such as violets, choose a unit of plant growth, such as a shoot, as an arbitrary unit of population size.

3. Choose an area of lawn for sampling in which this species is relatively common. Before taking any samples, observe physical features of the habitat such as shade, soils, or small dips or mounds affecting water runoff that might help you interpret the pattern you see. Develop hypotheses relating the reproductive history of your species and habitat features with the distribution you are measuring.

4. Next, you must select sites for quadrat samples within your study area. You can obtain a fairly unbiased sample by tossing the nail within the sample area without aiming for any particular spot, and then pushing it into the soil wherever it lands. Hold the string at the first knot and stretch it out taut from the nail.

5. Now move the string in a circle (Figure 7.9). The length marked by your closest knot becomes a radius of a circular quadrat with area 1 m^2. If this circle is too small to include several individuals, move out to the second knot for a 2-m^2 quadrat, the third knot for a 3-m^2 quadrat, or the fourth knot for a 4-m^2 quadrat, as needed. After you decide on the appropriate scale, use the same size quadrat for all your samples.

6. As you move the string in a circle, count how many individuals fall within this quadrat. When the circle is complete, record this number in the Data Table for Methods A, B, and C. On the third column of the table, make notes of any landscape features that may affect your plant. Pull out the nail, make another toss to relocate your circular plot, and repeat for a total of 20 samples. Your sampling is complete when you have recorded 20 quadrat counts.

7. Analyze results on the Calculations for Methods A, B, and C page following the data table. By comparing the variance of your 20 quadrat counts with the mean, you will determine whether the plants you sampled are aggregated, random, or uniformly dispersed.

METHOD B: DISPERSION OF HERBIVORES ON PLANT RESOURCE UNITS

[Indoors/outdoors, employs biological collections, year-round]

In this exercise, we make use of plant parts serving as resource units for herbivorous animals. Leaves are resource units for plant-eating insects. Seeds and fruits, such as acorns or crab apples, serve as resource units for seed predators and fruit-eaters. An even-aged stand of campus trees can be considered a series of resource units for foraging birds or nesting squirrels.

Rather than constructing quadrats to count individuals per unit of space (absolute density), we will count the number of animals per resource unit (ecological density). By counting animals or animal signs in each plant part in a collected sample, we can calculate a variance as well as a mean, and get a determination of the aggregation or dispersion of animals among their resource units. As seen with quadrat counts in the Introduction, analysis of pattern among resource units can yield interesting insights into the population biology of the species.

Research Question

When animals choose resources, or insects choose egg-laying sites, do they randomly disperse themselves (or their offspring) among resource units?

Preparation

Campus landscapes provide many options for collecting plant parts bearing evidence of associated animal species (Figure 7.10). Leaves of almost all temperate deciduous trees are riddled with leaf miner tunnels and insect galls in late summer. Close examination of nuts or acorns almost always reveals small, round emergence holes made by seed-eating weevils or bruchid beetles, which have matured inside the seed and escaped through the seed coat. Seeds are easy to preserve, so the instructor can make a large collection (on or off campus) to use repeatedly for this exercise. Leaf samples can be preserved in a plant press, and maintained as a permanent collection, provided the brittle specimens are handled carefully. (It is best not to use leaves from existing herbarium pages, since these tend to be biased toward perfect specimens without visible insect damage.)

a   b

Figure 7.10   Ecological density is measured as numbers of individuals per resource unit. Insects on plant parts provide good study organisms: (A) Galls on maple leaves, (B) Acorn weevil larvae.

Alternatively, students may venture out onto the campus to collect their own specimens as part of this laboratory. Fall is best for collecting leaves, but seeds can be found throughout the year. In temperate locations, winter is the best time to find squirrel nests, which look like balls of leaves, in the bare branches of large deciduous trees. Empty bird nests are also apparent in winter and early spring, and provide a good index of ecological density if enough of them can be found. In spring, the class may find aphids or scale insects, which can be counted under a dissecting microscope. Once the collection is brought back to the laboratory, this exercise requires little time for data collection.

Materials (per laboratory team)

Collection of 20 resource units, or data from 20 resource units collected outside

Dissecting microscope or hand lens for small organisms

Procedure

1. If making your own collection, observe or collect 20 plant parts of equivalent size. Decide on a sampling plan in advance, so that the specimens are chosen with minimal bias. If using a laboratory collection such as a box of acorns, select 20 plant parts in a random sampling plan.
2. Count animals, or evidence of animal activity, on each plant part. Observe any life history observations you can, to help interpret intraspecific interactions that result in a pattern of dispersion among resource units.
3. Record numbers of animals for each of the 20 resource units as a data column in the Data Table for Methods A, B, and C at the end of this chapter. Bear in mind that one plant resource unit is a sampling unit in your study. Complete the Calculations section following the table, and answer the questions to discuss your results.

METHOD C: DISPERSION OF ISOPODS IN AN ARTIFICIAL HABITAT

[Laboratory, year-round]

Research Question

Do terrestrial isopods seek out conspecifics when selecting a habitat, or do they disperse to partition resources more evenly by avoiding habitat spaces occupied by others?

Preparation

Terrestrial isopods (Figure 7.11) include "pill bugs" (genus *Armadillium*), which roll into a ball when threatened, and the flatter and less flexible "sow bugs" (genus *Oniscus*). These arthropods are widely distributed, and are easily collected from leaf litter, under rocks and logs, or in lumber piles around buildings in most of North America. They can also be obtained from laboratory supply companies. Isopods feed on decaying organic material and are not difficult to handle or maintain in the laboratory. Use a covered container with some air exchange, such as a terrarium, plastic food container with perforated lid, or plastic shoe box. You might find a spoon and a small paintbrush handy to "sweep" them up and transport them from one place to another. It is very important to keep damp leaf mold or moistened paper toweling in their habitat at all times, because isopods use gills for respiration, and require high humidity.

If you are turning over logs, rocks, or lumber piles in search of isopods, remember that more aggressive animals, such as scorpions and venomous snakes, frequently hide in the same places. It is always wise to lift the side of a rock farthest from you first, and to be careful where you place your hands. When you are finished searching in a natural setting, replace rocks and logs where you found them.

Figure 7.11   Terrestrial isopods, also called "pill bugs" (genus *Armadillium*), roll into a ball when threatened.

Materials (per laboratory team)

Plastic sweater box or other box-shaped container with a flat bottom, about 30 × 50 cm

5 medium potatoes

3–4 inch paring knife

60 map pins (with round heads, about 5 mm diameter)

20–30 live terrestrial isopods

Cutting board or cutting surface on lab bench

Procedure

1. Prepare "resource islands" from potatoes. First, cross-section a potato with parallel cuts evenly spaced 1 cm apart. You should get four or five round sections from each potato. Discard the ends, and cut more potatoes until you have 20 sections of approximately equal size and thickness. Then insert three map pins in a triangle formation on the bottom of each potato slice, so that the potato slice will stand on the map pins like a three-legged stool (Figure 7.12). The pins should leave about $\frac{1}{2}$ cm of space for the isopods to crawl underneath.

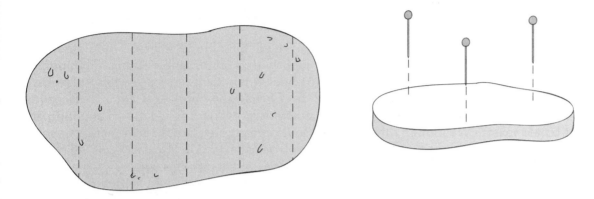

Figure 7.12   Slice potatoes and insert pins to make isopod shelters.

2. Arrange the 20 potato slices, pinheads down, in the bottom of the sweater box. Distribute the potato slices uniformly, with space separating each one. You now have an artificial habitat with 20 resource islands suitable for isopod colonization.
3. Release 20–30 isopods into the box. Take a few minutes to observe their behavior. How do they react to the box edges? to the potato "islands"? to one another?
4. Leave the isopods overnight. The underside of the potato slice islands provide a dark, humid microclimate that is preferred by isopods. They also feed on the potatoes. If lights are left on in the room, all isopods should select one of the potato slices to hide under.
5. After the isopods have had time to adjust to their habitat (12–48 hrs), count the number of isopods underneath each potato slice. You will need to pick up each "habitat island," because the isopods often invert themselves and cling to the potato. Record 0 if there are none in that sample. The 20 numbers you record should sum to the number of isopods you released into the box.
6. Complete the Data Table and calculation page at the end of this chapter. The mean number of isopods per potato slice ($\bar{x}$) is the ecological density, and the variance/mean ratio provides an index of aggregation for the captive isopod population.
7. Interpret your results by answering Questions for Methods A, B, and C.

Data Table for Methods A, B, and C

Sampling Unit	No. of individuals in the sample	Note any habitat features associated with this sample.
1		
2		
3		
4		
5		
6		
7		
8		
9		
10		
11		
12		
13		
14		
15		
16		
17		
18		
19		
20		

CALCULATIONS FOR METHODS A, B, AND C

1. Enter your 20 counts of organisms per sampling unit (x_i) in the second column of the table below.
2. Calculate a mean ($\bar{x}$) by summing all the counts and dividing by the sample size ($n = 20$).
3. Subtract the mean from each data value to obtain the deviation from the average (d_i).
4. Square each deviation (d^2)$_i$. (Note that this step takes care of the negative signs.)
5. Add up all the squared deviations (Σd^2).
6. Divide the sum of d^2 values by (sample size − 1) to calculate the sample variance (s^2).
7. Finally, divide the variance by the mean ($\bar{x}$) to compute the variance/mean ratio ($s^2/\bar{x}$).
8. Refer to the Introduction, and to the methods section you used, to interpret this ratio.

Sample number (i)	No. of organisms in sample i (x_i)		Deviations (d_i)		Squared deviations $(d^2)_i$
1		$- \bar{x} =$		$\char`\^2 =$	
2		$- \bar{x} =$		$\char`\^2 =$	
3		$- \bar{x} =$		$\char`\^2 =$	
4		$- \bar{x} =$		$\char`\^2 =$	
5		$- \bar{x} =$		$\char`\^2 =$	
6		$- \bar{x} =$		$\char`\^2 =$	
7		$- \bar{x} =$		$\char`\^2 =$	
8		$- \bar{x} =$		$\char`\^2 =$	
9		$- \bar{x} =$		$\char`\^2 =$	
10		$- \bar{x} =$		$\char`\^2 =$	
11		$- \bar{x} =$		$\char`\^2 =$	
12		$- \bar{x} =$		$\char`\^2 =$	
13		$- \bar{x} =$		$\char`\^2 =$	
14		$- \bar{x} =$		$\char`\^2 =$	
15		$- \bar{x} =$		$\char`\^2 =$	
16		$- \bar{x} =$		$\char`\^2 =$	
17		$- \bar{x} =$		$\char`\^2 =$	
18		$- \bar{x} =$		$\char`\^2 =$	
19		$- \bar{x} =$		$\char`\^2 =$	
n = 20		$- \bar{x} =$		$\char`\^2 =$	
Total # organisms $\sum (x_i) =$		Sum of Squared Deviations $\sum (d^2)_i =$			
Mean $\sum (x_i)/n =$		Sample Variance $s^2 = \sum (d^2)_i/(n-1) =$			
	Variance/Mean $(s^2/\bar{x}) =$				

Questions for Methods A, B, and C

1. Based on the variance/mean ratio, what can you conclude about the spatial pattern of your population? How might you explain this pattern, given observations you made as you were sampling?

_____

_____

_____

_____

_____

2. Random sampling is very important if the data you collected are meant to represent a larger population. In retrospect, do you have any questions or concerns about the validity of the sampling method? If bias exists, how might you alter your method to randomize your samples?

_____

_____

_____

_____

_____

3. An index of aggregation is maximized in patchy distributions if the size of the quadrat is the same as the size of the organism's aggregations. Might a larger or smaller sampling unit (or a different sized resource unit) have affected your results?

_____

_____

_____

_____

_____

4. Would you expect another organism from the same biological community to exhibit a similar index of dispersion? Is spatial pattern a property of the organism, or of its habitat?

_____

_____

_____

_____

5. Lesser horseshoe bats (Figure 7.6) and many other bat species are increasingly rare. How does a highly aggregated distribution, for at least part of the life cycle, influence the status of endangered species?

_____

_____

_____

_____

FOR FURTHER INVESTIGATION

1. In Method A, does spatial pattern vary among species of lawn plants, or in different parts of the campus? What does spatial pattern tell you about community interactions?
2. In Method B, how would you expect population density to affect the pattern of resource use by your herbivore? It might be possible to compare collections from different years or from different sites to address this question.
3. In Method C, how might the addition of predators such as centipedes affect patterns of settlement by an isopod population in the laboratory? If the same population is sampled repeatedly over a period of time, does its pattern of dispersion change? What environmental factors influence these changes?

FOR FURTHER READING

Eliason, E. A. and D. A. Potter. 2001. Spatial distribution and parasitism of leaf galls induced by *Callirhytis cornigera* (Hymenoptera: Cynipidae) on Pin Oak. *Environmental Entomology* 30 (2): 280–287.

Krebs, C. J. 1998. *Ecological Methods* (2nd edition). Menlo Park (Benjamin Cummings).

Chapter 8

The Niche

Figure 8.1 White trout lily, *Erythronium albidum*, is an Eastern forest species adapted to the spring ephemeral niche.

INTRODUCTION

Erythronium albidum, the white trout lily (Figure 8.1), is an Eastern North American wildflower whose form and life history are well suited for its deciduous forest environment. The plant stores carbohydrates in an underground bulb, accumulating energy from one year to the next, until it has amassed sufficient resources to bloom. Large white petals clearly advertise trout lily's presence in a shady habitat, and its flower shape and nectar composition are attractive to the early spring community of insect pollinators.

Ecologists have long noted the exquisite fit between organisms and their surroundings. When we speak of the organism's apt response to a particular environment, we use the term **biological adaptation**. Looking at the organism/environment fit from the other side, it is instructive to ask what kind of environment would accommodate an organism with a given set of physical, developmental, and behavioral characteristics. We refer to an organism's environmental requirements, including both living and nonliving factors, as its ecological **niche**. The niche is a biological metaphor borrowed from a much older architectural concept. In architecture, a niche is a recessed shelf built into a wall, just the right size and shape to accommodate a statue or religious icon (Figure 8.2). By analogy, the ecological niche describes a favorable set of conditions allowing an organism to take its place in the biological community. Although this analogy is useful in describing biological adaptation, it should not be carried too far. The architectural niche is a physical space measured along the axes of height, depth, and width, but the ecological niche transcends physical dimensions.

Figure 8.2 An architectural niche is just the right size and shape to accommodate a statue.

For example, white trout lilies occupy physical positions on the forest floor, but their niche also includes the dimension of time. Trout lilies and other early wildflowers are called **spring ephemerals** because they grow and bloom for only a few weeks in March or April, and then die back to the ground, remaining dormant for the rest of the summer. Since photosynthetic plants must have light, these little wildflowers could never compete directly with the giant trees shading the forest floor. To collect enough energy to survive and reproduce, ephemerals mobilize resources stored in their roots to develop spring leaves very quickly after the soil temperature rises above the freezing point. Small plants in temperate deciduous forests enjoy a brief advantage at this time of year, because solar radiation warms the surface of the ground faster than the air. Spring ephemerals take maximum advantage of their "head start" to collect and store a year's worth of solar energy before their larger competitors can open buds and unfurl their canopy of leaves overhead. Thus, the spring ephemeral niche is carved out of the photosynthetic opportunity temporarily available between spring thaw and the leafing out of canopy trees. Figure 8.3 illustrates the shape of this niche along a time axis.

In addition to the dimensions of physical location and time, the niche of an ephemeral wildflower is described by many other resource axes. Soil pH must be within limits not too acidic and not too alkaline. Air temperature must be above the minimum warmth needed for enzyme functioning, but below a maximum that interferes with carbon dioxide fixation in photosynthesis. Rainfall must keep the soil moist, but not so wet that the roots are deprived of oxygen. To follow the reasoning of ecologist G. Evelyn Hutchinson, if niche space in three dimensions could be described as a volume, an actual organism's niche described along a larger number (call it n) of resource dimensions would occupy an "n-dimensional hypervolume." N-dimensional objects are hard to draw, but you can imagine the niche stretching out to different lengths along a number of different axes representing all the environmental factors important to an organism's survival. We generally concentrate on one or two of these niche dimensions at a time, so the niche can be illustrated with a simplified graph, as seen in Figure 8.4.

Check your progress:

Sometimes ecologists say that the habitat is an organism's address, and the niche is its occupation. What is the "address" of the white trout lily? What is its "occupation"?

Hint: The organism's occupation is its way of using available resources.

A significant shortcoming of the niche analogy is that an architectural niche maintains the same size and shape, whether or not a statue sits within it. In contrast, ecological niches do not maintain a constant size and shape independent of their occupant. In fact, it is arguable whether the ecological niche even exists in the absence of an organism, because the bounds of the niche are established, at least in part, by interactions between the organism and other members of the biological community. For a more realistic representation of the dynamic nature of niches, think of the niche as a soap bubble. A bubble floating through the air by itself assumes a maximum volume and self-determined shape. Like the bubble, a biological niche extends outward to an organism's maximal tolerance limits along all resource dimensions as long as there are no complicating interactions with competing species. Maximum use of resources is demonstrated by spring ephemerals grown in open, sunny gardens, which achieve denser growth and stay in bloom longer without the limitations imposed by overshadowing trees. Ecologists use the term **fundamental niche** to describe the resource dimensions an organism could theoretically occupy in the absence of competition.

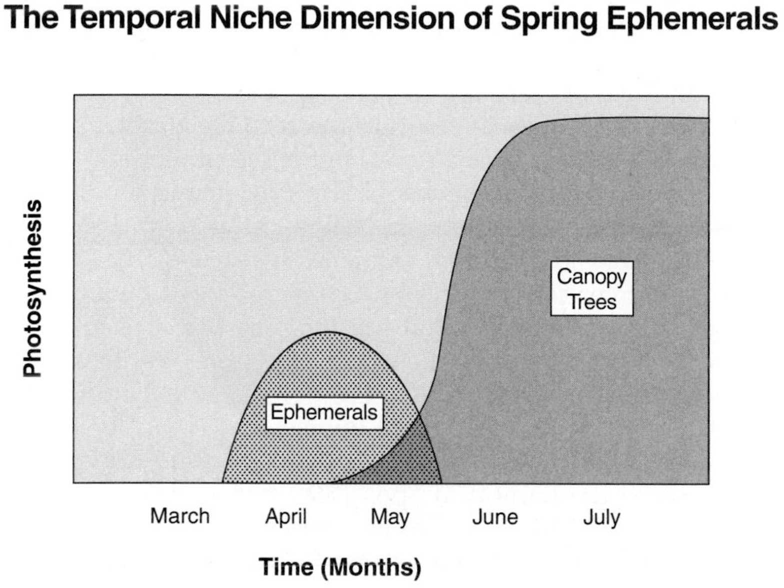

Figure 8.3 Spring ephemerals occupy a niche separated from that of canopy trees along a temporal resource axis.

But what happens in a community filled with competing species? Examine the close-up photograph of soap suds in Figure 8.5. When bubbles are packed closely together, each bubble's shape is influenced by the size, location, and pressure of other bubbles around it. This is a good way to visualize niche spaces in a community.

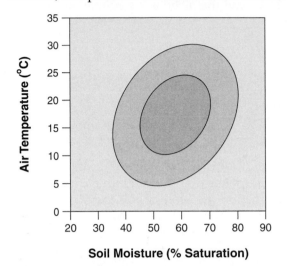

Figure 8.4 Two dimensions of the niche of a generalized spring ephemeral wildflower are moisture and temperature. The inner (darker colored) oval represents optimal conditions, under which the plant can grow and reproduce. The outer (lighter colored) oval represents marginal conditions under which the plant can survive, but without increasing its population size. The tilt in the axis of the oval indicates interaction between resource dimensions: when weather is warmer, plants need more soil moisture to replace water evaporating from their leaves. Although only two dimensions are illustrated here, a plant's niche includes other resource dimensions, such as soil texture, soil nutrients, pH, shade, pollinator numbers, and pressure from herbivores. Any two of these could be illustrated on a graph of this type.

In the presence of competition, the niche space of an organism shifts and contracts, just as crowded bubbles are squeezed and contorted by their neighbors. The altered shape of the niche in the presence of competition is called the **realized niche**. The soap bubble picture represents the dynamic nature of realized niches in a community, in which each species interacts with many others to claim its share of the total resource space. If another bubble forces its way into the foam (representing introduction of a new species), all the others must contract or move a little to make room. If one bubble pops (representing local extinction of a species) the entire niche space is rearranged, and neighboring bubbles expand to take up the vacancy. This is why "empty niches" are rarely observed in nature.

Close observation of natural communities can help identify how niches are divided among competing species. In 1958, Robert MacArthur documented this kind of **niche partitioning** by carefully observing small songbirds called warblers in Northeastern spruce forests. Five warbler species feed on insects in spruce trees, but MacArthur noticed that the species coexist in this habitat by specializing on different parts of the tree. For example, Yellow-rumped warblers feed on the trunk and bottom branches of spruce trees; Black-throated green warblers feed near the middle of the tree; and Cape May warblers feed at the top. Bay-breasted warblers feed on the bare branches near the tree's center, while Blackburnian warblers feed on new needles out at the perimeter. Like closely-packed bubbles, each warbler species occupies its own position in niche space, seeking food from a broader microhabitat if possible, but competing most effectively within the combination of conditions and resources that it exploits more effectively than any other bird.

Check your progress:

For the white trout lily, what factors determine the fundamental niche? What additional factors determine the realized niche? Why is the fundamental niche larger?

Hint: Consult Figure 8.3. What might happen to the ephemeral niche if trees are thinned by an ice storm?

Experimental studies clearly demonstrate the dynamic nature of niches. J. C. Holmes (1973) studied niches of worms that parasitize the intestines of rats. The mammalian intestine is a good experimental environment, since the chemistry of intestinal contents exhibits a gradient from anterior (nutrient rich) to posterior (nutrient poor). Holmes demonstrated that two types of parasites compete for attachment sites in the rat intestine. Tapeworms exploit a generalist niche, attaching themselves in a fairly broad range of environmental conditions from the anterior to the posterior end of the intestinal tract. Spiny headed worms, called acanthocephalans, were shown to be more specialized, concentrating their attachments in the anterior region. When a rat already infected with tapeworms was subsequently co-infected with acanthocephalans, the tapeworm niche contracted (Figure 8.6). The broad fundamental niche of the tapeworm shrinks to a significantly smaller realized niche in the presence of more specialized acanthocephalan competitors.

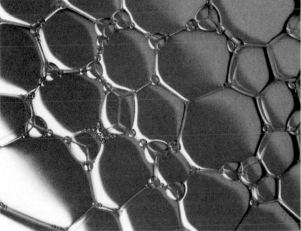

Figure 8.5   Soap bubbles packed tightly together provide a conceptual model for niches in nature.

Check your progress:

What prevents a warbler from feeding outside of its niche when other warbler species are present?

Hint: Aggressive interactions between species are sometimes involved, but not always.

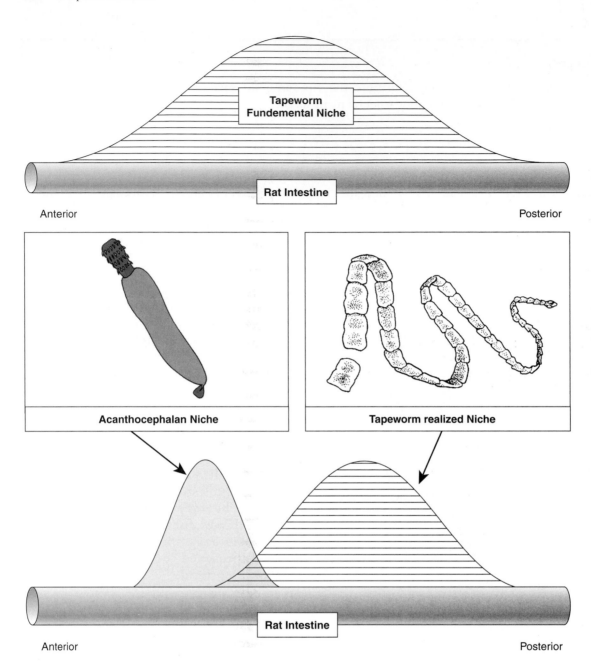

Figure 8.6 Holmes demonstrated a niche shift when two species of parasitic worms compete for attachment sites in the rat intestine (top figure). With a single-species infection, tapeworms exhibit a broad fundamental niche. When faced with competition from acanthocephalans (bottom figure), the tapeworm's realized niche contracts toward the posterior of the intestine. Peaks represent frequencies of parasite attachment.

In a study of the rocky intertidal zone on the coast of Scotland, Joseph Connell (1961) demonstrated that competition between two barnacle species causes niche contraction along an environmental gradient established by the tides. The smaller of these barnacles was in the genus *Chthamalus*, and the larger was *Balanus* (subsequently reclassified as *Semibalanus*). The intertidal zone is the part of the rocky shore extending from the extreme low tide mark to the extreme high tide mark. A gradient perpendicular to the shoreline runs from lower sites more frequently covered by the sea to higher sites seldom covered. Barnacles are easily observed subjects for experimentation because they disperse as larvae in seawater, and then attach permanently to the rocks to grow to adulthood. Because barnacles need to be covered by water to feed and bathe their gills, the higher parts of this intertidal gradient are more stressful for them. Through transplant experiments and protective cages, Connell showed that the smaller *Chthamalus* barnacle is able to establish itself anywhere from the extreme high tide zone to the mean tide zone of the tidal gradient. (Predatory snails removed them from the area below the mean tide mark.) Valves at the top of the shell allow *Chthamalus* to keep moisture inside during relatively long periods of exposure, so it can survive right up to the top of the tidal gradient. In ecological terms, we would say *Chthamalus* has a broad fundamental niche. However, because of competition with the larger and faster growing *Semibalanus* barnacles that dominate the bottom and center parts of the intertidal gradient, *Chthamalus*'s realized niche is confined to the high tide zone (Figure 8.7). *Semibalanus* cannot live in the upper reaches of the intertidal gradient, because it cannot withstand desiccation as well as *Chthamalus* can. It is often the case that harsh environmental conditions create a refuge for species less able to compete, but more able to withstand the rigors of a marginal habitat. To paraphrase the old Scottish song, *Chthamalus* takes the "high road" and *Semibalanus* takes the "low road," so each species persists in its own niche on the rocky shore.

Check your progress:

In the presence of competitors, niche contraction often occurs. How might a species change its niche over time?

Hint: Behavioral, developmental, and evolutionary responses can all occur, at different time scales.

Figure 8.7 Tidal zones on the coast of Scotland define a resource axis (white line) for barnacle niches. *Chthamalus* is able to survive anywhere from extreme high tide to extreme low tide in the absence of competition. However, when the larger barnacle *Semibalanus* is present, it crushes or overgrows the smaller *Chthamalus*. The smaller species is thus confined to a realized niche in the high tide zone. *Semibalanus* cannot live as far from the sea because its shell does not hold in moisture as well as *Chthamalus's* shell.

METHOD A: FEEDING NICHES IN THE COMMUNITY AQUARIUM

[Laboratory activity]

Research Question

Do common aquarium fish species exhibit different feeding niches?

Preparation

Several weeks in advance of this laboratory, establish a 10-gallon (or larger) freshwater community fish tank in a place convenient for student observations. A simple design, with under-gravel filter, about an inch of natural aquarium gravel, a few stones, aquarium plants, a heater to maintain water temperature between 72 and 78° F, and moderate aeration should provide good results. A hood that opens at the top for feeding is essential for this experiment. If you use tap water to fill the tank, let the water age for a few days to allow chlorine to evaporate from the water before adding fish. Don't overstock the tank; a conventional guideline is one inch-long fish per gallon of water in the tank. For this exercise, three to five species are sufficient. Put fish in the tank at least a week before this laboratory, to give them time to adjust to the new environment. Feed the fish once a day with flake food, the same kind to be used in the experiment.

If you would like to simulate a more natural fish community, choose fish originating from a common geographic region. For example, the following fish are all native to the Amazon basin:

Guppy	*Poecilia reticulate*
Molly	*Poecilia sphenops*
Cardinal tetra	*Paracheirodon axelrodi*
Black tetra	*Hyphessobrycon herbertaxelrodi*
Dwarf cichlid	*Apistogramma cacatuoides*
Angelfish	*Pterophyllum scalare*
Corydoras catfish	*Corydoras arcuatus*

Materials (per laboratory team)

A fish tank may be observed by several teams at once, but the feeding observation can only be conducted once per day.

Flake fish food in a Petri dish or other easily accessible container

A small spatula or forceps to pick up single pieces of food

Procedure

1. Note that fish can occupy three feeding niches within your tank. They can take food from the surface, from the water column as it sinks, or off the bottom. On the data page at the end of this chapter, label the three column headings representing these three feeding niches.
2. Consult your instructor to find out the names of the fish species in your tank. Make sure you can identify each kind of fish. Write the names of these fish on row labels in the Data Table for Methods A, B, or C at the end of this chapter.
3. Moving slowly so that you do not frighten the fish, drop a few flakes of fish food into the top of the tank. Note what kind of fish eats the food, and in which of the three niches the food is eaten. If more than one fish grabs at the same flake, note which fish swallows the largest portion. Make a tally mark in the appropriate cell in the data sheet to record this feeding event.
4. Continue to add fish food, a few flakes at a time, noting which kind of fish and in which feeding niche each piece is eaten. Fish feeding at the top may become sated before fish feeding in the water column or at the bottom get a chance to feed. Discontinue the experiment when all the fish stop eating the food.
5. Discuss with your lab partners the morphology and behavior of the fish in your tank. Note which species seem more aggressive and which are passive, which species feed first and which last, which move in schools and which feed alone, which tend to float easily in the water and which stay on the bottom. Observe how the morphology of their mouth and fins seems to adapt each species to a particular feeding niche.
6. Complete the calculation page to test the null hypothesis that all fish species are equally likely to exploit the three feeding niches. Use a Chi-square contingency test, explained on the Calculation Page for Methods A, B, and C, to find out if you can reject this hypothesis. Answer Questions for Method A on the following page to interpret your results.

Questions (Method A)

1. Interpret the results of your Chi-square test. If it shows significance, what does this mean, in terms of feeding niches in the aquarium?

   _____

   _____

   _____

   _____

   _____

2. How might interactions among fish in an aquarium differ from their behavior in the wild? Do dominance interactions, for example, seem more intense or less intense than you would expect in a tropical stream?

   _____

   _____

   _____

   _____

   _____

3. Part of a fish's adaptation is behavioral and part is morphological. For one of the fish species you observed, discuss how behavioral traits and physical traits work together to adapt this fish to a particular niche.

   _____

   _____

   _____

   _____

   _____

4. In this exercise, we examined niches along the dimension of the food's vertical position when it is eaten. What other niche dimensions inside a community fish tank or series of tanks might be identified and similarly examined?

   _____

   _____

   _____

   _____

   _____

5. If the most aggressive fish species were removed from the tank, would you expect other species to exploit their niches in the same way, or to expand their use of available resources? How does this question relate to the soap bubble analogy in the Introduction?

   _____

   _____

   _____

   _____

METHOD B: NICHES OF BIRDS AT A FEEDING STATION

[Laboratory/Outdoor activity–Winter season]

Research Question

Do birds at a feeder exhibit different feeding niches?

Preparation

Several weeks in advance of this laboratory, put up a bird feeder designed for sunflower seeds. An ideal location is out of the flow of traffic, with shrubs or trees nearby, and within view of a classroom or laboratory window. If squirrels are abundant on your campus, a feeder designed to limit squirrel access will attract more birds. Stock the feeder with sunflower seeds (black oil sunflower seeds are especially good for this purpose), adding food regularly for a week or two before the lab begins so that birds become habituated to the feeding station. Winter birds are generally not difficult to identify using field guides. For regional bird lists and additional help with bird identification, local ornithological societies are a good resource.

Materials (per laboratory team)

Binoculars, if needed to observe bird behavior

Field guide to birds, and a list, if available, of bird species most commonly seen on your campus

Procedure

1. Find a comfortable place to sit and observe the bird feeder established on your campus. If you sit outdoors, stay back at least 15 meters, and remain as still as possible so that you do not disturb birds at the feeder. Observing from inside through a window is ideal.
2. Identify the birds you see. A field guide is useful, and your instructor may have a local bird list to help narrow the possibilities. For each bird you identify, write the common name on a row label on the calculation page at the end of this exercise. Add new species as they appear.
3. Note that there are at least three feeding niches around a sunflower seed feeder. Birds may eat seeds while perched on the feeder, they may pick up seeds at the feeder and fly somewhere else to remove the hulls and eat the seeds, or they may feed on seeds that have fallen to the ground under the feeder. Write brief descriptors, "at feeder," "away from feeder," and "on ground" on the column headings in the calculation page.
4. For a period of 30 minutes or more, note each bird that visits the feeder. At one-minute intervals, note birds sitting at the feeder at this time, birds feeding on the ground at this time, and birds that have taken a seed and flown away in the past minute. Make a tally mark for each bird in the data table, indicating species and feeding niche for that feeding event. You may count the same bird twice if it is occupying the niche for more than one interval of time. All together, you should have at least 30 feeding events recorded. If you have not observed that many feeding events, extend your observation time or pool your data with other laboratory teams who have observed the feeder at a different time.
5. To test the null hypothesis that birds occupy feeding niches at random, use data from three to five of the most common bird species in your survey. Complete the data table at the end of this chapter, and follow the directions to perform a Chi-square **contingency test**. Interpret your results by answering the Questions for Method B on the following page.

Questions (Method B)

1. Interpret the results of your Chi-square test. If it shows significance, what does this mean, in terms of feeding niches for birds at your feeder?

_____

_____

_____

_____

_____

2. How does a bird's beak influence the way it eats sunflower seeds? Do differences in beak shape help to explain any differences in feeding niches you observe?

_____

_____

_____

_____

_____

3. Are some species more aggressive in defending a place at the feeder than others? How do less aggressive birds manage to obtain food?

_____

_____

_____

_____

_____

4. Position of birds eating sunflower seed is the niche dimension we explored in this exercise. In what other ways do birds seem to partition niche space at a bird feeder or series of feeders?

_____

_____

_____

_____

_____

5. Behavior can be instinctive (genetically programmed) or learned. Birds at a feeder exhibit both kinds. How could you determine whether a bird's exploitation of a particular feeding niche is learned or instinctive?

_____

_____

_____

_____

_____

METHOD C: PLANT NICHES ON A DISTURBANCE GRADIENT

[Outdoor activity–spring, summer, or fall]

Research Question

Do footpaths through campus lawns create niches for plants adapted to varying levels of disturbance?

Background

Paved campus walks often fail to cover all the common pedestrian routes. Shortcuts between buildings become dirt paths through campus lawns. Edges of footpaths exhibit a disturbance gradient, from highest frequency of disturbance at the center of the path to low disturbance beyond a transition zone of approximately a meter's width on each side (Figure 8.8). Unless your grounds staff regularly applies herbicides that reduce lawn biodiversity, plant species composition will typically exhibit change along this disturbance gradient.

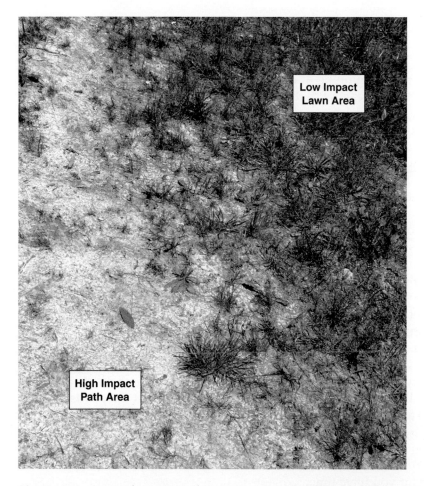

Figure 8.8 Disturbance gradient creates niches for plants resistant to trampling along a campus path.

Preparation

This exercise requires that students observe differences among plant species in the lawn. At least three species are needed for the Chi-square analysis, so make sure you can find at least three common plant species along a pathway before assigning this exercise. Taxonomic identification is desirable, but not necessary to complete the calculations. If possible, use a guide to common lawn weeds to learn botanical names of species at your site before the laboratory. You may want to develop a species list with brief descriptions to make naming the plants easier for students. If species are not identified, students can still record frequencies of occurrence for the unknown species. Students can describe or make a drawing of each species, calling their descriptions species 1, species 2, etc.

It would be best to collect data for this experiment at a time when foot traffic is low, to minimize inconvenience to ecologists and pedestrians.

Materials (per laboratory team)

Meter stick

Guide to common weedy plants in your region

Procedure

1. Select a position along the campus path. Beginning at the edge of green plant growth, position your meter stick perpendicular to the pathway, with the zero end of the scale toward the center of the path, and the 100-cm end out in the lawn.
2. Within the transect of your meter stick, the transitional zone can be arbitrarily divided into a 30-cm high-impact zone nearest the path, a 31–60-cm zone of intermediate impact, and a 61–90-cm zone of low impact. Instructions to follow assume the three zones are each 30 cm wide, adding up to a total transitional gradient of 90 cm. If these zones seem too wide or too narrow for your site, adjust your definitions of high impact, medium impact, and low impact zones by changing the distances to more appropriate widths for your path edge. Be sure to agree as a class on the definitions of your three zones, so that you can pool data later if needed.
3. Find the 3 cm mark on your meter stick and identify what kind of plant lies directly under this mark. Write the name (or description) of this plant as species 1 on the Data Table for Methods A, B, or C at the end of this chapter, and indicate that its position is in Niche A, which we will designate as the high-impact zone.
4. Repeat step 3 at positions 6, 9, 12, etc., for every third cm on your meter stick through position 30. This will yield 10 observations in the high-impact zone. Add species lines to the data table as needed. Each time you encounter a species, enter a tally mark in the appropriate cell of the data table.
5. Continue your observations at every third cm in positions 33 through 60. List these plants as occupants of Niche B, which we will call the intermediate-impact zone. Then list the plants you observe at positions 63 through 90 as occupants of Niche C, the low-impact zone.
6. Sample a second transect at a nearby point along the path, again placing your meter stick at right angles to the path edge and writing down what kinds of plants you find at each position from 3 to 90 cm, noting which plants fall into the three zones. Continue adding tally marks, pooling observations from several transects if you have time.
7. To test the null hypothesis that plants occupy the three zones (or niches) at random, use observations for three to five of the most common plant species in your survey. Complete the Calculation Page for Methods A, B, or C at the end of this chapter. Write "High impact" in the Niche A box, "Medium impact" in the Niche B box, and "Low impact" in the Niche C box. Also, write names of the most commonly observed plants (or your own brief descriptions) in the boxes labeled Species 1, Species 2, Species 3, etc. Follow the directions to perform a Chi-square contingency test. Interpret your results by answering the Questions for Method C on the following page.

Questions (Method C)

1. Interpret the results of your Chi-square test. If it shows significance, what does this mean in terms of niches for plants along a disturbance gradient?

    _____

    _____

    _____

    _____

    _____

2. The transition from high disturbance to low disturbance was arbitrarily divided into three zones of equal width in this study. Is this an adequate way to describe the way plants are partitioning the niche space? Did the changes you saw between Niches A and B seem equal to the changes you saw between Niches B and C?

    _____

    _____

    _____

    _____

    _____

3. Compare the results of this study to Joseph Connell's work on barnacle niches. How might the plants most tolerant of foot traffic be compared to *Chthamalus* barnacles?

    _____

    _____

    _____

    _____

    _____

4. If you assume plants found near the path cannot compete elsewhere in the lawn, how could you test your hypothesis? Briefly explain an experimental method.

    _____

    _____

    _____

    _____

    _____

5. Adaptation in plant form and growth habit results from evolutionary change over the history of a species. Since human footpaths represent a comparatively recent environmental influence, how can you explain apparent adaptations to an artificially introduced feature of the environment?

    _____

    _____

    _____

    _____

    _____

Data Table (Methods A, B, or C)

	Niche A	Niche B	Niche C
Species 1			
Species 2			
Species 3			
Species 4			
Species 5			

1. In the labeled boxes, describe or identify each species you observe, concentrating on the five most common types.
2. In the boxes heading each column, identify niches as directed in the procedure for this method.
3. Make tally marks as you observe species in each of the three niches.
4. Enter totals in the Calculation Page for Methods A, B, or C, and follow instructions to calculate a Chi-square contingency test.

Calculation Page (Methods A, B, or C)

Chi-Square Contingency Table

	Niche A	Niche B	Niche C	Row Totals
Species 1 observed:				
(expected):	( )	( )	( )	
Species 2 observed:				
(expected):	( )	( )	( )	
Species 3 observed:				
(expected):	( )	( )	( )	
Species 4 observed:				
(expected):	( )	( )	( )	
Species 5 observed:				
(expected):	( )	( )	( )	
Column Totals:				
				⇑ GRAND TOTAL

1. For this analysis, you will need records for at least three (and no more than five) of your most commonly observed species. Write the name of the first species in the box labeled "Species 1," the name of the second in the box labeled "Species 2," and so forth. If you have fewer than five species, leave the bottom boxes blank.
2. Inside the boxes marked "Niche A," "Niche B," and "Niche C," write a shorthand description of the three niches in which your observations were made.
3. For Species 1, record the number of times you observed this species in Niche A inside the cell at the top-left corner of the graph, *above the parentheses*. Going across the top row, enter the number of times you saw this species in Niche B and Niche C. Then enter observations for Species 2 in the second row, Species 3 in the third row, etc.
4. After entering your observations, add the three numbers in the top row and record the sum in the column labeled "Row Totals" at the right. Repeat for the other species, summing the total number of observations for each species as a row total.
5. Then add all the observations in the first column to determine the total number of individuals you saw occupying Niche A. Enter the column total at the bottom of the Niche A numbers. Calculate column totals for Niche B and Niche C in the same way.

6. Sum up all the row totals and enter the result in the cell labeled "Grand Total" at the bottom-right corner of the table. To check your math, sum all the column totals, and you should get the same result.

7. You should now have the table filled out as below. Four species are observed in this example.

	Niche A	Niche B	Niche C	Row Totals
Species 1	A_1	B_1	C_1	Row Total 1
Species 2	A_2	B_2	C_2	Row Total 2
Species 3	A_3	B_3	C_3	Row Total 3
Species 4	A_4	B_4	C_4	Row Total 4
Column Totals:	Column Total A	Column Total B	Column Total C	**GRAND TOTAL**

Expected values will be calculated from the **null hypothesis** that species assort randomly into niches. Using the laws of probability (see Chapter 6), we can calculate how many individuals we would expect in each niche, based on species abundance and total niche occupancy. If a pattern in the data does not conform to these expectations, we can reject the null hypothesis and follow the alternative reasoning that our species fill these niches in nonrandom patterns.

For each cell in the table, you will need to calculate expected values using a simple formula: (Expected value) = (Column Total)(Row Total)/(Grand Total). For example, the expected number of observations for Species 1 in Niche A would be calculated as follows:

$$\text{Expected number (A1)} = \frac{\textbf{(Column Total A)(Row Total 1)}}{\textbf{GRAND TOTAL}}$$

Perform this calculation, and *enter the expected number for Species 1 in Niche A within the parentheses* in the top-left cell of the data table.

8. For each of the other cells, multiply the cell's row and column totals, divided by the grand total, to get an expected number. Enter the expected value within the parentheses at the bottom of each cell in the contingency table.

9. For a Chi-square test, you will need to calculate the number of independently varying cells in the table. This is called **degrees of freedom**, abbreviated as **d.f.** For a contingency test, degrees of freedom (d.f.) is calculated as follows.

d.f. = (number of species − 1)(number of niches − 1) =

10. Calculate d.f. for your contingency table, and write the result in the box. The table has spaces for up to five species, but make sure to calculate d.f. based on the number of species you actually observed.

11. Enter a cell label for each cell you completed in your earlier table. For example, label the data for Niche B and Species 2 as "B2." Note that each cell in the earlier table gets its own row in this one.

12. From your entries in the first table, copy all observed and expected values in the columns labeled "O" and "E."

$$\text{Chi-square} = \Sigma\,(O - E)^2/E$$

where: Σ means summed over all cells
O = the observed value
E = the expected value

Chi-Square Calculation Table

CELL LABEL niche, species	OBSERVED NUMBER (O)	EXPECTED NUMBER (E)	$(O - E)$	$(O - E)^2$	$(O - E)^2/E$
				Chi-square value =	

13. Calculate the amount of deviation by subtracting the expected value from the observed value in each row. Enter the difference, with sign, in the column labeled "$(O - E)$." As a test of your math, check to make sure this $(O - E)$ column sums to zero.
14. Square each deviation. Note that positive and negative signs disappear when you square these values. Enter the squared value in the column labeled "$(O - E)^2$."
15. Divide each squared deviation by the original expected value. Enter the result in the last column, which is labeled "$(O - E)^2/E$."

16. Finally, add up all the $(O - E)^2/E$ values, and record the sum at the bottom-right corner of the calculation page. This is the **Chi-square** value for your data set. Because it is calculated from deviations of observed values from expected values, the greater the departure from null hypothesis expectations, the higher the value of Chi-square. A zero value of Chi-square would indicate an exact match with expectations, meaning that species are randomly selecting niches exactly as your null hypothesis predicted. Even if species were selecting niches at random, some degree of pattern could be expected due to sampling error, so we could anticipate a small Chi-square value most of the time, just due to chance.

As deviations from random expectations grow larger, Chi-square grows too. At some breaking point, called the **critical value** of Chi-square, we can no longer accept the null hypothesis as an explanation for data patterns generating these large deviations. At this point, we say an alternative, nonrandom explanation is warranted. By convention, scientists agree that the critical value of Chi-square has been reached if the odds of generating deviations of this magnitude by random sampling error alone are less than 5%. This 5% probability, written as **p = 0.05** by statisticians, means that our results can be considered **"statistically significant."** This does not mean the alternative explanation we pose is always right, but it does mean we have eliminated the null hypothesis with 95% confidence. An even higher value of Chi-square, corresponding to a p value of 0.01, is the threshold customarily designated as a "highly significant" departure from random expectations. In this case, our conclusion is the same, but our confidence in rejecting the null hypothesis has risen to 99%.

In summary, think of p as the error rate we can expect if we reject the null hypothesis based on a given Chi-square value. *A high value of Chi-square corresponds to a high level of statistical significance, which in turn corresponds to a low p value.*

17. Equating a Chi-square value (along with the degrees of freedom in your experimental design) with its corresponding p value is easily accomplished using Chi-square tables (Appendix 5). A simplified Chi-square table is included below, for your convenience. First, find the row equivalent to the **degrees of freedom** you calculated for this data set. Remember this is (Niches − 1)(Species − 1) so your degrees of freedom should be 4, 6, or 8 if you observed three niches and 3–5 species. Next, compare your Chi-square value with the two Chi-square numbers aligned with this number in the table.

Under the heading "p = 0.05" and in the row corresponding to your degrees of freedom, you will find a critical value of Chi square. *If your calculation of Chi-square yielded a number larger than the critical value, then your results can be considered significantly nonrandom.* In the context of your experiment, this supports the hypothesis that your species are partitioning themselves among niches in a nonrandom fashion.

Under the heading "p = 0.01," you will find a Chi-square value indicating a higher level of significance. If your calculation of Chi-square yielded a number larger than this, you can have even greater confidence that your data show a nonrandom pattern. The conclusion is the same as for p = 0.05, but your evidence is stronger in this case.

A Chi-square value lower than the number in the p = 0.05 column could have come from observations of species randomly selecting niches. This may or may not be the case, but low Chi-square values mean you do not have sufficient evidence to reject the null hypothesis of random resource use.

Simplified Chi-Square Table of Critical Values

DEGREES OF FREEDOM	CHI-SQUARE GREATER THAN THIS VALUE INDICATES SIGNIFICANCE $p = 0.05$	CHI-SQUARE GREATER THAN THIS VALUE INDICATES HIGHER SIGNIFICANCE $p = 0.01$
4	9.49	13.28
6	12.59	16.81
8	15.51	20.09

From Rohlf, F. J. and R. R. Sokal. 1995. *Statistical Tables*, 3rd ed. W.H. Freeman, San Francisco.

FOR FURTHER INVESTIGATION

1. Compare the feeding niches for fish in single-species tanks with feeding behavior in a community tank. For example, a tank containing a small population of a single species (e.g., black tetras) could be used to establish the fundamental feeding niche, expressed as the percentage of their feeding on the top, water column, and bottom of the tank. Then a second species (e.g., guppies) could be added and the realized niches determined in a second trial. Can you demonstrate a shift in feeding niches with the addition of competitors, as Holmes did for parasitic worms?

2. Establish different kinds of feeders at your campus feeding station to extend the niche space. A thistle seed feeder or mixed seed feeder will attract birds poorly adapted to eating sunflower seed. You can attract some insect-eating birds with suet (beef fat) suspended in a mesh bag of the type used by groceries for oranges and onions. Also attractive to insect-eating birds, a peanut butter feeder can be made by drilling large holes in a piece of wood and pushing peanut butter into the holes with a table knife. Observe the numbers and kinds of birds coming to the variety of feeders. Do niches you observed in Method B shift as new resources become available?

3. Find resource gradients other than path edges on your campus. If you live in a dry region, for example, downspouts around buildings may create wetter conditions near the downspout, with dryer zones farther away. Artificial watering systems also create resource gradients. Try taking transects from wetter to dryer zones, monitoring the plant composition along the transect as you did for Method C. Can you identify realized niches along this resource axis?

4. An instrument called a penetrometer measures the force required to push a small-diameter piston into the soil. This instrument provides a measure of soil compaction due to trampling along a foot path. Use a penetrometer to measure soil compaction on a transect from the center of a footpath through the gradient you observed in Method C. Does plant community composition correlate well with soil compaction?

FOR FURTHER READING

Connell, J. H. 1961. The influence of interspecific competition and other factors on the distribution of the barnacle. *Chthamalus stellatus*. *Ecology* 42:710–723.

Holmes, J. C. 1973. Site selection by parasitic helminths: interspecific interactions, site segregation, and their importance to the development of helminth communities. *Canadian Journal of Zoology* 51:333–347.

Hutchinson, G. E. 1957. Concluding remarks. *Cold Spring Symposia on Quantitative Biology*. 22:415–27.

MacArthur, R. H. 1958. Population ecology of some warblers of northeastern coniferous forests. *Ecology* 39:599–619.

Rohlf, F. J. and R. R. Sokal. 1995. *Statistical Tables*, 3rd ed. W. H. Freeman, San Francisco.

Chapter 9

The Community Concept

Figure 9.1 Mount Chimborazo, where Alexander von Humboldt's observations of vegetational change with altitude led to an ecological concept of the community.

INTRODUCTION

Traveling through South and Central America at the beginning of the 19th century, Prussian explorer Alexander von Humboldt risked thundering rapids, scorching deserts, deep caves, steep mountain trails, snakes, jaguars, and tropical diseases to survey vast unmapped regions of the American interior. Humboldt was a prolific and widely appreciated writer; his accounts of these adventures were acclaimed during his lifetime in both Europe and the Americas. He was also a physical scientist well schooled in the use of astronomical and meteorological instruments, a botanist who collected and named an immense catalog of previously unknown plant species, and a geologist constantly seeking causal explanations for physical characteristics of the earth. Although the term "ecology" was coined after his time, Humboldt's persistent questions about interactions between physical and biological features of the planet clearly distinguish him as a pioneer in the field.

An ecologically significant insight came to Humboldt as he climbed Mount Chimborazo near Quito, Ecuador, in 1802. Impressive even among Andean peaks at 6315 meters, this volcanic cone was thought in Humboldt's day to be the highest mountain in the world (Figure 9.1). Without adequate clothing or any real climbing equipment to make his way through the ice at the mountaintop, Humboldt was forced to stop short of the summit. There his altimeter recorded 5878 meters, which was higher than any climber had reached on any mountain before that time.

As amazing as his conquest of Mt. Chimborazo was, Humboldt's significant discovery came from careful observations of the plant life, coupled with meticulous altimeter readings he recorded as a matter of routine along the way. As he descended from the bare rocks and snow fields of the high country, one imagines him mentally reviewing the pattern of plant life he had observed on the way up. At 4200 m, he reached a zone of grasses and lichens. Descending to 3600 m, he walked over broad-leaved herbaceous plants. At 3000 m, he entered a zone of evergreen shrubs. Between 2700 m and 1500 m he passed through an oak forest, and at lower elevations he saw ferns and palms. Hardly pausing for rest after his climb, Humboldt began writing a treatise on plant geography, identifying associations of plants found within predictable altitudinal zones across South America (Helferich, 2004). Due in no small part to Humboldt's mountaintop inspiration, biologists who came after him began asking not only "what species are in this place," but also, "why are these species found together in this environment?"

Check your progress:

List in order, from low elevation to high, the vegetational zones Humboldt observed in the Andes.

Hint: The effects of altitude on plant communities resemble changes seen as one travels from the equator towards the poles.

Building on Humboldt's integration of geology and climatology to explain plant associations, American biologist Steven Forbes contributed the idea that every species in nature is influenced not only by its physical environment, but also by all the other species feeding it or preying on it. Forbes thought the species of a community could be studied as parts of a great living machine, with a complete understanding of the whole accessible only through an exhaustive inventory of its components. Near the end of the 19th century, Forbes and a team of researchers worked for many years in biologically productive lakes along the Illinois River floodplain, collecting the insect life, sampling the plankton, dredging the bottom, and seining the fish to document a network of **trophic** relationships we now call a food web (Schneider, 2000). Subsequent studies of food webs in freshwater, marine, and terrestrial systems demonstrated that complex interdependence among species is a typical feature of biological communities (Figure 9.2).

In the early years of the 20th century, botanist Frederic Clements pushed the idea of biological interdependence even further. He compared the species within a living community to organs in a great natural body, which Clements called the **super-organism**. Just as organisms grow and develop from birth to maturity, Clements noted that ecosystems go through stages of development. (See Chapter 14 for a discussion of this kind of community change.) If a community is like a super-organism, Clements argued, then removal of some of its component parts could result in the death of the whole system. Ecologists influenced by Clements began to think of communities as well-defined biological entities, each with a characteristic list of component species. For several decades, ecologists turned to the task of identifying communities, locating them on maps, and giving them names based on their dominant plant forms. A good example of this sort of community analysis is demonstrated by the intricate maps of A. W. Kuchler, who identified no fewer than 116 types of plant associations within the boundaries of the continental United States (Figure 9.3).

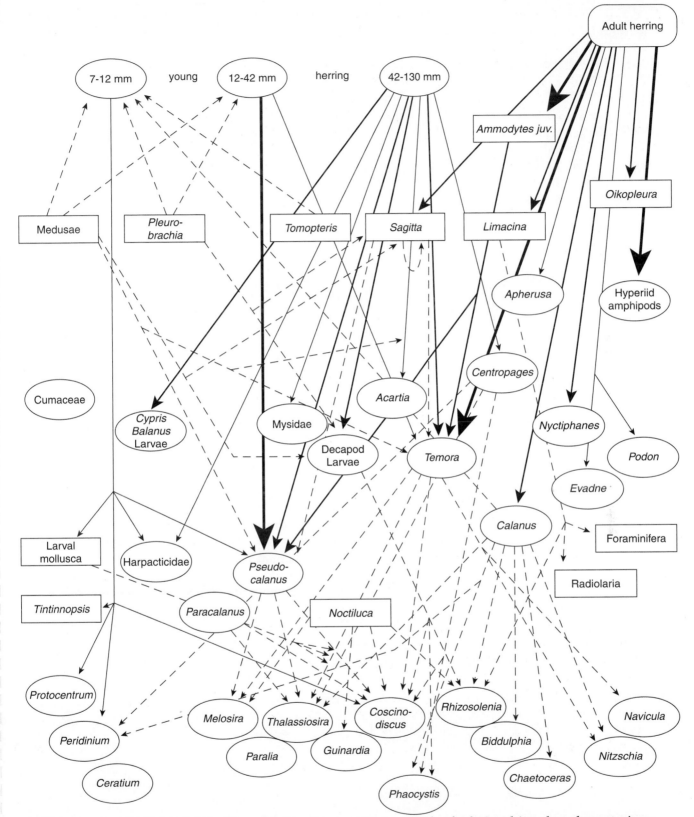

Figure 9.2 A marine food web demonstrates the complex network of relationships that characterizes natural communities. Arrows connect predator to prey species relevant to herring populations. (From Hardy, A. C. 1924. The herring in relation to its animate environment. Part I. The food and feeding habits of the herring with special reference to the east coast of England. *Fishery Investigations, Series* 27(3): 53.)

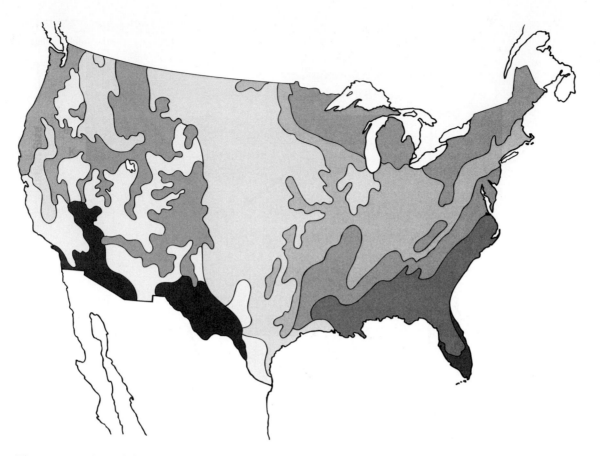

Figure 9.3   Simplified Kuchler map of potential natural vegetation in the United States. Kuchler's original work illustrated 116 types of plant communities, each plotted in a different color.

Not every ecologist accepted the idea of the community as a super-organism. As early as 1926, Henry Allen Gleason began promoting the idea that species adapt individually. Gleason doubted the premise that community members are so interdependent they cannot survive separately. Rather, he viewed a plant association as a loose collection of species whose individual habitat requirements happen to intersect in a particular region. Although he recognized that communities could be limited by abrupt discontinuities in the physical environment, such as a lake shore, he did not believe that transitions between community types are always so easy to define. Gleason called his idea the **continuum** model, because he thought communities generally have fuzzy boundaries. In Gleason's view, a traveler walking from one type of environment to another would see some species dropping out of the prevailing community as others appeared, according to the distinct habitat requirements of each species (Gleason, 1939).

Check your progress:

Contrast Clements's and Gleason's views on the community.

Hint: The "super-organism" idea was embraced by one but not the other.

Clements and Gleason stimulated lively debates among ecologists for many years. "Are biological communities simply equal to the sum of their parts," they asked, "or do special properties emerge above the level of the organism which define and control the community's existence?" An elegant comparative approach to this question was developed by Robert Whittaker (1956) in a study of tree species distributions in the Great Smokey mountains. Just as Humboldt had described in the Andes, Whittaker recognized that plant community composition changes with altitude. Whittaker tested the super-organism theory by close examination of species distributions along an environmental gradient from the warm, moist habitats at lower elevations to the cooler, drier habitats at higher elevations. If Clements was right about communities as super-organisms, then all the species in a particular plant association should occupy the same zone on the mountain slopes. If conditions become too harsh for some of its component species, survival of the entire super-organism should be limited at that point. Clements's model predicts sharp boundaries between communities, with every species' range falling within one community or another, but never straddling the border. Gleason, on the other hand, predicted gradual transitions among plant associations, with species ranges more or less independent of one another (Figure 9.4).

Whittaker sampled the trees at regular intervals along an altitudinal gradient and plotted the relative abundance of each species as a function of altitude. By comparing changes in species abundance along the gradient, Whittaker was able to test Clements's prediction of clear dividing lines between communities against Gleason's null model of independent species distributions. Your task in this laboratory exercise will be to replicate Whittaker's method on a smaller scale. To maintain our objectivity, we will not review Whittaker's results prior to our own experiment, but his 1956 paper is cited at the end of the chapter if you are curious.

Check your progress:

Why did an altitudinal gradient better suit Whittaker's purpose than some other environmental variable, such as soil type or herbivore prevalence?

Hint: If the environment changed abruptly, then how might an ecologist interpret rapid changes in community structure?

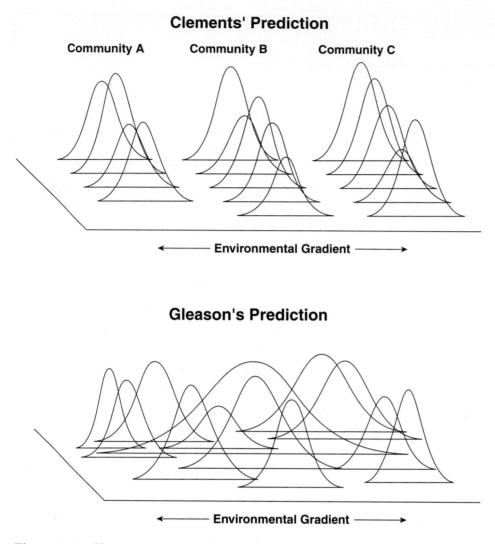

Figure 9.4   Clements envisioned communities as distinct entities (top), while Gleason thought that species distributions would be independently arrayed along a continuum (bottom). Each curve represents a species distribution, with height representing species abundance and width representing tolerance limits along an environmental gradient.

METHOD A: GEOGRAPHICAL DISTRIBUTION OF TREE SPECIES
[Laboratory or classroom activity]

Research Question
Do local plant communities have well-defined boundaries?

Preparation
Before the laboratory, students will need a list of 10 native trees or shrubs common to a forest community in your area. Local parks, State Forestry divisions, and native plant societies are good sources of species lists. If you live in an area without native forests, then grassland or desert species also work well for the purpose of this exercise.

Materials (per laboratory team)
List of 10 trees (or woody plant species) native to a forest community in your region

Field guide to native trees (e.g., R. J. Preston Jr., 1989) with North American range maps for each species

Colored pencils, in at least 10 different colors

Procedure
1. Look up the geographic range of the first tree on your species list, making sure it is shown to be native to your region. From your collection of colored pencils, select a color to represent this tree. On the blank map of North America (Figure 9.5), draw a line illustrating the outer boundary of this species' natural range. Color in the first box in the figure legend, and label this box with the species name.
2. Repeat step one for all 10 trees on your list, using a different color to show the outer boundary of each species' range. Make sure to color in boxes and add all species names to the legend under the map.
3. Answer the Questions for Method A to interpret your composite range map.

Figure 9.5 Outline the distribution of each tree species on this map of North America.

Questions (Method A)

1. Your analysis is based on a two-dimensional depiction of species distributions rather than the single dimension of altitude that Whittaker examined. However, you can still examine the correlation or independence of species ranges to test the "super-organism" theory of the community. Based on Figure 9.4 what prediction would the Clements model make for the ranges of your 10 tree species? What prediction would the Gleason model make?

    _____

    _____

    _____

    _____

2. Which model seems to apply to your data? Do all 10 trees conform to a common community boundary, or do some species ranges extend far beyond the limits of other species?

    _____

    _____

    _____

    _____

3. Do you feel confident testing the Clements vs. Gleason question based on inspection of the map? If not, what other kinds of information would be useful?

    _____

    _____

    _____

    _____

4. If several species seem to share a common boundary, is biological interdependence the only explanation? What geological or climatic factors may explain why all these trees are limited at the same boundary line?

    _____

    _____

    _____

    _____

5. If you find the ranges of some trees to be much larger than others, does this mean the community is a useless concept? How exactly would you define a plant community, based on what you have learned in this exercise?

    _____

    _____

    _____

    _____

METHOD B: DIRECTION OF EXPOSURE AND PLANT COMMUNITIES
[Outdoor activity, spring, summer, or fall]

Research Question
Do north-facing and south-facing sides of buildings support distinct plant communities?

Preparation
This laboratory examines plants that grow along building foundations. Unless your grounds-keeping staff uses trimmers or herbicides to eliminate all vegetation along the foundations, plant diversity is often remarkably rich in these microhabitats, even if lawns are maintained as near-monocultures. Look for maintenance buildings, monuments, or other structures less intensively landscaped for the best sites. It is always a good idea to check with your facilities management director to ensure that pesticides have not been recently applied before the class begins sampling.

A critical requirement for this exercise is that vegetation be compared at the bases of walls facing different compass directions. In the Northern Hemisphere, vegetation at the base of a north-facing wall receives little or no direct sunlight through most of the year, so conditions tend to be much cooler and moister than they are at the base of a south-facing wall. As a transition zone, east-facing and west-facing walls should be sampled for comparison. The walls you choose can be on different buildings (for example, the four sides of a quad area would work nicely), but all should be free of plantings and share similar microhabitat characteristics, aside from the direction of exposure. Dividing the sampling work among laboratory teams and pooling data from the class will be necessary to complete this exercise within a typical 2–3 hour laboratory period.

Materials (per laboratory team)
A 10-meter tape line, marked in cm

A magnetic compass, for determining which direction walls are facing

A local plant list, or guide to weedy vegetation (if available)

Clipboard or notebook, with lined notebook paper for preliminary data

Procedure
1. Locate the wall of a building with vegetation, but no landscaping, at its foundation. Use a compass to measure which direction (in degrees) the wall is facing (Figure 9.6).
2. Stretch out your tape line along the foundation, spaced about 10 cm from the edge of the building. This line will provide a **transect** for surveying vegetation at this site. Stretch the tape out to include at least 5 meters of linear habitat space for your sample. Place a pencil on the 10-cm mark of the tape line, pointed toward the building and extending 2 cm inward toward the building. Place a mark on the side of the pencil where it crosses the inside edge of the tape line. Every measurement will be made by sliding the pencil sideways along the tape, and positioning it with the mark at the tape's edge, so your pencil extends the same distance in from the tape line each time (Figure 9.7).

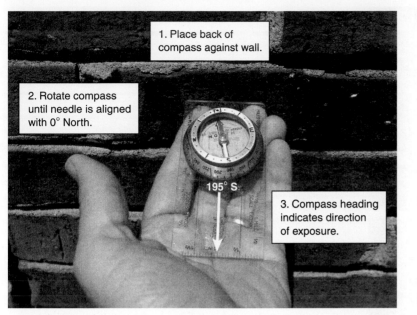

1. Place back of compass against wall.

2. Rotate compass until needle is aligned with 0° North.

195° S

3. Compass heading indicates direction of exposure.

Figure 9.6 Using a compass to determine direction of exposure of a wall.

3. On the 10-cm mark, identify the plant that lies closest to the point of the pencil. Write the name of this plant on a piece of notebook paper, and make one tally mark beside it to indicate you have observed this species. If you do not know the name of the plant, call it "Species 1" and record a short description with drawings in your notebook so that you can identify it again.

4. Slide the pencil to the 20-cm mark and note the second plant in your survey. If it is the same species, make a second tally mark. If it is a different species, enter a new line in your preliminary data table.

5. Repeat step 4 every 10 cm for a total of 50 to 100 observations at this site.

6. Identify the five most abundant plant species, along with the number of times each species was found in your survey. Add up all the numbers to get a total number of sample points for this transect. Then divide the number for each species by the total to get a relative frequency of this species in the community.

7. On the polar coordinate Graph for Method B (Table 9.1), find the direction (in degrees) that matches your compass reading for the first site. Draw a line from the center point through this compass reading. You now have recorded an axis of orientation for the wall. Along this axis, mark the relative frequencies for each of your five species, using the concentric circles on the graph as a scale. Consider the center of the polar graph the 0% point, and count each ring crossed by your axis as 5% on the frequency scale. Place a point on the axis for each of the five most abundant plants in your survey. Label these points with species numbers, and fill in the legend to indicate which species each number represents.

Figure 9.7   Sampling vegetation along a transect at a building foundation.

8. Repeat steps 1–7 on a wall that faces a different direction. As you find new species, continue using a new number for each type of plant so that you can number all the species on your composite data display. If you have time, try to sample transects at the base of four walls facing roughly North, East, South, and West. Share data with other laboratory teams if the sampling takes more time than you are allotted for this experiment.
9. After you have summarized all your data on the polar coordinate graph, examine the vegetational patterns it represents. If you have transect data from walls facing a number of different directions (eight or more), you may want to draw lines connecting points representing the same plant type to show more clearly how individual species are distributed around campus buildings. Answer the Questions for Method B to interpret your results.

Table 9.1

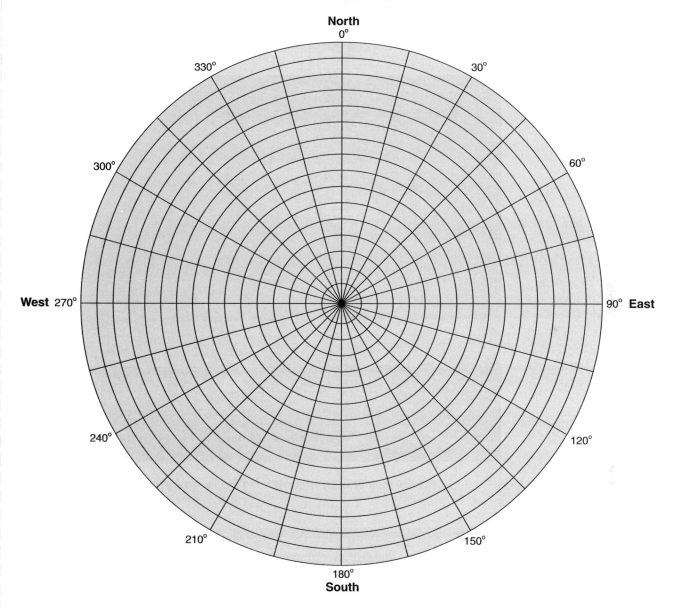

Table 9.1 - Data Table for Method B.

Questions (Method B)

1. Compare the plant communities on the north-facing vs. south-facing sites. How would you characterize these two community types? Could you give these two communities names, based on their dominant species?

    _____

    _____

    _____

    _____

2. Look at the east-facing and/or west-facing site data. Do these intermediate microhabitats support distinct plant associations, or a blend of species from the north-facing and south-facing sites?

    _____

    _____

    _____

    _____

3. What do your data from this experiment indicate about the super-organism concept of Clements? Are species distributed around buildings within distinct communities with well-defined limits, or do species seem to have individual distributions on a directional continuum of plant associations?

    _____

    _____

    _____

    _____

4. How do you think the plants in your survey got to these sites? How might distributions of colonizing plants compare with distributions of species that do not disperse so easily?

    _____

    _____

    _____

    _____

5. Cultivated plants are derived from wild species adapted to different kinds of habitats such as forests, deserts, or prairies. In selecting plants for foundation plantings around buildings, would you take into consideration the kind of environment each plant originally came from? Which side of the building would best approximate a forest niche? a prairie niche?

    _____

    _____

    _____

    _____

METHOD C: MICROBIAL COMMUNITIES ON A SALT CONCENTRATION GRADIENT

[Laboratory activity]

Research Question

Do members of a microbial community exhibit overlapping distributions on a salt concentration gradient?

Preparation

A hay infusion is a simple way to culture a community of bacteria and protozoans in the laboratory. Although cultures will not take much time to set up, students will need to monitor the development of their hay infusion communities over a period of 3–5 weeks. If you have compound microscopes readily accessible in the laboratory, a 15-minute observation period once a week should be enough for meaningful data collection. Since this exercise also involves community development over time, the class can maintain the most productive cultures for a longer period as a demonstration of ecological succession (Chapter 14.)

Hay for an infusion can be ordered from supply houses, but dried grass collected from a wet meadow will work. Loose hay from a livestock barn or petting zoo is also a good source for hay infusions. Many types of bacteria and resting stages of protozoa are present in the hay itself. Water from an aquarium or pond is best for this exercise, since it will also contain bacteria, protozoa, and invertebrates. Alternatively, rain water or spring water will work. Do not use tap water, as it contains chlorine which kills microorganisms. For culture tubes, use new tubes or carefully rinsed tubes. Traces of soap or chemicals will prevent growth of protozoans.

Although the typical flora of hay infusions are benign microorganisms, students should wash their hands carefully after taking samples as a precaution. Be aware that early stages of infusions often produce strong odors. If you share a lab with other classes, you may want to incubate infusions in a hood or well-ventilated prep room.

Materials (per laboratory team)

6 1-liter volumetric flasks for stock solutions (can be shared among four teams)

6 large test tubes (potato tubes) with slip-on caps for aerobic culture

Test-tube rack

Hay (a handful) for infusions

Pond or aquarium water

Sodium chloride, reagent grade

Electronic or triple-beam balance, accurate to 0.01 g

Weighing papers

Scissors

Grease pencil or marker for labeling test tubes

6 Pasteur pipettes, with bulbs

Compound microscope, magnification up to 400×

Glass slides and cover slips

Identification guide for protozoa

Drawing paper, pencils

Procedure

1.   Prepare a series of salt concentrations for your cultures as shown in the following table. Label each volumetric flask with the parts per million (ppm) number to show how many grams of salt will be added to a liter of water. Grams of salt needed for each concentration are indicated in the second row of the table. Weigh the salt first, add it to the flask, and then fill the flask up to the 1-liter mark with aquarium or pond water. These stock cultures will be used for your series of hay infusions.

Table 9.2   Salt Concentration Series for Hay Infusion Communities

Concentration (parts per million)	Fresh water 0 ppm	Trace of salt 1000 ppm	Brackish 2000 ppm	Saline 5000 ppm	Highly saline 10,000 ppm	Sea water 35,000 ppm
Grams of salt added per liter of water	No added salt	1 gram per liter	2 grams per liter	5 grams per liter	10 grams per liter	35 grams per liter

2.   Mark a series of six large test tubes (potato tubes) with the ppm numbers shown in the top row of Table 9.2. Use scissors to trim hay to 1/2 the length of your tubes, and weigh out 10 grams of hay to place in each tube.

3.   Pour the salt solutions matching your test-tube labels to fill each tube in the series 3/4 full of a different concentration of salt water. Cover the tubes with slip-on caps, to allow some air into the culture.

4.   Incubate the tubes at room temperature for a week. If possible, place your cultures under a fluorescent lamp or in a window so that photosynthetic organisms can grow.

5.   After one week, use a Pasteur pipette to remove a small amount of fluid from near the top of the fresh water (0 ppm) infusion. Place a drop of this fluid on a glass slide, and place a cover slip carefully on top. Examine the infusion community under a compound microscope, beginning at low power and then at high magnification. Draw and describe what you see. Repeat this microscopic observation for the other five salt concentrations, using a different pipette for each tube to avoid cross-contamination.

6.   Return the infusions to the place you have been incubating them, and let them sit a second week. Examine each culture at week 2, and compare your results with week 1. Continue observing for a third and fourth week, if necessary, until you see several protozoan species in large numbers in at least some of the cultures.

7.   As soon as you have at least one of your hay infusions with three or more common species, fill in the Data Table for Method C at the end of this section. Draw or identify the most common species, and then check boxes to indicate which of the six cultures contain this species. Answer the Questions for Method C to interpret your data.

Data Table (Method C)
Microbial Communities Along a Gradient of Salt Concentrations

Salt Concentration (parts per million)	Fresh water 0 ppm	Trace of salt 1000 ppm	Brackish 2000 ppm	Saline 5,000 ppm	Highly saline 10,000 ppm	Sea water 35,000 ppm
Species 1						
Species 2						
Species 3						
Species 4						
Species 5						

Questions (Method C)

1. If you consider the series of increasingly saline solutions as a habitat gradient, how does your experiment compare with Whittaker's study of trees in the Great Smokey Mountains?

    _____

    _____

    _____

    _____

2. Somewhere between no salt and the salt concentration of sea water, a freshwater protozoan will reach its tolerance limit, and the experimenter would not expect to see that species in more saline solutions. What would the Clements model predict for all the ranges of species in a microbial community sampled along this salt gradient? What would Gleason predict? (See Figure 9.4.)

    _____

    _____

    _____

    _____

3. Which model best describes your data? Explain.

    _____

    _____

    _____

    _____

4. If you had collected dried grass from a salt marsh near the sea as your source of culture medium and resting protozoans for this experiment, would you expect a different result from your series of infusions? Why?

    _____

    _____

    _____

    _____

5. Early stages of hay infusions are dominated by bacteria; then protists that feed on bacteria tend to reduce bacterial populations. From the appearance and odor of your cultures, could you tell which tubes had the most robust bacterial populations? Although you cannot see bacteria well under a compound microscope, do you suspect different types may grow in different salt concentrations? Explain.

    _____

    _____

    _____

    _____

FOR FURTHER INVESTIGATION

1. Try Method A with 10 mammal species or 10 nonmigratory bird species. Do you achieve similar results?
2. In Method B, you sampled every 10 cm along a transect line. This systematic approach guarantees an even distribution of data points within the sample area. An alternative approach is to use a random number table to select 50 points between the values of 0.00 m and 10.00 m along the transect line. You may want to compare results from the two transect methods to assess the relative merits of systematic vs. random sampling.
3. From your study sites in Method B, take soil samples with a core sampler. Measure soil moisture by weighing the soil sample, then drying it in an oven, then weighing it again. The percentage soil moisture is [(wet wt. − dry wt.)/(dry wt.)] × 100%. You can also measure soil temperature on site with a soil thermometer. Do soil moisture and soil temperature explain some of your plant distribution data?
4. Sample the bottoms of your culture tubes from Method C and compare the community near the top of the culture with the community near the bottom. Does the oxygen gradient within a tube compare to the salt gradient among the series of tubes in your experiment?
5. Continue monitoring your most productive culture for an extended period of time. Observations of a hay infusion over longer periods conform to many of the same principles as ecological succession in a forest community (see Chapter 14).

FOR FURTHER READING

Clements, Frederic E. 1936. Nature and structure of the climax. *J. Ecology* 24:252–284.

Gleason, Henry A. 1939. The individualistic concept of the plant association. *American Midland Naturalist* 21:92–110.

Hardy, A. C. 1924. The herring in relation to its animate environment. Part I. The food and feeding habits of the herring with special reference to the east coast of England. *Fishery Investigations* 27(3):53.

Helferich, Gerard. 2004. *Humboldt's Cosmos: Alexander von Humboldt and the Latin American Journey That Changed the Way We See the World*. Gotham Books, New York.

Preston, R. J., Jr. 1989. *North American Trees*. Iowa State U. Press, Ames.

Schneider, Daniel W. 2000. Local knowledge, environmental politics, and the founding of ecology in the United States: Stephen Forbes and "The Lake as a Microcosm." *Isis* 91(4):681–705.

Whittaker, Robert H. 1956. Vegetation of the Great Smokey Mountains. *Ecological Monographs* 26(1):1–80.

Chapter 10
Competition

Figure 10.1 Honeypot ant (left) and harvester ant (right) compete for resources in the Chiricahua Mountains of Arizona.

INTRODUCTION

Harvester ants (Figure 10.1, right) are ecologically important members of desert communities. These ants range far and wide to collect small seeds for food, hoarding their substantial harvest in huge underground nests. Honeypot ants (Figure 10.1, left) often feed on nectar, so they have a somewhat different **niche**. (For a discussion of the niche, see Chapter 8.) However, honeypot ants do compete with harvester ants for dead insects and other scavenged food. In fact, honeypot ants frequently attack harvester ants to rob the larger ants of food they are carrying. This kind of negative community interaction is called **interference competition**, because one species actively interferes with the success of the other. Interference can be behavioral, as seen in warring ants, or it can be chemical, as seen in plants and fungi. A *Penicillium* fungus growing through rotting fruit produces antibiotics that interfere with the growth of its bacterial competitors.

Creosote bush (Figure 10.2) releases chemical herbicides that interfere with the growth of grasses and other plants. This sort of chemical warfare against competitors, called **allelopathy**, has been proposed as an explanation for creosote's dominance in dry scrub environments across the American southwest. Although allelopathy can be clearly demonstrated in controlled laboratory experiments, field researchers have identified many complicating factors, including herbivory by small mammals, drought, fire, and soil microorganisms, all potentially influencing the way plant chemicals affect the outcome of competition in nature (Halsey, 2005).

Figure 10.2 Creosote bush (*Larrea tridentata*) releases allelopathic chemicals that can retard the growth of competing plants around its root zone.

Competitors do not always face off in head-to-head contests. Hawks and owls hunting for the same rodents in a meadow may not hunt at the same time, but the presence of one predator reduces food for the other. Two tree species in a forest, though in physical proximity, compete primarily by extracting nutrients, water, and sunshine from their habitat to the exclusion of other trees, and not necessarily through active interference. These more passive forms of mutually detrimental interaction are called **exploitative competition**.

It is important to recognize that individuals pursuing limited resources are generally faced with two sources of competition. Competitive interaction among members of the same species is called **intraspecific competition**. Competitive interaction among members of different species is called **interspecific competition**. (Similar prefixes are used to distinguish intramural sports, among members of your own school, from intermural sports, between rival schools.)

Laboratory Studies

Key insights into the nature of competition came from the microbiology laboratory of Georgyi Frantsevitch Gause, who worked in Russia during the 1930s. Gause understood that the complexity of ecological systems makes identification of cause and effect very difficult, so he used his training in microbiology to develop the simplest model ecosystem he possibly could. Gause cultured *Paramecium*, a genus of ciliated protozoans, in small culture tubes on a diet of the rod-shaped bacterium *Bacillus subtilis*. To eliminate bacterial population growth as a confounding variable, Gause cultured the bacteria separately on solid media, scraped mature colonies off the culture plates, and resuspended the cells in a nutrient-free

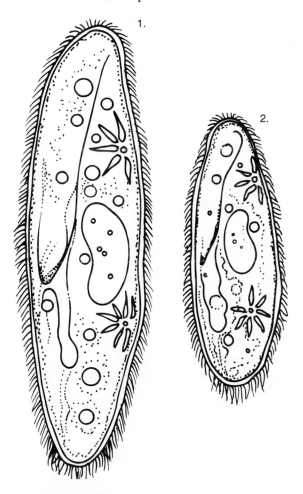

solution of physiological salts. Paramecia could then be introduced to feed on bacteria in these suspensions, but the bacteria could not reproduce. Every day, Gause put his protozoan cultures in a centrifuge at low speed to push all Paramecia to the bottom without killing them. He then used a pipette to remove the old fluid on top, and replaced it with a fresh bacterial suspension. He then gently stirred the Paramecia up from the bottom. This method guaranteed a constant food supply and constant waste removal for the ciliate. With bacterial food supply and the chemical environment held constant, growth of Paramecia was the only variable Gause needed to monitor in his experiments.

Gause's most famous experiment involved two closely related species: *Paramecium caudatum*, which is the larger, and *Paramecium aurelia*, which is the smaller of the two organisms illustrated in Figure 10.3. Species in the same genus, which are called **congeners**, were a wise choice for Gause's competition studies, because taxonomically related species would be expected to use similar resources in similar ways. To determine growth characteristics of each species, Gause first cultured each species separately. These isolated populations followed **logistic growth** curves, reaching similar **carrying capacities** when cell volume was taken into account. When Gause introduced both *Paramecium* species into the same culture tube, *P. aurelia* always persisted, and *P. caudatum* always died out. Although smaller, *P. aurelia* was consistently the stronger competitor (Figure 10.4).

Figure 10.3 *Paramecium caudatum* (1) and *Parame-cium aurelia* (2), from Gause's classic publication on competition.

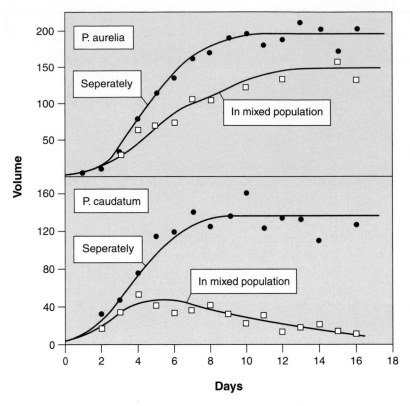

Further experiments with other pairs of protozoans produced similar results; when grown in mixed culture, even slight advantages in efficiency or stress tolerance allowed one species to survive and grow while the other species declined to extinction. The only way Gause was able to keep two species alive for any length of time was to create a more complex environment, such as a culture tube with glass wool in the bottom, which allowed each species to dominate in a different microhabitat. These observations led to Gause's **competitive exclusion principle**. Stated simply, Gause (1934) determined that *competing species cannot coexist within the same ecological niche.*

Figure 10.4　Gause's results for *Paramecium aurelia* (top graph) and *Paramecium caudatum* (bottom graph). Each species was grown separately (solid data points) and then in mixed population (open data points). Results show total biomass accumulation (numbers of individuals multiplied by cell size) over a 16-day period. Under competitive pressure in mixed cultures, *P. aurelia* growth is only slowed down, but *P. caudatum* is driven to extinction.

Check your progress:

If the competitive exclusion principle applies to all of nature, how can we account for so many species still existing in the world?

Hint: Natural ecosystems contain a great variety of niches.

Intraspecific vs. Interspecific Competion in Plants

In plants, the number of individuals is an indicator of the success of a population, but size of individuals is important as well. Large individuals tend to consume more resources and produce more offspring than small ones. For example, if sagebrush and bunchgrass are competing in a Great Basin habitat, counting individuals may not be the best way to determine which species is more successful, since individual size at maturity varies so much in plants of this type (Figure 10.5). A better measure of competitive outcomes in plants is **biomass** accumulation over time. To measure biomass, ecologists harvest all plants from a sampled area, separate the harvest by species, dry the plant material, and weigh the samples to determine which species has amassed more grams of tissue per unit area over the time of the study.

Dutch biologist C. T. de Wit developed a method for comparing the effects of intraspecific and interspecific competition in plants. His experimental design, called a **replacement series,** begins with seeds of two species planted in a series of pots. Let's call these plants Species A and Species B. An equal number of seeds are started in each pot. However, the proportions of the two species vary from all Species A to combinations of A and B, to all Species B (Figure 10.6). This is called a replacement series because Species B individuals are gradually replaced by Species A individuals, illustrated in the figure from left to right. If interspecific competition is less severe than competition within the species, results resemble Figure 10.7A. The two bottom curves on the graph show dry weights of Species A and Species B, separately, for each combination of planting frequencies. The top curved line shows the biomass of both species summed together. The straight line connecting the 100% A biomass and the 100% B biomass represents a null hypothesis: that the total yield at each frequency along the replacement series is predicted by the proportions of A and B planted in that pot. This is what we would expect if competition between species is exactly as intense as competition within the species. For example, for a mixed planting begun with 50% species A seeds and 50% species B seeds, the null hypothesis projects total biomass exactly halfway between the maxima for Species A and Species B.

Figure 10.5   Sagebrush and bunchgrass compete in a Great Basin habitat.

Seedlings Planted in a Replacement Series

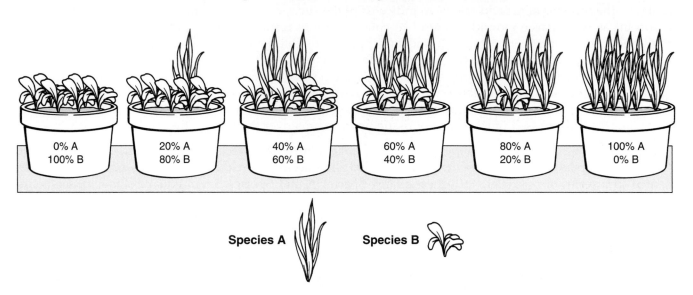

Figure 10.6 Replacement series of two-species plantings for de Wit competition experiment. Each pot in this example has five plants, but frequency of the narrow-leaved species A ranges from 0% at the left to 100% at the right.

The actual biomass totals can be compared with the null model to determine whether interspecific competition compares with intraspecific competition. If the two plants produce a higher combined biomass in mixed planting than in pure stands, then the total biomass line curves above the straight line predicted by the null model. This would be expected if the two plants are extracting different nutrients from the soil, or are in other ways exploiting different niches. Sometimes the de Wit experiment indicates high interspecific competition (Figure 10.7B). If one species interferes with growth of the other, as expected in allelopathic plants, then the combined biomass line sags below the straight line of the null model. Agricultural scientists conduct this kind of experiment to find out whether mixed species plantings could improve productivity of cropland or rangeland. For ecologists, the de Wit method can give insights into the extent of **niche overlap** between two competing plant species. The more closely two species compete in the same niche, the more likely combined yield in mixed plantings will be depressed below null expectations.

Check your progress:

What is the expected shape of the total yield curve in a de Wit diagram if interspecific competition is equivalent to intraspecific competition?

Answer: A straight line connecting the two 100% points

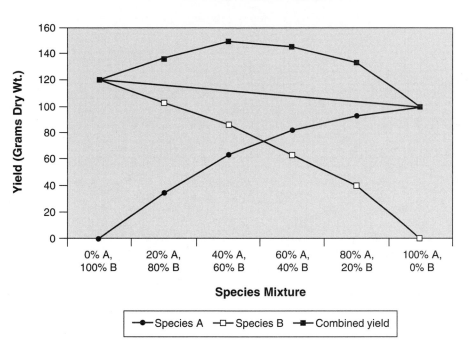

**de Wit Diagram
Low Interspecific Competition**

Species Mixture

Yield (Grams Dry Wt.)

●— Species A   □— Species B   ■— Combined yield

Figure 10.7A   de Wit diagram showing competition between species less severe than competition within species. The combined yield (top curved line) is above null expectations based on averaging individual species yields (straight line).

Modeling Competition

Although the de Wit diagram is a good way to describe the results of competition among plants, it does not forecast the outcome of competition in population terms. The challenge of predicting population growth in the presence of competitors was taken up by two researchers on different continents at the beginning of the 20th century. American biophysicist Alfred Lotka and Italian mathematician Vito Volterra independently discovered that well-known population growth equations could be modified to incorporate the effects of competition. Their simple mathematical models, called the **Lotka-Volterra competition equations**, have generated many useful questions and experimental investigations of competition in nature. (Lotka and Volterra also followed parallel approaches to the study of predator-prey relationships. You can find a discussion of those equations in Chapter 11.)

Lotka and Volterra's model of competition begins with the assumption that two species are competing for the same resources, and that their rates of population growth are described by the **logistic growth** equation. (For an introduction to the logistic growth model, see Chapter 4.) To demonstrate how this model works, let's name our two competitors Species 1 and Species 2. To keep track of the numbers of both species, we will identify their population sizes as N_1 and N_2. We can assume each species has a maximum **intrinsic rate of reproduction** (r), so we can designate the population growth rates of the two species r_1 and r_2. Since resources are finite, each species has a **carrying capacity** (K) within the shared environment. Even if both species are limited by the same resource, their carrying capacities may not be the same. For example, if Species 1 is a bison and Species 2 is a prairie dog, we would not expect a patch of grassland to support equal numbers of bison or prairie dogs! We therefore need to designate K_1 as the

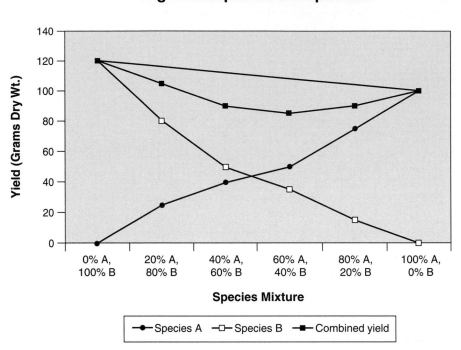

**de Wit Diagram
High Interspecific Competition**

Figure 10.7B   de Wit diagram showing competition between species greater than competition within species. The combined yield (concave curve) lies below null expectations based on averaging individual species yields (straight line).

carrying capacity for Species 1, and K_2 as the carrying capacity for Species 2. Finally, we will need symbols to show the negative impact of these two species on each other. The impacts are not always equal. To go back to our bison and prairie dog example, we would expect a single bison to have a greater negative influence on the colony of prairie dogs than a single prairie dog would have on a herd of bison. We will therefore define α as the per-capita effect of Species 2 on Species 1, and β as the per-capita effect of Species 1 on Species 2. The constants α and β are called **competition coefficients**.

Check your progress:

Identify each of these symbols from the Lotka-Volterra equations.

$N_1 =$ $K_1 =$

$N_2 =$ $K_2 =$

$r_1 =$ $\alpha =$

$r_2 =$ $\beta =$

Hint: Remember that each species must have a population size, a growth rate, a carrying capacity, and a competitive effect on the other.

To begin as simply as possible, let's imagine that Species 1 is enjoying population growth by itself in the absence of any competitors. Its numbers are predicted by the logistic model:

Look at the fraction at the end of the equation. The numerator of this fraction is the carrying capacity of the environment (measured in numbers of individuals the environment can hold) minus the number of individuals already there. In other words, the numerator shows the number of "empty slots" available for colonization. Since this number of openings is divided by total carrying capacity, $(K_1 - N_1)/K_1$ represents the proportion of the environment that remains unfilled.

Growth Equation for Species 1 in isolation:

$$\Delta N_1 = (r_1)(N_1)\frac{K_1 - N_1}{K_1}$$

ΔN_1 = change in population size from one time interval to the next

r_1 = population growth rate of Species 1

N_1 = population size of Species 1

K_1 = carrying capacity of Species 1

Now considering the whole equation, this mathematical statement says the change in the number of individuals of Species 1 is equal to Species 1's intrinsic rate of increase multiplied by the current population, multiplied by the portion of the environment that is still open for colonization. If numbers of Species 1 are low, then N_1 is near 0, and $(K_1 - N_1)/K_1$ approximates (K_1/K_1), which would be equivalent to multiplying by 1. This means that the third term in the equation has little impact on population growth as long as population size is low. However, as N_1 approaches K_1, the third term approaches zero, and multiplying by zero results in no change in the population. In other words, as the environment fills up, the population approaches zero population growth so that numbers are stabilized at carrying capacity. This is the meaning of logistic growth.

Check your progress:

How does the size of N_1 relative to K_1 determine the growth rate of species 1 in the absence of competition? Explain what $(K_1 - N_1)/K_1$ means, in biological terms.

Hint: Consider the effects of multiplying by the third term in the growth equation when $N_1 = 0$. Compare this answer to your result if $N_1 = K_1$.

Next, we will assume that Species 2 enters the habitat. As Species 2 occupies space and consumes resources, the number of "empty slots" for growth of Species 1 in the environment will decline. How much will they decline? We can show carrying capacity being consumed by both species as follows:

Number of "empty slots" for Species 1 = $K_1 - N_1 - \alpha N_2$

From the total carrying capacity for Species 1, we subtract the number of Species 1 individuals, and we also subtract the number of Species 2 individuals multiplied by the competition coefficient α. Think of α as a conversion factor that transforms each individual of Species 2 to an equivalent number of Species 1, for the purpose of calculating resource consumption. For example, α would be quite large if we were converting bison to an equivalent number of prairie dogs, but quite small if we were converting prairie dogs to an equivalent number of bison.

Having incorporated both species into the model, we can now calculate the growth rate of Species 1 in the presence of its competitor as follows:

Note that this is just the logistic growth equation, with Species 2 thrown into the numerator of the third term. As long as the carrying capacity K_1 exceeds the total numbers of Species 1 and Species 2 (converted by α to Species 1 equivalents), then populations of Species 1 will grow. The actual rate of growth depends on the numbers already present (N_1) and the population growth rate (r_1). Population growth is indicated by a positive value of ΔN_1, stable population size is indicated by a zero value of ΔN_1, and population decline is indicated by a negative value of ΔN_1.

Competition Equation for Species 1:

$$\Delta N_1 = (r_1)(N_1)\frac{K_1 - N_1 - \alpha N_2}{K_1}$$

ΔN_1 = change in population size from one time interval to the next

r_1 = population growth rate of Species 1

N_1 = population size of Species 1

N_2 = population size of Species 2

α = per-capita effect of Species 2 on Species 1

K_1 = carrying capacity of Species 1

Check your progress:

Use the competition equation to calculate the number of new individuals expected in the population if $N_1 = 43$, $N_2 = 86$, $\alpha = 0.4$, $r_1 = 0.75$, and $K_1 = 110$.

Answer: 9.6 new individuals added, so we project a total of 52.6 individuals of Species 1 in the next time period.

To simplify interpretations of the model, it is useful to ask, under what conditions does the population have zero population growth? Examine the competition equation for Species 1. There are three terms multiplied together in the equation. If any of these terms is zero, then ΔN_1 will be zero. However, the first two terms give trivial solutions. In the case of the first term, population growth is obviously zero if N_1 is zero, because the species cannot grow if it is already extinct. In the case of the second term, the population cannot grow if $r_1 = 0$, because that would indicate a zero intrinsic rate of increase, with no biological capacity for reproduction beyond population replacement. The third term is therefore the interesting case. As we saw in our earlier discussions, ΔN_1 will be zero whenever $K_1 - N_1 - \alpha N_2 = 0$. In biological terms, the population stops growing whenever the two species fill up the carrying capacity of the environment. Solving this equality for N_1 gives us the following zero-growth equation conditions for Species 1:

Zero-growth Equation for Species 1:

$\Delta N_1 = 0$ whenever:

$$K_1 - N_1 - \alpha N_2 = 0$$

or:

$$N_1 = K_1 - \alpha N_2$$

ΔN_1 = change in population size from one time interval to the next
N_1 = population size of Species 1
N_2 = population size of Species 2
α = per-capita effect of Species 2 on Species 1
K_1 = carrying capacity of Species 1

Note that the zero-growth equation is in the form of a straight line with a slope equal to $-\alpha$. The best way to illustrate the conditions for zero growth is with a **joint abundance graph** (Figure 10.8). This graph shows numbers of Species 1 on the x-axis and numbers of Species 2 on the y-axis. A point on this kind of graph shows both population sizes simultaneously. Changes in the joint abundance of the two populations can be shown by moving this point around within the graph space. The figure shows the zero growth equation for Species 1 plotted as a line on the graph. This line is called the **zero growth isocline** for Species 1. Remember that the zero growth isocline represents the combined numbers of Species 1 and Species 2 that equal the carrying capacity for Species 1. Anywhere below this line (closer to the graph's origin) is a set of conditions allowing growth of Species 1. Anywhere above and to the right of this line represents an overloaded carrying capacity, which will result in decline of Species 1. Arrows on the graph show the direction the joint population point will move, depending on its position relative to the isocline.

Check your progress:

Beginning with the zero-growth equation for Species 1,

$$N_1 = K_1 - \alpha N_2$$

Solve for N_1 when $N_2 = 0$

Then solve for N_2 when $N_1 = 0$

Answer: Look at the isocline on the graph.
When $N_2 = 0$, the line intersects the x-axis. What value of N_1 do you see at this point?
When $N_1 = 0$, the line intersects the y-axis. What value of N_2 do you see at this point?

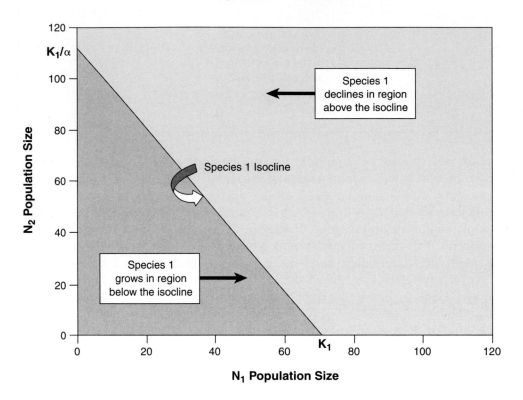

Figure 10.8   Joint abundance graph showing the Species 1 zero-growth isocline. The N_1 population, shown on the horizontal axis, grows whenever the joint abundance lies in the region to the left of the isocline (shaded) and declines whenever the joint abundance lies in the region to the right of the isocline (unshaded).

Now we need to consider Species 2. By the same arguments we used for Species 1, we can generate the following complementary equations:

Competition Equation for Species 2:

$$\Delta N_2 = (r_2)\,(N_2)\,\frac{K_2 - N_2 - \beta N_1}{K_2}$$

ΔN_2 = change in population size from one time interval to the next
r_2 = population growth rate of Species 2
N_2 = population size of Species 2
N_1 = population size of Species 1
β = per-capita effect of Species 1 on Species 2
K_2 = carrying capacity of Species 2

And by similar logic, the following equation can be used to draw a zero growth isocline for Species 2 (Figure 10.9).

Zero-growth equation for Species 2:

$\Delta N_2 = 0$ whenever:

$$N_2 = K_2 - \beta \, N_1$$

ΔN_2 = change in Species 2 population size from one time interval to the next
K_2 = carrying capacity of Species 2
N_1 = population size of Species 1
β = per-capita effect of Species 1 on Species 2

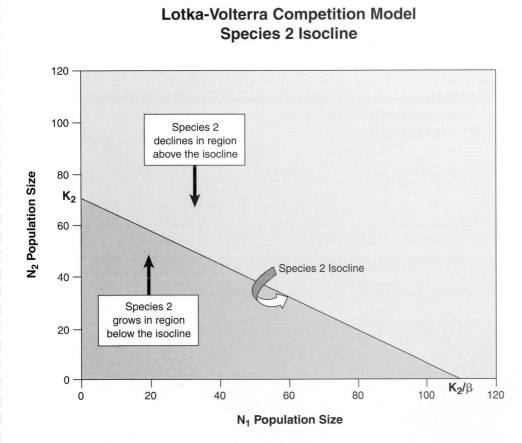

Lotka-Volterra Competition Model
Species 2 Isocline

Figure 10.9   Species 2 isocline. The N_2 population, shown on the vertical axis, grows whenever the joint abundance lies in the region below the isocline (shaded) and declines whenever the joint abundance lies in the region above the isocline (unshaded).

Note that numbers of Species 2 are represented vertically on the graph, so the labels on the isocline are inverted relative to the isocline for Species 1.

Check your progress:

By inspecting the zero-growth isocline for Species 2 in Figure 10.9, can you estimate the value of K_2 in this example? Can you estimate the value of β?

Answer: $K_2 = 70$
$\beta = 0.64$

The Lotka-Volterra model gets interesting when both species isoclines are shown on the same graph (Figure 10.10). Since the isoclines cross in the illustrated example, the graph space is divided into four regions. Near the origin is a region below both isoclines, so both species can grow. The black arrow in the figure shows the direction of change on the joint abundance graph within this region. Above and to the left is a region below the Species 1 isocline, but above the Species 2 isocline. Species 1 can continue to grow below its isocline, so the joint abundance moves toward the right, but Species 2 is declining so the joint abundance moves down. The arrow is a vector summing both of these trends, indicating a change down and toward the right. At the lower right is a region above the Species 1 isocline but below the Species 2 isocline, so the joint abundance moves up and to the left as Species 1 declines and Species 2 grows. Finally, at the upper right is a region above both isoclines, so the joint abundance declines down and to the left in this region.

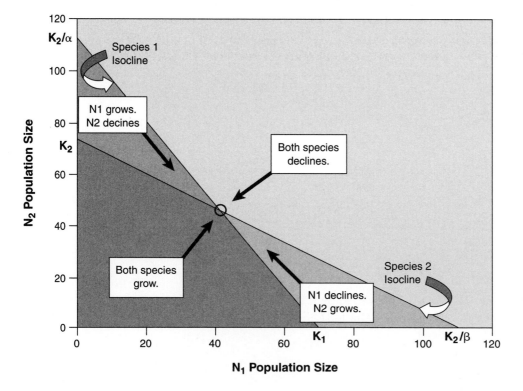

Lotka-Volterra Competition Model
N1 and N2 Isoclines

Figure 10.10

Note that wherever the two populations begin, their joint abundance moves toward the point of intersection of the two isoclines. This point, indicated by a circled dot on the graph, is a point of equilibrium because it is the only place on the graph in which ΔN_1 and ΔN_2 both equal zero. The model predicts coexistence of Species 1 and Species 2, with stable populations determined by the intersection of the isoclines. If environmental perturbations move the joint equilibrium off this stable equilibrium point, the system returns to equilibrium. You can think of this model like a round-bottomed bowl holding a marble: whatever happens to move the marble away from equilibrium, it tends to return to the same point.

The particular outcome predicted in Figure 10.10 occurs only when $K_2/\beta > K_1$ and $K_1/\alpha > K_2$. This tends to happen when the competition coefficients are low, making K_2/β and K_1/α large in relation to carrying capacity. Biologically, this makes sense. Coexistence between competitors would be expected if their competitive effects on one another are small. As α and β become larger, the value of the fractions decreases, and it is less likely the isoclines will cross in this direction.

Figure 10.11 shows four possible outcomes for Lotka-Volterra competition, depending on the orientation of the two zero-growth isoclines. Graph A predicts coexistence, as you have already seen. In Graph B, the isoclines cross in the opposite direction, with $K_2/\beta < K_1$ and $K_1/\alpha < K_2$. There is still an equilibrium point at the intersection, but note the direction of the vectors of change pointing away from the center of the graph. Any small perturbation moving the point away from the center will result in one species or the other driving its competitor to extinction. You can think of this condition as an upside-down bowl with a hemispherical bottom and a marble balanced carefully on top. A slight push destroys the unstable equilibrium, and the marble falls to rest at a more stable equilibrium point. This model predicts either one species or the other will win, but the outcome depends on initial conditions.

Graphs C and D in Figure 10.11 show what happens when the isoclines do not cross. In these cases, the species whose isocline lies farthest from the origin will continue to grow when its competitor is above capacity, so the same species always drives the other to extinction. A marble on an inclined plane is a good way to visualize these graphs.

There is no question that the Lotka-Volterra competition models are oversimplified, and fail to incorporate all the factors influencing competition in nature. However, we should remember that models are supposed to be simplifications of a complex reality. The key to a good model is retaining a few key elements of the system so that we can better grasp how it works. For nearly a century, the Lotka-Volterra competition equations have helped ecologists to explain community interaction, and to generate more interesting questions for their studies of competition.

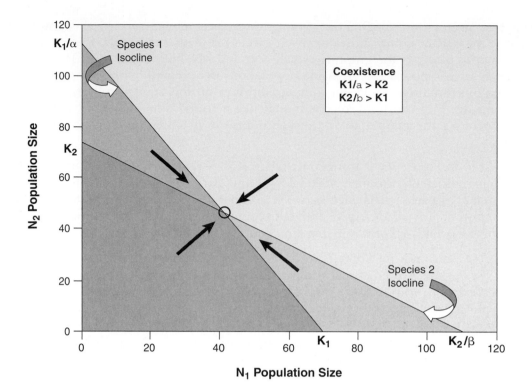

Figure 10.11A

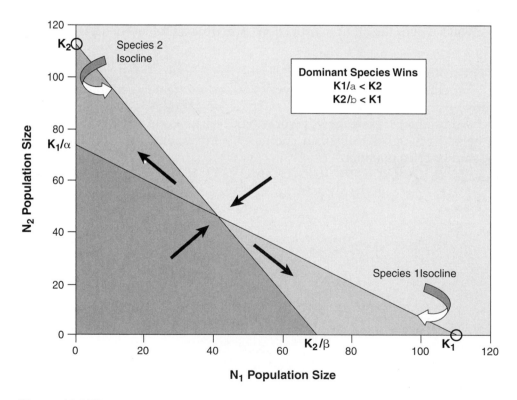

Figure 10.11B

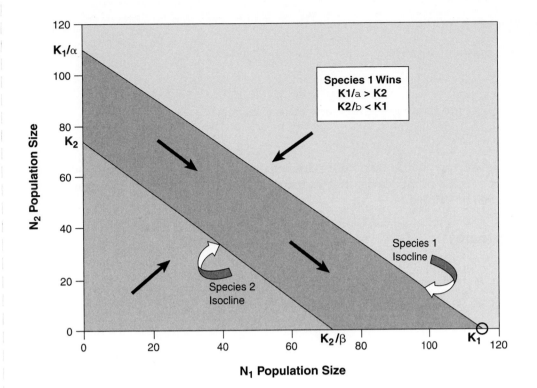

Figure 10.11C

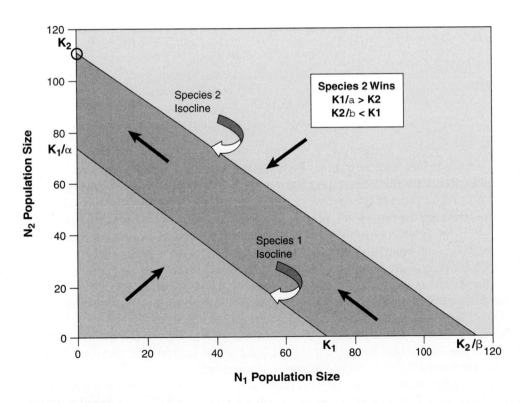

Figure 10.11D

METHOD A: ALLELOPATHY

[Laboratory activity, spring, summer, or fall]

Research Question

Is lettuce sensitive to allelopathic compounds found in black walnut leaves?

Preparation

Seeds can be obtained at garden supply stores, but plant parts as a source of allelopathic compounds are seasonal. The exercise calls for black walnut leaves, but radish leaves, sweet potato plants grown from cuttings, or alfalfa sprouts can be substituted.

Materials (per laboratory team)

3 Petri dishes, 10 cm diameter

9 sheets of round filter paper, 9 cm diameter

Grease pencil or marker to label glassware

Lettuce seeds (packet of 75 or more seeds)

Lettuce leaves, 100 g

Black walnut leaves, 100 g

Distilled water (1 liter)

Blender

2 large funnels

3 flasks, 500-ml each

Paper toweling or cheesecloth

Access to refrigerator

Procedure

1. Weigh out 50 grams of lettuce leaves and 50 grams of walnut leaves.
2. Place 50 g lettuce leaves in blender with 300 ml distilled water. Blend long enough to disrupt cells.
3. Push the center of a paper towel (or piece of cheesecloth) into a large funnel. Place the funnel in a large flask. Pour your leaf extract through the funnel, filtering out leaf fragments to get a clear solution of plant compounds. Collect about 250 ml of lettuce leaf extract, then label the flask and place it in a refrigerator.
4. Wash the blender thoroughly with soap, and rinse several times.
5. Repeat steps 2 and 3 with black walnut leaves to make a walnut leaf extract.
6. Place three sheets of filter paper in each of the three Petri dishes. Add 5 ml of lettuce extract, or enough to saturate the filter paper, to one of the dishes. Add an equal amount of walnut extract to the paper in a second dish, and an equal volume of distilled water to the third dish.
7. Place 25 lettuce seeds on top of the wet filter paper in each of the dishes. Cover the dishes, and label them on the sides of the lids so that the tops transmit light. Place both under constant light from a growth lamp or fluorescent light. Incubate at room temperature.
8. Check dishes every couple of days to ensure that the filter paper stays moist. Add equal amounts of distilled water to all dishes as needed to keep paper moist. Seedlings should be fully developed within a week.
9. Compare germination success and seedling size in the three dishes. If you wish, use a Chi-square contingency test (see Chapter 8) to test the null hypothesis that germination is equal in the three treatments. If seedlings do emerge, measure the length of the root and calculate a mean and variance for each treatment group (see Chapter 1). Answer the following questions to analyze your data:

Questions (Method A)

1. Do black walnut leaves seem to have an allelopathic effect on germination of lettuce seedlings? on growth rates after germination?

_____

_____

_____

_____

_____

2. Why was lettuce extract used in one of the treatments? Why was plain water used in the third treatment? Were there any differences between lettuce extract and plain water treatments? Why do you think this might be so?

_____

_____

_____

_____

_____

3. What does this experiment imply about the way black walnut might compete with other plants? How might you test this hypothesis in a field setting?

_____

_____

_____

_____

_____

4. If other trees react in the same way as lettuce to black walnut compounds, how would results of a replacement series involving black walnut appear in a de Wit diagram? Draw and label a sketch to indicate your conclusion.

_____

_____

_____

_____

5. With reference to the Lotka-Volterra models, what sort of competition coefficient would you expect for the effects of black walnut on a broad-leaved competitor?

_____

_____

_____

_____

_____

METHOD B: COMPETITION WITHIN AND BETWEEN SPECIES

[Laboratory activity, spring, summer, or fall]

Research Question

How do intra- and interspecific competition compare among legumes and grains?

Preparation

This experiment requires bench space in a greenhouse, growth chamber, or extensive growth lights in a laboratory. Small peat pots of the type that expand when wet work well for this experiment. Mung beans and wheat seeds can be ordered from supply houses, but are also available from health food stores.

Materials (per laboratory team)

6 seed-starting pots (2–3" diameter) in trays, with potting soil

Mung bean seeds

Wheat seeds

Tap water, aged in a watering can or large flask for watering

Electronic balance, accurate to .01

12 large plastic weighing dishes

Procedure

1. Use label tape to identify each pot as shown in the first row of the following table. Fill pots with potting soil. Place the pots in a tray to catch excess water and water all pots liberally. To plant seeds, push a pencil point into the wet soil to make a hole and push in the seed until it is just below the soil surface. Plant *initial numbers of seeds* as shown in the second and fourth rows of the following table:

Label	0% B 100% W	20% B 80% W	40% B 60% W	60% B 40% W	80% B 20% W	100% B 0% W
Initial # bean seeds	0	4	6	8	10	12
Final # bean seeds	0	2	4	6	8	10
Initial # wheat seeds	12	10	8	6	4	0
Final # wheat seeds	10	8	6	4	2	0

2. Note that two extra seeds of each species are placed in each pot to account for any failures in germination. After seeds germinate, pull out excess seedlings to achieve the final number of seeds in each pot, as shown in the second and fourth rows of the table.
3. Place all pots in a randomized pattern under growth lights or on a greenhouse bench. Grow at room temperature (20–25° C) for 3–4 weeks. Check plants frequently and add water to the underlying tray to ensure all pots are equally moist.
4. After 3–4 weeks, cut all plants off flush with the soil, and separate plants from each pot into two weighing dishes; one for wheat and one for beans. You should have 10 weighing dishes for six pots, because the 100% pots contain only one species and the others have two. Expose all your harvested plant material to the air in a fairly dry place for a week.

5. Weigh the dried plant material in their weighing dishes, then dump out the plant material and reweigh the dish itself. Subtract the weights to determine the dry weight of beans (or wheat) in that pot. Enter data in the following data table.
6. Calculate class means for each species in each pot, and use the class averages to complete a de Wit diagram for this experiment. Answer the questions for Method B to interpret your results.

Results for Method B
Yields from Plant Competition Replacement Series
(Dry Weight in Grams)

TREATMENT GROUP:	0% B 100% W	20% B 80% W	40% B 60% W	60% B 40% W	80% B 20% W	100% B 0% W
Bean weight (Individual data)	X					
Wheat dry wt. (Individual data)						X
Bean weight (Group average)	X					
Wheat weight (Group average)						X

de Wit Diagram for Group Data

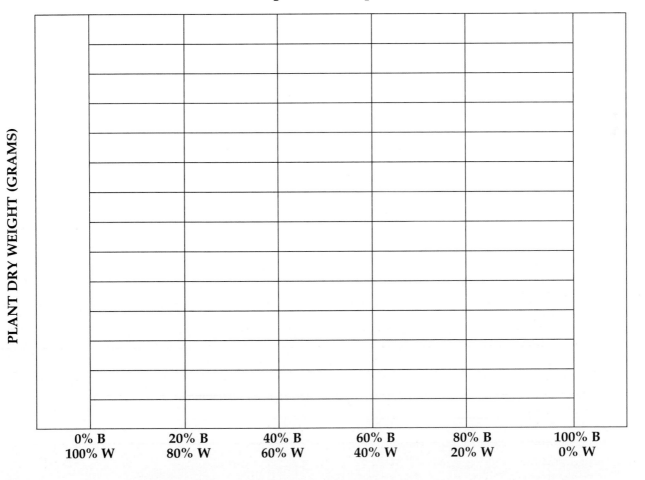

PLANT DRY WEIGHT (GRAMS)

| 0% B 100% W | 20% B 80% W | 40% B 60% W | 60% B 40% W | 80% B 20% W | 100% B 0% W |

Questions (Method B)

1. Explain the null hypothesis in this experiment. What is the value of a null hypothesis?

_____

_____

_____

_____

_____

2. For mung beans growing with wheat, is interspecific competition stronger or weaker than intraspecific competition? Is the same true of wheat?

_____

_____

_____

_____

_____

3. Do mung beans and wheat seem to have substantial niche overlap, or do they occupy distinct niches? How can you tell from the de Wit diagram?

_____

_____

_____

_____

_____

4. Would beans and grain be good candidates for mixed plantings in the same field to increase agricultural yields, or not? Explain.

_____

_____

_____

_____

_____

5. Referencing the Lotka-Volterra model of competition, would you suspect the two competition coefficients α and β to be large or small in this system? Are they likely to be equal or unequal in size? Explain.

_____

_____

_____

_____

METHOD C: COMPETITION PROBLEMS

[Classroom or homework activity]

1. Observe the population dynamics of the two *Paramecium* species illustrated in Figure 10.4. Assume the carrying capacities of each species (in volumetric units as shown on the graphs) are indicated by the maximum size of the populations when cultured in isolation. Draw a joint abundance graph, with Lotka-Volterra isoclines consistent with the results of competition in this experiment. (Consult Figure 10.11 for possible arrangements of the isoclines. You will not have values for α and β, so the important consideration is the relative positions of the isoclines.)

2. Assume the maximum density of harvester ants in a desert region is eight colonies per hectare in the absence of honeypot ants. Five colonies of honeypot ants move into the area, and as a result of interference competition, the number of harvester ant colonies falls to six colonies per hectare. If α is a competition coefficient representing the effect of honeypot colonies on harvester colonies, what is the value of α?

3. Assume a prairie reserve in Kansas has sufficient grassland to maintain a herd of up to 180 bison. Both bison and elk are introduced to the reserve. From past experience, assume that α, the effect of elk on bison, is 0.3 and that β, the effect of bison on elk, is 0.45. Carrying capacity for elk is 230. Use the Lotka-Volterra equations to predict the equilibrium population sizes of these two competing species. Draw a joint abundance graph with isoclines to illustrate your answer.

4. In the Kansas preserve described in problem 3, if the bison herd is culled every year and maintained at a constant number of 50 individuals by the park staff, predict the equilibrium number of elk. Explain your answer by referencing isoclines on the joint abundance graph.

5. Assume that Species 1 (cattails) and Species 2 (rushes) compete for space around a pond in such a way that $K_1 = 80$ shoots per m^2 and $K_2 = 115$ shoots per m^2. Competition coefficients are $\alpha = 0.8$ and $\beta = 1.6$, indicating a high level of niche overlap. Draw a joint abundance graph with isoclines, adding arrows to indicate the direction of change within each area of your graph. In theory, if the populations were at the point of intersection of the two isoclines in your drawing, both cattails and rushes would demonstrate zero growth, and the system would be in equilibrium. Is the population likely to remain at this point? What biological factors might determine the actual outcome for this pond?

FOR FURTHER INVESTIGATION

1. Locate ant colonies on your campus. Sidewalks and paving stones often shelter nests. Map colony sites, measuring distances between colonies. Use small dabs of acrylic paint to mark individual ant workers, and follow them as they forage. Do ants forage near other colony sites, or do the colonies seem to have territorial boundaries?

2. Many weed species, such as crabgrass and ragweed, have been shown to release biologically active compounds. Try collecting weeds of different types from your campus or weedy lots or fields to extend the allelopathy investigation outlined in Method A.

3. Try the replacement series experiment for seeds collected from wild plants or trees in your area. Does the de Wit diagram you produce explain distribution patterns of these species?

4. Use a spreadsheet program to model the Lotka-Volterra equations. Because the growth equations are in the discrete form, you can use the solution from one year as the population size input for the next. Plot results on a joint abundance graph to see what happens as populations approach the zero-growth isoclines.

FOR FURTHER READING

de Wit, C. T. 1961. Space relationship within populations of one or more species. *Society of Experimental Biology Symposium* 15:314–329.

Gause, G. F. 1934. *The Struggle for Existence.* Accessible at: http://www.ggause.com/Contgau.htm

Halsey, Richard W. 2005. In search of allelopathy: an eco-historical view of the investigation of chemical inhibition in California coastal sage scrub and chamise chaparral. *Journal of the Torrey Botanical Society* 131:343–367.

Holldobler, Bert. 1986. Food robbing in ants: a form of interference competition. *Oecologia* 69(1):12–15.

Chapter 11

Predators and Prey

Figure 11.1 Starfish are important predators in marine ecosystems.

INTRODUCTION

The sea star *Asterias* may seem slow and benign to human observers, but it is a formidable predator on mollusks (Figure 11.1). Moving on flexible arms and hundreds of tiny tube feet, the sea star actively seeks its prey on the sea floor. When it encounters a clam or mussel, the sea star grasps the bivalve, pries open the shell, and everts its stomach into the opening to digest the soft body of the mollusk inside. Although the bivalve's shell seems impenetrable, the sea star has evolved a body plan and predatory behavior that circumvents the prey's defenses. This kind of evolutionary arms race is typical of predators and prey; each adaptation of the prey leads to **co-adaptation** by its predators.

On an ecological time scale, studies of sea stars have taught ecologists a lot about the influence of predators on natural communities. In a classic field experiment, Robert T. Paine (1966) artificially removed sea stars from tide pools and monitored the resulting changes in diversity of invertebrate animals in a rocky intertidal environment. Released from predation, mollusk populations grew and competition among species increased. As a result, some species drove others to extinction, and overall **biodiversity** decreased. The so-called "**top-down control**" of ecological community structure by predators has been demonstrated in many kinds of biological communities. If a single species at the top of the food chain seems especially crucial to maintaining the character of a biological community, the top carnivore is called a **keystone** predator. The term "keystone" is borrowed from architecture, and refers to a structural element that holds an entire system together (Figure 11.2). The supposition is that the entire structure collapses if the keystone is removed. Prey can function as keystone species as well. Population dynamics of plankton in freshwater lakes, for example, can significantly impact kinds and numbers of fish in the community. This is called "**bottom-up control**."

Figure 11.2   The keystone (A) holds other parts of an arch in place. By analogy, a keystone species is essential to maintaining community structure. Removal of the keystone (or extinction of a keystone species) results in drastic community change.

Not all types of sea stars enhance biodiversity. In Pacific waters from Australia to the South American coast, a spiny sea star called the crown of thorns "starfish" has increased both its distribution and its numbers, inflicting terrible damage to coral reefs throughout its range (Figure 11.3). Hard corals may seem unlikely food for a predator without teeth, but the crown of thorns feeds by extruding its stomach over the thin layer of living polyps that cover a coral's calcareous skeleton. As it moves across the reef, the crown of thorns leaves behind a white swath of dead coral. Without its living foundation, the reef deteriorates and many other species of fish and invertebrates are lost. Why are crown of thorns sea stars increasing in numbers? One widely accepted theory is that population growth of the sea star is linked to a decline in large predatory marine snails called tritons. In a reverse of the *Asterias*/clam relationship, the triton is a

Figure 11.3 Crown of thorns starfish feeding on corals off the coast of Hawaii.

mollusk that feeds on crown of thorns sea stars (Figure 11.4). Unfortunately, the triton shell is highly sought by collectors, and its populations have been depleted by divers taking living tritons from reefs throughout the Pacific. Released from predation by the increasingly rare triton, populations of the crown of thorns sea star have grown exponentially, with predictable consequences for reefs. Whether sea stars eat mollusks or mollusks eat sea stars, the importance of predator-prey relationships in maintaining a stable community structure emerges as a common ecological theme. To understand why some predator/prey systems are more stable than others, we need a more quantitative understanding of predator and prey populations.

Check your progress:

Based on the two predator/prey cases listed above, what ecological consequences would you expect from overfishing large predatory fishes from the world's oceans?

Answer: Removal of top predators often causes loss of species diversity due to competitive interactions among smaller species released from top-down control.

Figure 11.4   A giant triton snail attacks a crown of thorns sea star.

When ecologists speak of stability in biological communities, we should not imagine a static environment with no year-to-year changes. Populations of predators and prey rise and fall, often demonstrating cyclic changes in numbers over many generations. An instructive example of fluctuations in mammalian predator/prey populations is found on Isle Royale, a large forested island 24 km off the north shore of Lake Superior. Here moose and wolf populations have existed together for over 50 years, but many smaller mammals otherwise expected in woodland habitats at this latitude are absent because of the great distance across very cold water to the mainland. Hunting is prohibited, as the entire island has been preserved as a national park. Movement of animals on or off the island is practically nonexistent. Isle Royale wolves have few other prey than moose, and moose have no other predators. For ecologists, this kind of simplified natural system provides an opportunity to study interactions between one predator and one prey without the confounding variables encountered in more complex communities. Results of 50 years of population census data are shown in Figure 11.5. Note that wolves and moose are shown on separate scales; it is typical for prey to exist in much larger numbers than their predators. Like many systems dominated by one predator and one prey species, these populations fluctuate in cycles. Although longer monitoring will test the generality of the trend, a period of roughly 20 years between peak numbers seems to characterize both populations, with the predator peak lagging several years behind the prey peak. Our understanding of this process involves mutual feedback as follows: 1) If prey and predator populations are both low, prey numbers increase exponentially. 2) When predators have more food, their survival and reproductive success increases, and their population growth follows that of their prey. 3) As predator populations increase, the death rate for prey exceeds the birth rate, and prey populations decline. 4) After prey numbers decline, predators deprived of food face population collapse. 5) The cycle begins again with low numbers of both species.

Wolf and Moose Populations on Isle Royale

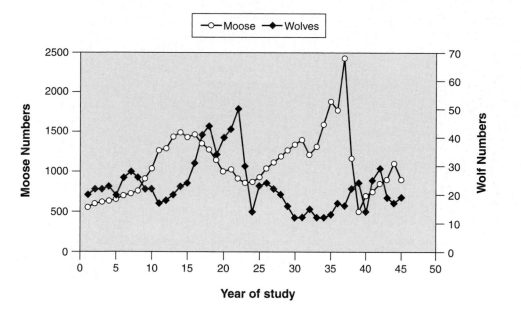

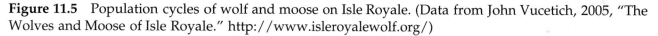

Figure 11.5   Population cycles of wolf and moose on Isle Royale. (Data from John Vucetich, 2005, "The Wolves and Moose of Isle Royale." http://www.isleroyalewolf.org/)

This is a common scenario in two-species systems, though the period of predator-prey cycles depends on the life spans and reproductive rates of the organisms. For predatory mites feeding on herbivorous mites in laboratory experiments, population peaks occur every three months. Snowshoe hare and Canada lynx cycles, documented over 200 years through fir trading records, show a predictable pattern of peaks every nine years (Figure 11.6). Clearly, any model we develop to explain predator-prey interactions must incorporate predator and prey population cycles.

Check your progress:

When predator and prey populations are plotted over several generations, why do predator population peaks usually lag behind prey population peaks?

Hint: Predators usually have longer lives and slower reproduction than their prey.

Ecology on Campus

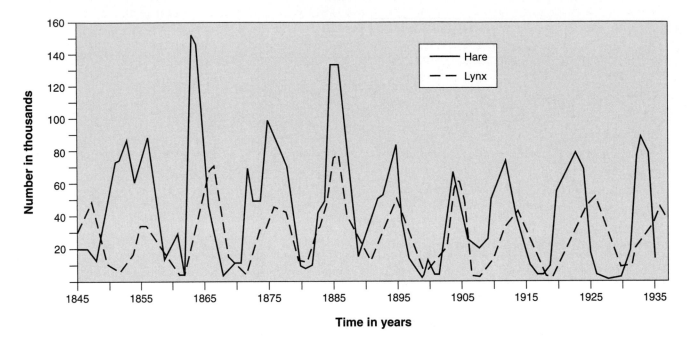

Figure 11.6   Snowshoe hare and Canada lynx populations rise and fall on a 9-year cycle. (After Bolen, E. G. and W. L. Robinson. 1999. *Wildlife ecology and management.* Prentice Hall, N. J.)

Modeling Predator and Prey Populations

Alfred Lotka, in the United States, and Vito Volterra, in Italy, independently modeled not only mathematical descriptions of competition (see Chapter 10), but also developed equations that describe the cyclic patterns of change in predator and prey numbers. To follow their reasoning, let's first define some terms. Let **P** be our symbol for the numbers of predators in our system, and let **H** stand for the numbers of prey (herbivores). Prey grow exponentially in this model with an intrinsic rate of growth equal to **r**. (For a more complete explanation of exponential growth, see Chapter 4.) We also need to quantify the rate of capture when predators encounter prey; let's call this the predation constant **p**.

As for predators, Lotka and Volterra assumed a constant death rate **d**. Since predator birth rate depends in part on prey consumed, they used the constant *a* as the predator birth constant. Think of *a* as the rate at which a predator converts food into offspring.

Check your progress:

Identify each of these symbols from the Lotka-Volterra predator-prey equations.

P =   **d** =

H =   **p** =

r =   *a* =

Hint: P, d, and *a* describe the predators.
H and r describe the prey.
The value of p affects both.

To model prey populations, we assume that changes in population size result from births minus deaths. Births are assumed to follow an exponential curve, with changes in prey numbers equal to the prey population size times the intrinsic rate of increase, or $r \cdot H$. Death rate of prey is assumed to be due to predation, which is modeled as the predation constant times the number of prey times the number of predators. In other words, negative change in population numbers due to predators killing prey is equal to $p \cdot H \cdot P$. Note that by multiplying the prey and predator populations together, we get a number proportional to the probability that a predator meets its prey during a given time interval. If both populations are high, the product of their numbers is very high. If one or both populations are low, the product of their numbers is low, and therefore the frequency of predation is low. Multiplying by p simply tells us how often an encounter becomes a successful kill for the predator. The population size for prey can therefore be stated as follows:

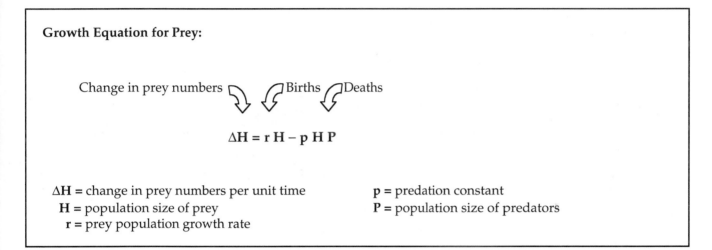

Growth Equation for Prey:

Change in prey numbers   Births   Deaths

$$\Delta H = r\,H - p\,H\,P$$

ΔH = change in prey numbers per unit time p = predation constant
H = population size of prey P = population size of predators
r = prey population growth rate

Any time the constants are known and census data indicate predators and prey numbers, we can use this equation to project the number of prey added to or subtracted from the population. In other words, $H + \Delta H$ = the herbivore population for the next time period.

Check your progress:

For a population of sea stars feeding on clams in the shallows around a small island, assume that $H = 1200$, $p = 0.004$, $r = 0.7$, and $P = 130$. What is the projected population size for clams?

Answer: 216 new clams added, so the prey population will grow to 1416.

To examine conditions critical to prey survival, it is instructive to ask, under what conditions can prey populations maintain a stable population size? If the population is not growing under these conditions, then $\Delta H = 0$. The equation can be simplified as follows:

Zero-Growth Condition for Prey:

$$0 = rH - pHP$$

Adding pHP to both sides: $$pHP = rH$$

Dividing both sides by **H**,
this simplifies to: $$P = r/p$$

H = population size of prey   p = predation constant
  r = prey population growth rate   P = population size of predators

---

Since both **r** and **p** are constants, the value **r/p** represents a constant number of predators needed to hold prey populations in check. Let's look at the biological meaning of the fraction on the right side of the equation. When **r**, the growth rate of the prey, is large, the number of predators needed to control the prey will be large. It also makes sense that the predation constant **p** is in the denominator, because if the predator is highly efficient at capturing prey, **p** will be large and thus the number of predators needed to control the prey will be relatively small.

Think of r/p as a threshold number of predators. Whenever P is lower than the threshold number, prey populations grow. Whenever P is higher than the threshold number, prey populations decline. A joint abundance graph (Figure 11.7) is a good way to illustrate both populations at the same time. Prey numbers (H) are shown on the horizontal axis of the graph, and predator numbers (P) are shown on the vertical axis. A point on this kind of graph represents both populations at the same time. Movement of this point within the graph space shows how both populations change over time.

Zero-growth conditions for the prey are represented by the line at the top of the shaded area, representing the number of predators (r/p) that would create zero-growth conditions for prey. On a graph, we call this an **isocline**. It is a graphic representation of the equation for zero-growth conditions for prey explained above. Note that the joint abundance point will move to the right (indicating prey growth) in the area under the isocline, and to the left (indicating prey decline) in the area above the isocline.

Check your progress:

For the predator-prey system illustrated in Figure 11.7, assume the predation constant is .004. Estimate the reproductive rate of the prey.

Answer: estimating r/p = 12.5 yields r = .05

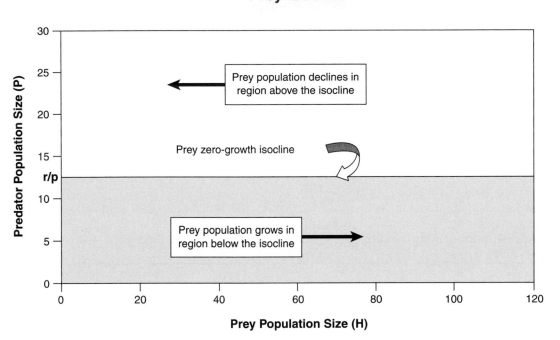

Lotka-Volterra Predation Model
Prey Isocline

Figure 11.7   Zero-growth isocline for prey. The position of the isocline is determined by the number of predators (r/p) required to keep prey numbers in check. Lower predator numbers allow prey population growth; higher predator numbers cause prey population decline.

Next we need to look at an equation to model predator population growth. Lotka and Volterra assumed that predators suffer a constant death rate (**d**), and that their growth was based on the amount of prey they are able to capture (**p**) multiplied by the reproductive rate per captured prey (**a**). As we saw for prey, changes in predator populations are equal to births minus deaths, as shown in the following equation:

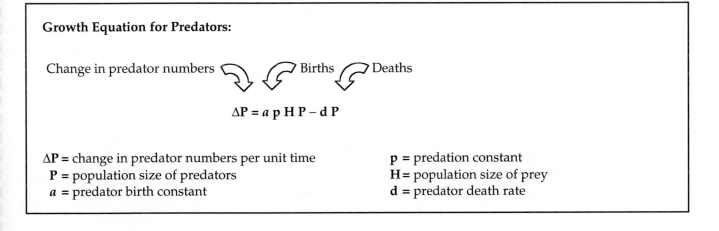

Growth Equation for Predators:

Change in predator numbers   Births   Deaths

$$\Delta P = a\, p\, H\, P - d\, P$$

ΔP = change in predator numbers per unit time   **p** = predation constant
P = population size of predators   **H** = population size of prey
a = predator birth constant   **d** = predator death rate

It is interesting to note that the three terms determining numbers of prey captured by predators once again appear as **p H P**, which is the capture rate multiplied by the two population sizes. However, number of captures determines deaths of prey, whereas it contributes to births of predators.

Check your progress:

To return to our sea star/clam example, assume that **H** = 1200, **p** = 0.004, *a* = 0.025, **d** = 0.12, and **P** = 130. What is the projected population size for sea stars?

Answer: Δ**P** = **0**, so the sea star population retains its current size of 130.

To calculate a zero-growth condition for predators, we again set Δ**P** = 0, as shown below:

Zero-Growth Condition for Predators:

$$0 = a \, p \, H \, P - d \, P$$

Adding dP to both sides: $d \, P = a \, p \, H \, P$

Dividing both sides by P
and rearranging, the
equation simplifies to: $H = d \, / \, (ap)$

a = predator birth constant   **P** = population size of predators
p = predation constant   **d** = predator death rate
H = population size of prey

What does this equation say biologically? Predator growth will be zero whenever the number of prey is equal to a value determined by **d** / (*a***p**). This is the threshold number of prey needed to maintain the predator population. The **d** in the numerator makes sense, because the higher the death rate of the predator, the more births (fueled by prey) it will take to replace those dying each generation. It also makes good biological sense for the predation constant (**p**) and the predator birth rate (*a*) to occupy the denominator, since these two terms measure the effectiveness of the predator at catching prey and converting them to offspring. The more offspring the predator can produce out of each prey encountered, the fewer prey are needed to maintain the predator population.

On a joint abundance graph, we can show this threshold number of prey as a zero-growth isocline for predators (Figure 11.8). This time the position of the isocline depends on a number of prey on the H axis, so the line runs vertically. Everywhere to the right of this line, the predators have sufficient food for population growth, so the joint abundance point moves upward. Everywhere to the left of this line, the predators are starving, so the joint abundance moves downward.

Check your progress:

From inspection of Figure 11.8, what is the minimum number of prey needed to maintain a population of predators in this system?

Answer: at least 50

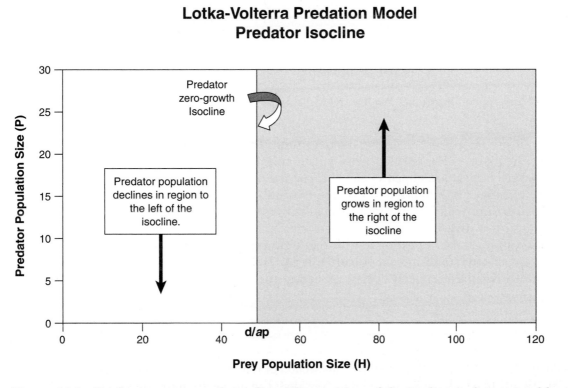

Figure 11.8 Predator zero-growth isocline. The position of the isocline is determined by the number of prey (d/ap) needed to maintain the predator population. Larger numbers of prey allow predator population growth; smaller numbers of prey cause predator population decline.

To see how the Lotka-Volterra predation model generates population cycles, let's look at both isoclines on the same graph (Figure 11.9). The isocline for zero prey growth intersects the isocline for zero predator growth near the center of the graph. If the two species were introduced to an environment at precisely this joint abundance, the model predicts equilibrium. At the intersection, predators are just sufficient to keep prey in check, and prey are just sufficient to feed the predators. If the joint abundance begins a bit off center, the populations will enter a cycle, as shown by the curved arrows in the graph. With two simple equations, Lotka and Volterra provide a quantitative theory for the periodic fluctuations often observed in predator and prey abundance.

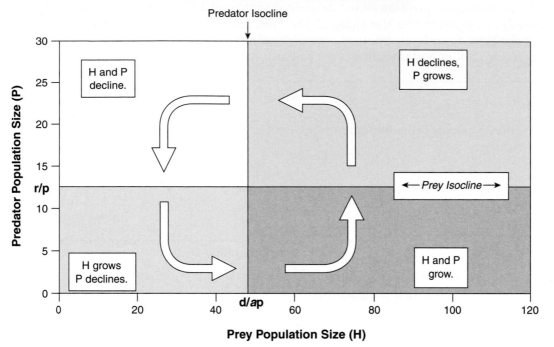

Figure 11.9   Predator/prey joint abundance cycles around the intersection of the zero-growth isocline.

Though common, not all predator-prey cycles persist over time. The joint abundance plot may produce a closed circle as shown in Figure 11.8, but it often forms a spiral. If the joint abundance plot spirals in toward the center, ecologists call it a damped cycle, because the oscillations get smaller and smaller until the populations come to rest at the intersection of the isoclines. The intersection of the isoclines thus represents a **stable equilibrium** which is resistant to small environmental disturbances. Alternatively, the two populations may spiral outward in an exploding cycle, oscillating at greater and greater extremes until one species or the other crashes to extinction. In this example, the isoclines intersect at an **unstable equilibrium** point, because any small movement off this joint abundance results in major irreversible changes in the system. To compare damped cycles, repeating (or limit) cycles, and exploding cycles, see Figure 11.10.

Check your progress:

In which kind(s) of cycle on Figure 11.10 would the intersection of the zero-growth isoclines be considered a stable equilibrium point? In which kind(s) would it be considered an unstable equilibrium point?

Hint: When disturbed, a system quickly returns to a stable equilibrium point. A disturbed system rarely returns to an unstable equilibrium point.

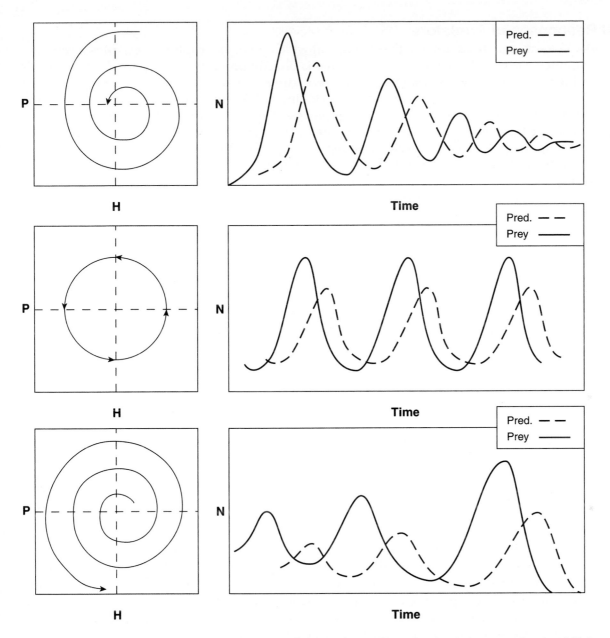

Figure 11.10 Predator-prey dynamics can produce damped cycles (top), limit cycles (middle), or exploding cycles (bottom). On the left are joint abundance graphs (see Figure 11.9). On the right, numbers (N) of predators and prey are plotted against time.

Functional Responses of Predators

The simple Lotka-Volterra equations assume that the rate at which predators capture prey (**p**) is a constant, so numbers of captures each predator makes per unit time is directly proportional to prey density. This is sometimes called a **type 1 functional response**. Can we assume predators always take prey in direct proportion to their numbers? An ecologist named C. S. Holling (1959) studying small mammals feeding on sawfly cocoons observed that predation is a two-step process: first the prey must be found, then it must be "handled." **Handling time** includes all activities required before the next capture, including removal of inedible parts, carrying the prey back to a nest, or eating and digesting the food item. If handling time is long in comparison to searching time, then it can become a limiting factor for a predator's consumption rate as prey density increases. When plotted against prey density, capture rate levels off in the type 2 response (Figure 11.11). Holling's disk equation, so-called because he did simulations in the laboratory involving sandpaper disks, recognizes that predators spend most of their time hunting for food when prey are scarce, but a greater proportion of their time handling food when prey become more abundant. The equation for the type 2 functional response is as follows:

Holling's Disk Equation:

$$C = A\,N/(1 + T_h\,A\,N)$$

C = number of prey captured
T_h = handling time
A = attack rate of predators
N = population density of prey

For vertebrate predators capable of learning from experience, the most abundant prey may be selected at an even higher frequency than their prevalence in the environment. The rarest prey may be ignored altogether. Vertebrate predators often develop a **search image** for prey with which they have experience. (This is not unlike a phenomenon you may observe in a university cafeteria. If unusual looking food items are presented on a buffet line, they are selected at a lower frequency than food items more familiar to students.) If predators form a search image, the functional response curve is S-shaped, due to disproportionately low predation of rare prey. This is called a **type 3 functional response** (Figure 11.11). At intermediate prey density, the rate of capture increases rapidly with enhanced predator experience, and then handling time slows the capture rate at very high density. Invertebrate predators such as hunting wasps are less likely to illustrate the S-shaped functional response curve, because most of their predatory behavior is genetically programmed, and not dependent on learning.

In conclusion, predators adjust to their prey over three time scales: 1) Over evolutionary time, predators develop adaptations in morphology and behavior to help them capture their prey. 2) Over generations, predator populations fluctuate up and down in response to prey numbers. 3) Over the lifetime of a single predator, handling time and learning affect rates of capture.

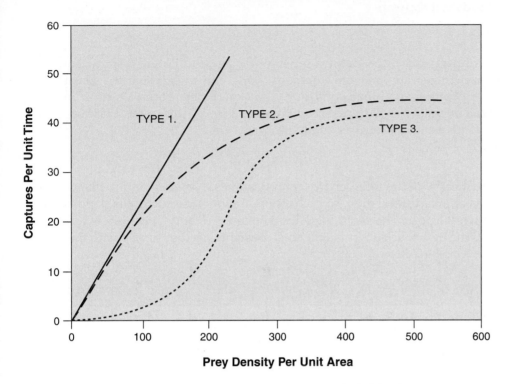

Figure 11.11   Functional Response Curves. Type 1 (solid line) assumes prey consumption is directly proportional to availability. Type 2 (dashed line) levels off at high prey density as the capture rate becomes increasingly limited by "handling time" required to process and digest prey. Type 3 (dotted line) is seen in vertebrate predators developing a "search image" for the most common prey. In this model, rare prey are underrepresented in predator diets in comparison to their frequency, and common prey are overrepresented.

METHOD A: PREDATOR-PREY SIMULATION

[Computer activity]

Research Question
How do the Lotka-Volterra predator-prey equations generate population cycles?

Preparation
Because the Lotka-Volterra equations are built on ideas about population growth, it is highly recommended that students complete readings and exercises from Chapter 4 before attempting exercises in this chapter. Instructions for this simulation are appropriate for Microsoft Excel software. If you use other spreadsheet software, instructions for writing cell formulas given in this procedure may require modification.

Materials (per laboratory team)

A computer station, with spreadsheet software

Background

This exercise simulates growth of prey and of predator populations using the Lotka-Volterra equations. The purpose of the exercise is to demonstrate how the equations work, and to examine the sensitivity of predator-prey systems to small changes in initial conditions. Although the Lotka-Volterra model is simplistic, and omits many factors operating in natural communities, it help us understand how even two-species interactions can generate interesting patterns of abundance over time.

Procedure

1. Set up a computer spreadsheet with the following column headings. Numbers in the first column (Column A) indicate the year in our simulation, so each row will show the state of the populations in another year as we go down the page. The second and third columns (B and C) will show numbers of prey and of predators. The next four columns will hold constants needed to calculate the Lotka-Volterra equations.

	A	B	C	D	E	F	G
1	Year	Prey	Predators	r	p	A	d
2							

2. The first entry in cell A2, under "year," should be 0, the starting point of our simulation. Go on down this column, numbering years 1, 2, 3, 4, etc. If you do not want to type all those numbers in, just enter 0 in cell A2, and then in cell A3 type the formula **= A2+1**. If you copy and paste this formula down the column to the one hundred fifty second row, you will get a series of numbered years from 0 through 150. Format all the cells below the first row in columns 1, 2, and 3 to display integer values, with no decimal places. The top of your spreadsheet should now look like this:

	A	B	C	D	E	F	G
1	Year	Prey	Predators	r	p	A	d
2	0						
3	1						
4	2						
5	3						
6	4						

3. Then you will need to enter starting numbers of prey and predators, along with values for the four Lotka-Volterra constants. To start off, try entering 100 prey and 15 predators. For the four constants, try the following values: Intrinsic rate of increase of the prey (r) = 0.1; the predation constant (p) = 0.01; the predator's birth constant (a) = 0.3; and the predator death rate (d) = 0.3. The top of your spreadsheet should now look like this:

	A	B	C	D	E	F	G
1	Year	Prey	Predators	r	p	A	d
2	0	100	15	0.1	0.01	0.3	0.3
3	1						
4	2						
5	3						
6	4						

4. Now you are ready to enter the Lotka-Volterra population growth equations for prey. In cell B3, which is the first empty cell under the Prey column, type the following formula:

$$=B2+(\$D\$2*B2)-(\$E\$2*B2*C2)$$

This formula calculates the number of prey in year 1 by starting with the number that were present in the previous year, then adding births and subtracting deaths. (This corresponds to the $\Delta H = r\,H - p\,H\,P$ equation from the Introduction.) Note that cell references in the formula include dollar signs in front of the letter and number of the cell address. This transfers the value within the cell, and not the position of the cell, which is necessary for constants. COPY cell B3 and then PASTE the formula into cells B4 through B152. Don't worry if you get an error message in the target cells at this point.

5. Next, you need to enter the Lotka-Volterra population growth equation for predators. In cell C3, in the first empty cell under the Predators column, type the following formula:

$$=C2+(\$F\$2*\$E\$2*B2*C2)-(\$G\$2*C2)$$

This formula calculates the number of predators in year 1 by starting with the number that were present in the previous year, then adding births and subtracting deaths. (This corresponds to the $\Delta P = a\,p\,H\,P - d\,P$ equation from the Introduction.) COPY cell C3 and then PASTE the formula into cells C4 through C152. You should see a cascade of numbers go down the page as the spreadsheet calculates numbers of predators and prey for each year of the simulation. The top of your spreadsheet should now look like this:

	A	B	C	D	E	F	G
1	Year	Prey	Predators	r	p	A	d
2	0	100	15	0.1	0.01	0.3	0.3
3	1	95	15				
4	2	90	15				
5	3	86	14				
6	4	82	14				

6. If your spreadsheet can display graphs, try highlighting the first three columns, including the headings, all the way down to row 152. Then click on the graph option on the menu bar, and select a scatter plot for your data display. You should be able to produce a graph with Year as the x-axis and Prey and Predator numbers as Series 1 and Series 2 on the y-axis. With the simulation parameters you have set by following these instructions, you should see an exploding cycle, with prey and predator populations cresting in three peaks of increasing size during the 150-year simulation. Ideally, you can display the graph within the spreadsheet just under the constants on the right, so that you can change initial

simulation conditions and observe results as the graph changes. You can also produce a joint abundance graph by highlighting columns 2 and 3, and using the same graphing options. For simulation conditions given above, your graphs should resemble those in Figure 11.12.

7. Try to stabilize the exploding cycles by adjusting the numbers of prey and predators you entered in Year 0. Small changes in initial numbers can make large differences in the simulation.

8. Adjust the four constants, one at a time, by replacing the numerical values under r, p, a, and d by numbers just a little larger or smaller. After each adjustment, look at the numbers output from the spreadsheet and the graphs. Take notes on your findings.

9. Produce large plots, both of numbers vs. time, and of joint abundance, of the most stable cycles you were able to produce in the simulation. If you cannot get computer printouts to show the graphs, draw them by plotting points by hand on graph paper. On the joint abundance plot, use the input parameters for the simulation to calculate the positions of the isoclines, and draw them in, as seen in Figure 11.9.

10. Use your experience with the model and your graphs to answer the Questions for Method A.

Predator-Prey Simulation Results

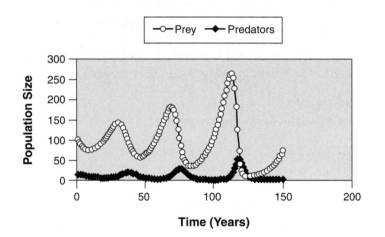

Joint Abundance Plot

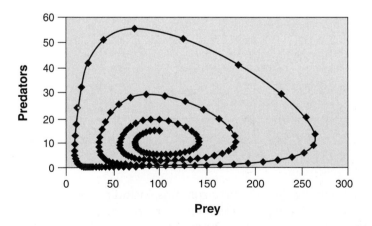

Figure 11.12 Sample output from predator-prey spreadsheet simulation.

Questions for Method A

1. On your joint abundance graph, use the values of the constants to calculate the positions of the isoclines (as shown in Figure 11.9) and draw the isoclines with a pencil and ruler on top of your population curve. Does the curve seem to spiral around the point of intersection? Explain.

    _____

    _____

    _____

    _____

2. On a joint abundance plot of cycles near the stable limit condition, are the circles symmetrical around the intersection of isoclines, as diagrammed in Figure 11.10, or not? Describe the shape of the cycles on the joint abundance graph, and reflect on why the increasing and decreasing parts of each cycle might be asymmetrical. (Hint: what controls predator births? deaths?)

    _____

    _____

    _____

    _____

3. During the 1920s, U.S. government policy encouraged extermination of predators such as wolves, mountain lions, and coyotes from large areas in the American West in order to maintain more abundant deer and antelope populations for hunters. Based on your model, what outcome would you expect from this effort to maintain prey in the absence of predators?

    _____

    _____

    _____

    _____

4. If you adjust each of the four constants, one at a time, by 10%, which seems to have the greatest impact on the simulation output? Examine the Lotka-Volterra equations to explain why this constant makes such a big difference in the predator-prey system. (Hint: This constant affects both the birth rate of predators and the death rate of prey.)

    _____

    _____

    _____

    _____

5. Predator-prey systems are difficult to balance when there is only one predator and only one prey. Would you expect a system with multiple prey species to be more stable? Why?

    _____

    _____

    _____

    _____

METHOD B: PREDATION BY *DAPHNIA* ON *EUGLENA*

[Laboratory activity]

Research Question

Do *Daphnia* feeding on *Euglena* experience population cycles?

Preparation

This laboratory works best if students are able to come in and check on their cultures frequently between laboratory periods. If student access to the lab is limited to weekly class meetings, it is very hard to monitor rapid changes in the populations. *Daphnia pulex* and *Euglena gracilis* are available from supply houses. Order large, mixed-age *Daphnia* cultures if possible. Make sure your *Euglena gracilis* strain has chloroplasts. Water quality is essential for culturing both *Daphnia* and protozoans. Traces of chlorine or detergent will kill protozoans and *Daphnia*, so all glassware should be carefully rinsed and water must be aged to let chlorine evaporate. Avoid deionized or distilled water, as it does not have sufficient trace minerals to support *Euglena*. Spring water can be purchased, but rainwater or tap water can be used if it is filtered or boiled for 15 minutes to remove contaminating algae. If boiled, cover container of water loosely with aluminum foil and allow to reoxygenate for several days before using it in cultures.

Euglena are photosynthetic, so they can grow without added food. *Daphnia* are notoriously difficult to culture for long periods, but most cultures go through at least one full cycle before they crash. While students are beginning their cultures with *Euglena*, keep *Daphnia* in the dark, and feed them a few drops of Baker's yeast suspended in water to avoid introducing algae along with the *Daphnia*. After inoculation with algae, cultures should be kept under fluorescent lights or grow lights at room temperature to support *Euglena* growth.

Materials (per laboratory team)

Access to stock cultures of *Daphnia pulex* and *Euglena gracilis*

Water for *Daphnia* culture

Glass culture dish, $4\frac{1}{2}$" diameter $\times$ 3" high

Glass plate (or extra stacking dish) to cover culture dish

Plastic dropper, 6" (one-piece plastic pipette with tapered tip)

Scissors

Pasteur pipette

Glass slides

Cover slips

Compound microscope

Dissecting microscope (optional)

Background

Daphnia are very small crustaceans that swim in the water column and feed on *Euglena* and many other kinds of plankton in freshwater lakes and ponds. *Daphnia* are able to reproduce by **parthenogenesis** (a form of asexual reproduction), which allows rapid population growth when food is abundant. Because *Daphnia* have a short life span, their populations can also decline quickly when prey are scarce. Since *Euglena* also have rapid growth rates and very short life spans, predator-prey cycles in *Daphnia*/*Euglena* populations occur quickly, over a few days or weeks.

Procedure

1. Fill culture dish ¾ full of culture water. With a Pasteur pipette, transfer one or two drops of *Euglena gracilis* stock culture to your dish. Cover your dish with another dish or glass plate, and place it under a light source. About three feet below a fluorescent lamp is enough light. Incubate at room temperature (20° to 23° C) for one week.

2. After one week, stir the culture well with the end of a pipette and then draw up a sample. Put one drop on a glass slide, cover with a cover slip, and examine the culture under a compound microscope, using the 10× objective (100× total magnification). You should see green, cigar-shaped *Euglena* cells swimming through the water. After you have learned what they look like, census the population as follows: move the slide slightly to a new location. Look through the eyepiece and count the number of *Euglena* cells you see in this field of view. If you see no *Euglena*, record zero as the number in this field. Repeat ten times. Take an average of the number of cells per field, and record in the calculation page for Method B.

3. When you have established a *Euglena* culture, it is time to add *Daphnia*. First look at the tip of your large plastic pipette. If the hole in the end is smaller than 2 mm in diameter, use scissors or a sharp blade to trim some of the end off the pipette so that *Daphnia* can pass through unharmed. Use your modified pipette to capture *Daphnia* from the stock culture. Transfer three large and three smaller *Daphnia* into your culture dish.

4. Record the date, the mean number of *Euglena* per field, and the number of *Daphnia* you placed in your culture on the first line of the Method B Data Table. Every few days, count the number of *Daphnia* by placing your culture dish on a black surface under a bright light. If numbers are hard to count, estimate by counting a wedge-shaped part of the culture dish and multiplying by the number of parts in the whole. Count the number of *Euglena* as described in step 2. Record the date, the age of the culture (as number of days elapsed since Day 0), the number of *Daphnia* per culture, and the number of *Euglena* per field in the next line of the data table.

5. Continue monitoring your *Daphnia/Euglena* system for several weeks. You should be able to see the two populations peak and then decline for at least one cycle. Plot the data on a joint abundance graph, with *Euglena* per field on the x-axis and *Daphnia* per culture on the y-axis. Answer the Questions for Method B.

Data Table (Method B)

DATE	AGE OF CULTURE No. days since beginning	DAPHNIA Number per culture	EUGLENA Mean number per field
	0	6	

Joint Abundance Plot of Results (Method B)

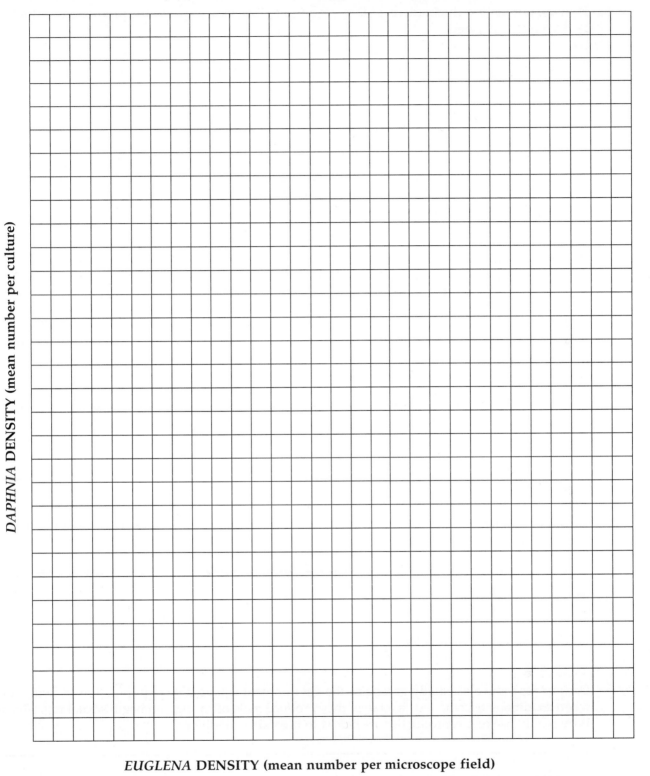

DAPHNIA DENSITY (mean number per culture)

EUGLENA DENSITY (mean number per microscope field)

Questions for Method B

1. How do population fluctuations of *Daphnia* and *Euglena* compare with the predictions of the Lotka-Volterra model?

_____

_____

_____

_____

_____

2. When plotted on a joint abundance graph, did the joint abundance move in a counterclockwise direction? If so, how does the shape of the curve compare with the simulation output in Figure 11.11?

_____

_____

_____

_____

_____

3. Based on your experiment, would you say this two-species community is very stable, or fairly fragile? What lessons might we apply to agricultural systems that replace natural communities with one or a few domesticated species occupying a very large landmass?

_____

_____

_____

_____

_____

4. Why was it important to introduce some large and some small *Daphnia* in your initial population? (Hint: See Chapter 5.)

_____

_____

_____

_____

_____

5. *Daphnia* reproduce asexually as long as the environment is favorable to population growth, but when conditions become less favorable, they produce male offspring and begin sexual reproduction. Why would asexual reproduction be linked to favorable conditions?

_____

_____

_____

_____

METHOD C: SIMULATING FUNCTIONAL RESPONSE OF A PREDATOR
[Laboratory activity]

Research Question
How does the functional response of a predator depend on handling time?

Preparation
Bolts and nuts used to represent organisms can be purchased at a hardware store. A 3/8" diameter is suggested, but other sizes are fine. Threads should go all the way to the head of the bolt, because a key variable is the number of turns necessary to remove a bolt—this represents handling time in the simulation.

Materials (per laboratory team)
25 2½" bolts, with nuts

1 sheet poster board (22" × 28" or larger)

Stopwatch or laboratory timer

Background
In this simulation, a student with eyes closed represents a predator searching for prey. A piece of poster board lying flat on a table or lab bench represents the habitat. Bolts represent prey, and nuts represent inedible parts of the prey (such as a clam shell or insect wings) that must be removed before the prey can be consumed. Time spent removing the nut from the bolt therefore serves as handling time in this experiment. The "predator" must search the board with only one finger, pointing straight down, moving across the "habitat" of the board. When a prey item is encountered, it is picked up and "handled" by removing the bolt. Bolt and nut are placed off the board and the search resumes, all without looking at the board. Other students in the lab group will time the simulations, count the number of prey "consumed," and replace the bolts with nuts reattached on the board to maintain a stable prey density.

Procedure
1. Designate one student in your group as the "predator." With eyes closed, hand the "predator" a bolt with the nut screwed all the way up to the head. Ask the student to remove the bolt from the nut (without spinning the nut around the bolt). After allowing once or twice for practice, conduct a time trial and measure the number of seconds it takes to remove the nut. Then divide the number of seconds by 60 to calculate handling time in minutes. Record this number, to the nearest hundredth of a minute, in the Data Table for Method C.
2. Lay the poster board flat on a table or lab bench. While the "predator's" eyes are still closed, scatter five "prey" (nut and bolt assemblies) onto the poster-board habitat. Your task is to determine how many prey can be captured and handled in a three-minute period. The "predator" searches for prey with one finger touching the board. The "predator" should try to move at a steady pace, and be careful not to touch prey with the whole hand. When a prey item is found, the "predator" picks up the prey, removes the nut from the bolt, places the "eaten" parts off the board, and then continues searching. Student observers should screw the nut back onto the bolt and quietly replace the "prey" item back on the board so that the prey density does not change during the experiment. Record the number of prey taken within a three-minute period. If a prey item has been found but not fully processed when time is up, count that bolt as an additional ½ prey consumed.

3. Calculate the "predator's" attack rate as follows:
   a) multiply the handling time per prey (in minutes) by the number of prey consumed during the three-minute interval. This is total handling time.
   b) subtract total handling time (in minutes) from the total predation time (three minutes) to determine total searching time (T_s). For example, if total handling time is 1.2 minutes, then T_s = 3 minutes – 1.2 minutes = 1.8 minutes.
   c) Use the following equation to calculate the attack rate for the "predator" in the first trial, which used five prey on the board. The total search time in the denominator is multiplied by 5 because that is the initial prey density.

   Attack rate = (# prey consumed)/(5 T_s)

   d) Record the attack rate (A) for this "predator" in the data table.

4. Repeat step 2 with the same predator, but with 10 bolts on the board. Then repeat with densities 15, 20, and 25. Record results for each prey density in the Data Table for Method C.
5. Allow a second lab group member to take a turn as predator. Measure handling time, record the number of bolts the "predator" is able to "consume" in three minutes at densities 5, 10, 15, 20, and 25. Calculate the attack rate after the first trial, and record all data for the second predator in the second row of the data table. If time allows, allow every member of the laboratory group to take a turn as "predator." If you do not have time for five predators, simply leave the bottom of the table blank, and compute column means based on the number of predators who were able to do the simulation.
6. Calculate average numbers of bolts "eaten" at each of the five densities for your group, and write the means in the data table. Also calculate a mean handling time, in minutes, for the entire group, and record that grand mean as well.
7. Plot results on the Graph for Method C. Assume 0 prey will be caught when prey density is 0. For densities 5–25, plot your mean number of captures per three-minute trial as points on the graph, and connect your points to make a curve. Compare your curve to the Functional Response curves in Figure 11.11.
8. Use the following equation to calculate an expected number of captures for trials at each of the five prey densities. Draw a dotted line (or use a different color) to show the expected functional response on the Graph for Method C.

Holling's Disk Equation:

$$C = A\,N/(1 + T_h\,A\,N)$$

C = number of prey captured per trial
T_h = handling time
A = attack rate of predators
N = population density of prey

---

9. Refer to the data table and the graph when answering Questions for Method C.

Data Table for Method C

PREDATOR #	PREY DENSITY 5	PREY DENSITY 10	PREY DENSITY 15	PREY DENSITY 20	PREY DENSITY 25	HANDLING TIME	ATTACK RATE
1	C =	C =	C =	C =	C =	$T_h =$	A =
2	C =	C =	C =	C =	C =	$T_h =$	A =
3	C =	C =	C =	C =	C =	$T_h =$	A =
4	C =	C =	C =	C =	C =	$T_h =$	A =
5	C =	C =	C =	C =	C =	$T_h =$	A =
Mean Observed values:							
Expected values:							

Functional Response of Predators to Prey Abundance

NUMBERS OF PREY CONSUMED (per three-minute simulation)

PREY DENSITY

Questions for Method C

1. Which of the three types of functional response curves shown in Figure 11.11 does your data plot most resemble? Does this make sense, given the rules of the simulation?

_____

_____

_____

_____

_____

2. How did your expected functional response curve match the graph of your observations? Propose hypotheses to explain any discrepancies between the two.

_____

_____

_____

_____

_____

3. If two nuts were placed on each bolt to simulate a prey harder to process and consume, how would you expect the functional response curve to change? Illustrate your answer with a small drawing.

_____

_____

_____

_____

_____

4. Outline a method for an experiment with real predators (such as preying mantises fed with crickets) to test Holling's model for predator functional responses.

_____

_____

_____

_____

_____

5. Why does a strong search image produce an S-shaped functional response curve? If one predator were feeding on many prey species, would this kind of functional response tend to increase or decrease biodiversity? Explain.

_____

_____

_____

_____

_____

FOR FURTHER INVESTIGATION

1. Use the spreadsheet model you developed in Method A to simulate the predator/prey dynamics you observed in Method B. Values of r and d for *Daphnia* are reported in the literature. The positions of the isoclines crossing at the center of observed predator/prey cycles may then allow estimation of *a* and p. How valuable are the Lotka-Volterra equations for describing *Daphnia/Euglena* population dynamics?

2. Try culturing *Daphnia* on more than one species of algae. Does a more complex food web lend greater stability to the system?

3. In late spring or summer, observe songbirds nesting on your campus. During their nesting season, birds such as American robins tend to spend most of the day searching for food or feeding young. Since robins tend to patrol territories on lawns, searching time can be measured as you observe a robin hopping across the grass looking for prey. Time spent flying to and from the nest, or in feeding young, can be considered handling time. With binoculars and a stopwatch, you can measure search time and handling time for nesting birds. Although prey density may be hard to measure, you can compare the proportions of time spent in searching and handling food in different kinds of habitat.

FOR FURTHER READING

Holling, C. S. 1959. Some characteristics of simple types of predation and parasitism. *Canadian Entomologist* 91:385–398.

Lubchenco, Jane and Bruce A. Menge. 1978. Community development and persistence in a low rocky intertidal zone. *Ecological Monographs* 48(1):67–94.

Lubchenco web site: http://lucile.science.oregonstate.edu/?q=node/view/131

Paine, Robert T. 1966. Food web complexity and biodiversity. *American Naturalist* 100: 6575.

Peterson, Ralph O. and John A. Vucetich. 2005. Ecological Studies of Wolves 2004–05 Annual Report. School of Forest Resources and Environmental Science. Michigan Technological University, Houghton, Michigan, USA, 49931–1295 http://www.isleroyalewolf.org/

Chapter 12

Mutualism

Figure 12.1   A ruby-throated hummingbird pollinates the red tubular flower of a trumpet creeper vine.

INTRODUCTION

The evolution of flowering plants over 100 million years ago was one of the greatest revolutions in the history of life. Flowers gave plants the capacity for faster gene exchange over longer distances, which in turn generated the variety of botanical forms that dominate our world today. Flowering plants owe much of their success to partnerships with pollinators. Early flowering plants may have been wind-pollinated, with insects visiting flowers primarily to feed on the energy-rich pollen, but plant/insect interactions quickly evolved as mutually beneficial relationships. Insects are still the most ecologically significant pollinators, but birds, bats, and even nonflying mammals are known to pollinate flowers. By attracting pollinators with strong scents, showy petals, and sugary nectar, flowers exploit the power of animal locomotion to overcome the disadvantages of a rooted existence. In return for a meal, the pollinator carries pollen cells clinging to its body to other flowers of the same species, facilitating sexual reproduction. This kind of relationship is called **mutualism** because the two organisms mutually benefit from their interactions.

Because pollen carried to a flower of the wrong species is of no value to the pollen donor or receiver, pollination is more efficient when the flower attracts a limited number of pollinators specializing in a single flower type. Flower color is an effective way to narrow the field of potential pollinators. Because insect vision is sensitive to ultraviolet wavelengths, but not to red light, a red flower is more likely to attract birds than insects. With the further adaptation of a trumpet-shaped flower more easily probed by a hummingbird's beak than a bee's tongue, the trumpet creeper eliminates many pollinators while powerfully signaling its preferred hummingbird pollinator (Figure 12.1). Hummingbirds in turn have an incentive to visit trumpet creeper, since its nectar resources are not likely to have been taken by insects prior to the bird's arrival.

Some mutualisms are so highly coevolved that the partners can no longer survive alone. *Yucca glauca*, a showy perennial flower of the Great Plains, has intrigued ecologists ever since its unique life history was first described over a century ago. Its pollination by the yucca moth *Tegeticula yuccasella* (Figure 12.2) presents a classic case of **obligate mutualism**. The relationship is obligate because the moth gets its only food from the seeds of the plant, and the plant has no other pollinator. As often seen in cases of obligate mutualism, each species has developed adaptive strategies to maximize its benefits from close association with the other organism.

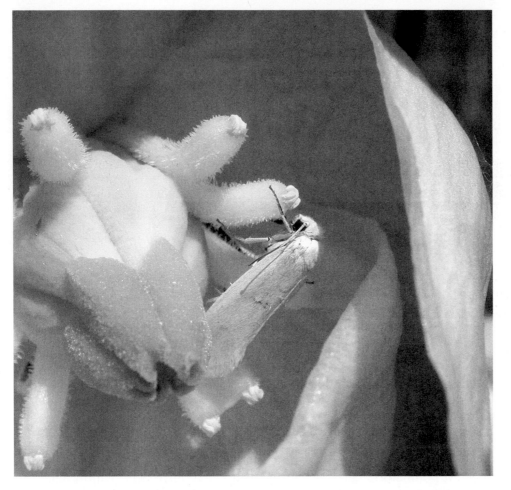

Figure 12.2   A yucca moth (*Tegeticula yuccasella*) pollinates a *Yucca* flower. The plant and moth are obligate mutualists; the plant has no other pollinator and the insect has no other food source.

Adult yucca moths normally live only a day or two, because their mouthparts are not adapted for feeding. Instead, the insect's specialized oral appendages are adapted for a totally different task. The female moth uses these mouthparts to gather sticky yucca pollen into a ball, which she carries under her chin. At sunset, the moth flies to another plant with her pollen load, settling on a fresh yucca flower. Stabbing into the base of the flower with a sharp abdominal appendage, she injects an egg into a seed chamber within the flower's ovary. Then, exhibiting behavior that looks quite intentional to human observers, the moth climbs to the receptive tip of the flower and rubs her pollen ball up and down on the stigma, ensuring pollination of the flower. Her eggs hatch inside the developing yucca fruit, and the young caterpillars chew their way through a row of immature seeds. Since there are generally more seeds in the chamber than one larva needs to complete its development, some of the seeds survive to propagate the yucca plant. Although nineteenth-century biologists credited the "wise little moth" with restraint and foresight in preserving the reproduction of her ally in this relationship, contemporary ecologists recognize that the plant's evolutionary adjustment of seed capsule size and of flower numbers contribute to a more biologically coherent explanation for the plant's reproductive success.

Check your progress:

Why have so many flowering plants evolved shapes and colors that limit, rather than expand, the number of pollinator species visiting their flowers?

Hint: Generalist pollinators are less likely to visit the same floral species twice in a row.

Not all mutualistic relationships are as completely codependent as *Yucca* and its moth. **Facultative mutualists** can survive on their own, but grow faster or produce more offspring in the presence of a coevolved species. Plants in the **legume** family, which includes peas and beans, exhibit facultative mutualism with bacteria of the genus *Rhizobium*. The bacterium has the rare ability to convert N_2 gas from the atmosphere into organic nitrogen (NH_4^+), which provides raw material for protein synthesis in both bacteria and plants. *Rhizobium* can grow alone in the soil, but when it encounters a legume root, the bacterium stimulates the growth of root nodules, in which the microbe takes up residence (Figure 12.3). Oxygen levels, which must be kept low to facilitate nitrogen fixation, are controlled inside the nodules by a layer of proteins similar to blood hemoglobin. Energy needed to drive the nitrogen-fixing reactions also comes from nutrients supplied by the plant. Legumes can grow without *Rhizobium*, but in unfertile soils, the plants have better access to growth-enhancing organic nitrogen with the help of their microbial **symbionts**.

Mutualistic associations frequently involve organisms from different kingdoms. Since mutualistic interactions are most likely to develop when the capabilities or by-products of one organism complement the needs of the other, this observation makes biological sense. Tree roots develop beneficial relationships with soil fungi, large-fruited plants rely on birds and mammals that eat their fruits to disperse seeds, cnidarian animals such as corals maintain living algae within their tissues for photosynthesis, and many herbivores rely on microbes in their guts to facilitate digestion. In the last example, mutualism is necessary because cellulose, a polymer of sugar, is the primary structural molecule in plant fibers. Although cellulose contains as

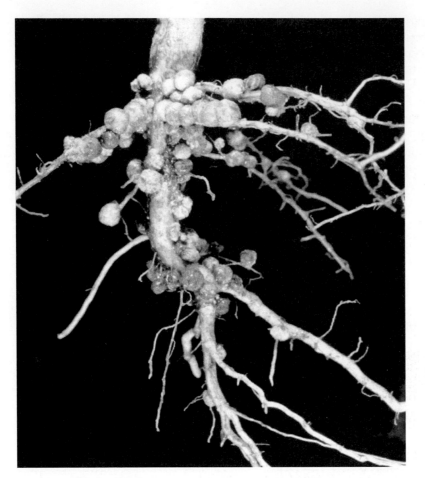

much chemical energy as starch, this food energy is unavailable to most mammals, including humans, because we do not have the enzyme required to cleave the sugar units from the polymer. This enzyme, called cellulase, is synthesized by a number of soil fungi, bacteria, and some protists. How can a cow, feeding on vegetation rich in plant fibers, derive energy from cellulose? The cow's rumen, a sac off the alimentary canal anterior to the stomach, is full of microorganisms that make the digestive enzyme for the cow, and in turn receive a warm, dark environment ideal for microbial growth. Termites (Figure 12.4), feeding almost exclusively on the cellulose in wood, have pockets off the hind gut that house a diverse community of cellulase-secreting microorganisms. One of these is a multi-flagellated protist called *Trichonympha* (Figure 12.5), an obligate symbiont that maintains its own mutualistic relationships with other microbes in the termite gut community.

Figure 12.3   *Rhizobium* bacteria induce the formation of nodules on the roots of legumes, which provide a hospitable environment for the microorganism. In return, *Rhizobium* fixes nitrogen in a chemical form useful to the plant.

Another two-kingdom partnership of great ecological significance is the lichen association. Lichens result from a partnership between algae (or in some cases, blue-green bacteria) and fungi. There are many lichen forms; some produce crusty splotches on rocks or tree bark, some develop leaf-like sheets, and others form spongy mats (Figure 12.6). The lichen body, called a thallus, is constructed like a sandwich. It has a thin layer of fungus on top, a layer of algae in the middle, and a thicker and looser layer of fungus filaments on the bottom (Figure 12.7). Laboratory studies of lichen reproduction suggest that the balance of power in this relationship is not equal. The fungus seems to control the growth and reproduction of its algal symbionts to such an extent that some lichenologists have questioned whether this association should be compared to a partnership or a kidnapping.

Figure 12.4   Termites digest cellulose in wood with the help of microbial symbionts.

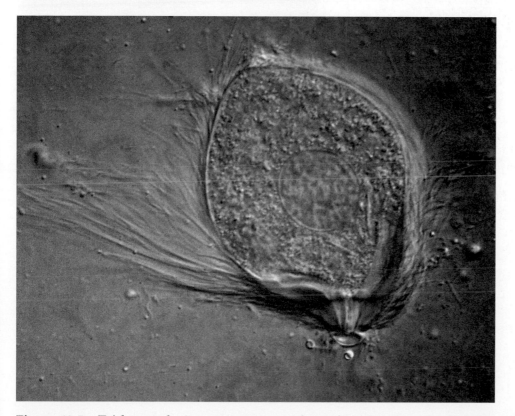

Figure 12.5   *Trichonympha* is a protozoan symbiont living in the gut of termites. The enzyme cellulase, provided by *Trichonympha*, breaks cellulose down into its component sugars and allows the termite to eat wood it could not otherwise digest.

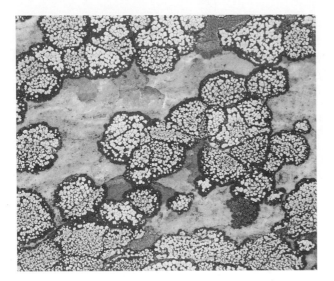

Figure 12.6A   Crustose lichen (adheres tightly to substrate in a single layer).

Figure 12.6B   Foliose lichen (leafy growth form).

Figure 12.6C   Fruticose lichen (bushy growth form).

Figure 12.6D   Squamulose lichen (with scale-like parts).

However controlled, a typical lichen association can tolerate drought and nutrient-poor conditions that would kill either component species. Algae inside the lichen produce food by photosynthesis, and the fungal layers protect the algae from drying out. The fungus can secrete acids to extract minerals from a rock substrate, so many lichens play an important role in soil formation in the early stages of ecological **succession**. The fungal elements of lichens also extract ions and organic molecules from water moving across their surfaces, so they help trap nutrients that would otherwise escape the ecosystem after a rain. This tendency to extract chemicals from the environment also makes lichens very susceptible to air pollution. Pollutants picked up by raindrops from the atmosphere are concentrated in lichens, and tend to kill them before other forms of life show ill effects. Where air quality is poor, lichens become less common, and some may disappear completely (Figure 12.8). Lichens are therefore useful as **bioindicators**, demonstrating long-term trends in air quality through changes in their diversity and distribution.

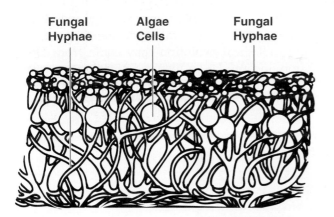

Fungal   Algae   Fungal
Hyphae   Cells   Hyphae

Figure 12.7   A lichen thallus is made up of fungal hyphae, with a layer of algae cells embedded near the upper surface.

Check your progress:

Name the mutualist associated with each of these organisms, and explain what each partner receives from the association.

cow
legume
yucca
lichen-forming fungi

Hint: Mutualism tends to involve species from different kingdoms.

Although these and many other cases of mutually beneficial interactions among species have been studied by generations of biologists, only recently have we realized the full significance of mutualism in the natural world. Microbial ecologist Lynn Margulis (1999) has expanded our understanding of mutualism's importance with her work on the **endosymbiotic theory** of cell origins. This idea, since corroborated with genetic and biochemical evidence, is that mitochondria and chloroplasts originated as free-living prokaryotes that infected, and subsequently became obligate mutualists with, a larger cell. This explains why chloroplasts and mitochondria have circular DNA, binary division, and ribosomal and cell membrane characteristics similar to those of prokaryotes. Probing further into our general understanding of life, Margulis proposes that much of the genetic variation driving biological evolution may come from DNA exchanges among cooperating symbionts. Although competition and predation inspired nineteenth-century biologists to describe a natural world "red in tooth and claw," an appreciation for the essential role of mutualism gives us a more sophisticated understanding of community relationships and the history of life.

Figure 12.8 Lichens indicate air quality. The Red Alder tree on the left is in a clean air environment, and is covered with gray lichens. The Red Alder on the right, impacted by air pollution from a pulp mill, is devoid of lichens.

METHOD A: EFFECTS OF *RHIZOBIUM* ON LEGUMES

[Laboratory activity]

Research Question

Does association with *Rhizobium* affect the morphology and growth of legumes?

Preparation

Plan at least four weeks between planting and harvest of pea plants for this experiment. If previously exposed to soil, plastic pots for pea culture should be washed in 15% chlorine bleach solution and allowed to dry before use. Vermiculite, garden pea seeds, and the symbiont, *Rhizobium leguminosarum*, can be purchased in garden stores or through biological suppliers. Seeds should be untreated. *Rhizobium* is available as an inoculant designed for application to soil in granular form. Make sure package directions specify its use on peas, or list *R. leguminosarum* as the bacterial species.

Nitrogen-free or low-nitrogen fertilizer is sold under the names "Bloom Booster" or "Mor Bloom." Look for a fertilizer formulation with zero as the first number, such as 0-12-12, or 0-10-10. (The second and third numbers signify available phosphorus and potassium.) Apply according to package directions. If you wish to make up your own nitrogen-free fertilizer, mix the following reagents with four liters of water. Tap water is adequate, but let it stand to allow chlorine to evaporate before use.

Nitrogen-free fertilizer, per 4 liters of water:

3.2 g potassium monohydrogen phosphate

0.8 g magnesium sulfate

0.8 g dihydrogen phosphate

0.4 g calcium sulfate

0.04 g ferric sulfate

Soak peas overnight in lukewarm water the night before the laboratory begins.

Materials (per laboratory team)

For the Planting Lab:

8 plastic flower pots, 4" diam.

Labeling tape

Paper towels

2 trays (or plastic shoebox lids), one for treated pots and one for untreated

Vermiculite potting medium, enough to fill pots

Small amount of pea inoculant (a packet can be shared with other groups)

40 garden pea seeds (previously soaked)

For the Harvesting Lab:

Metric ruler

Glass slides

Single-edge razor blade or scalpel

Dissecting microscope

Calculator or computer for computation of t-test

Procedure

Planting Lab:

1. Label four flower pots "*Rhizobium* treated," and four pots "Control." Fill all eight flower pots to within 1" of the rim with clean vermiculite potting medium. Water pots thoroughly, and allow to drain.
2. Press five previously soaked pea seeds onto the vermiculite in each of the eight pots, but do not bury the seeds.
3. In the "*Rhizobium* treated" pots, introduce the bacterium by sprinkling a very small amount of granular inoculant over the peas, making sure a few grains are touching each of the seeds. Be very careful not to contaminate the control pots with the bacterium. Washing hands before handling control pots is a good precaution.
4. Sprinkle a thin layer of Vermiculite over the peas in both treatment and control pots, covering them to a depth of 1/2". Cut a circle out of paper toweling 4" in diameter, and cover the surface of the planting medium in each pot with a layer of paper. Water the peas once more through the toweling to ensure the top layer is damp.
5. Place pots in a growth chamber at 70–75° F, with 16 hr light per day, or in a cool greenhouse, maintaining the treated and control pots in different trays. The entire class can randomize and rotate tray positions as the seedlings mature to control for any position effects on plant growth. Check for sprouts within a few days, and remove the paper toweling as soon as shoots emerge.
6. Let peas mature for 4–5 weeks before harvesting the plants. Once a week, water the pots with a low-nitrogen fertilizer solution. Add water to trays between fertilizer treatments as needed to keep vermiculite moist, making sure treated and untreated plants receive the same watering and fertilizer regime.

Harvesting Lab:

1. Ideal harvest time is when flower buds appear. First carefully unwind pea plants from each other and measure the height of each pea from the surface of the potting medium to the highest shoot tip, recording your results in the Data Table for Method A. Record any observations you can about the plants' condition, color, leaf size, and other details. Sample size could be as high as 20 for the treatment group and 20 for the control group, but may be less if some of the seeds failed to germinate.
2. After measuring heights of control peas and *Rhizobium*-treated peas, carefully invert the pots and remove the plants from the vermiculite. Shake and rinse off plant roots, keeping treated and untreated plants separated. Count the number of nodules on the roots of each plant, recording your numbers and noting sizes of nodules.
3. Remove one of the root nodules and place it on a glass slide. Use a single-edged razor blade or sharp scalpel to cross-section the nodule. Look for a rust-colored layer near the outside of the nodule. Make a drawing to record your observations.
4. Use a t-test to test the null hypothesis that *Rhizobium*-treated plants grow to the same mean height as controls. Complete the calculations in the box below the data table to determine significance for the difference between the two means. (Refer to Appendix 2 for more information about performing a t-test.)
5. Use your data and observations to answer Questions for Method A.

Data Table for Method A

RHIZOBIUM-TREATED SEEDLINGS				CONTROL SEEDLINGS		
Plant #	Height (cm)	# Nodules		Plant #	Height (cm)	# Nodules
1				1		
2				2		
3				3		
4				4		
5				5		
6				6		
7				7		
8				8		
9				9		
10				10		
11				11		
12				12		
13				13		
14				14		
15				15		
16				16		
17				17		
18				18		
19				19		
20				20		
Group Mean	$\bar{X}_1 =$			Group Mean	$\bar{X}_2 =$	
Standard Deviation	$S_1 =$			Standard Deviation	$S_2 =$	

t-TEST FOR PLANT HEIGHT DATA:

Null hypothesis: mean height of *Rhizobium*-treated plants is equal to that of controls.

t = test variable associated with data significance   d.f. = degrees of freedom
$\bar{X}_1$ = mean height of treated plants $\bar{X}_2$ = mean height of control plants
n_1 = sample size of treated plants n_2 = sample size of control plants
S_1 = standard deviation, treatment S_2 = standard deviation, control

$$t = \frac{(\bar{X}_1 - \bar{X}_2)}{\sqrt{(S_1^2/n_1) + (S_2^2/n_2)}} = \boxed{}$$

$$\text{d.f.} = n_1 + n_2 - 2 = \boxed{}$$

From t-table, the t value equivalent to a significance level of p = 0.05 = ☐

Conclusion: ☐

Questions (Method A)

1. Compare the appearance of the plants treated with *Rhizobium* and the control plants. Did you notice any differences in vegetation? color? root growth?

   _____

   _____

   _____

   _____

   _____

2. Did your t-test produce the expected result, based on information in the Introduction? Explain.

   _____

   _____

   _____

   _____

3. If you had added nitrogen fertilizer to all pots, how would you expect the outcome to have been different? Would categorization of this mutualism as facultative or obligate depend on the environment? Explain.

   _____

   _____

   _____

   _____

   _____

4. Why do vegetarian diets rely heavily on foods made from legumes such as peanut butter, beans, and tofu? Why can't other kinds of plants produce storage compounds in their seeds equivalent to the dietary value of legumes?

   _____

   _____

   _____

   _____

5. Describe what you saw inside the root nodule. What makes the nodule a good site for nitrogen fixation?

   _____

   _____

   _____

   _____

METHOD B: MICROBIAL SYMBIONTS IN TERMITE GUT AND LICHEN THALLUS

[Laboratory activity]

Research Question

Do termites have symbiotic microorganisms?

Preparation

Termites can be ordered from biological supply houses, but a USDA permit is required for shipping to Arizona and Maine. The genus *Zootermopsis* is desirable for observing protists; some genera in the family Termitidae rely on bacteria rather than protozoans for cellulose digestion.

Termites can also be collected from rotting wood or old lumber piles. (Always exercise caution when turning over rocks, logs, or lumber—snakes and scorpions also occupy these sites.) Most termites are fairly easy to maintain in the laboratory. A battery jar or large glass container with air holes in the lid, half-filled with sandy soil, and a piece of wet sponge to maintain humidity provide good culture conditions for termites. Unlike ants, worker termites function well in the absence of a queen. Feed termites on brown paper, bark, or pieces of rotten wood dropped onto the surface of the soil. Avoid using pieces of treated lumber, as it contains chemicals toxic to termites.

A 0.7% saline solution will protect protozoa from osmotic shock. Add 0.7 grams NaCl or iodine-free table salt to 100 ml deionized water, and dispense into dropper bottles for each laboratory team. Termites can be distributed in Petri dishes, but put a small piece of damp paper towel in the dish if termites are to be kept this way for more than an hour or two.

For lichen observations, local lichens can be collected. Prepared slides of the lichen thallus in cross section are readily available.

Materials (per laboratory team)

Compound microscope, magnification to 400×

1 small Syracuse watch glass

Fine forceps

2 dissecting needles

Dropper bottle of 0.7% saline solution

Glass slides

Cover slips

Petri dish containing live termites

Lichen specimens

Prepared slide of lichen thallus cross section

Procedure 1 (Termite Symbionts)

1. Place a termite in the watch glass. With forceps and a dissecting needle, remove the head of the termite.
2. Cover the termite abdomen with about 10 drops of 0.7% saline. It is important to free protists from the gut under saline, because these microorganisms are anaerobes, and air is toxic to them. Tease the termite abdomen apart with dissecting needles. When the gut is opened, the saline should become milky.
3. With a Pasteur pipette, remove a few drops of fluid and place on a glass slide. Cover with a cover slip and observe with the compound microscope, first under medium power, and then on high.
4. Make observations of protozoans in the suspension, make drawings, and answer Questions for Method B.

Results for Method B: Drawings of Organisms Found in Termite Gut

Draw, label, and describe locomotion of organisms you found.

Questions for Method B on Termite Symbionts

1. Did you find *Trichonympha* in the termite gut (see Figure 12.5)? Describe these and any other types of protists you found in the termite gut community.

   _____

   _____

   _____

   _____

2. *Trichonympha* and other microbes supply cellulase, the enzyme needed to break down cellulose, which is the primary chemical constituent of wood and other fibrous plant parts. What ecological consequences would ensue if we did not have *Trichonympha* and other organisms producing this enzyme?

   _____

   _____

   _____

   _____

3. Young termites engage in proctodeal feeding, which involves ingesting secretions from the anus of older termites. Do your observations help to explain this behavior?

   _____

   _____

   _____

   _____

4. How could antibiotics designed to stop narrowly defined categories of infectious microbes be used to discover which organisms are mutualists and which are **parasites** or **commensals** in the termite gut?

   _____

   _____

   _____

   _____

5. Human beings carry a diverse community of bacterial species in their large intestine, including *Eschericia coli* strains that rarely cause disease in healthy people. The normal bacterial flora outcompete any foreign bacterial species passing through the system. Based on these facts, is it understandable that antibiotics taken for a sore throat might cause intestinal discomfort?

   _____

   _____

   _____

   _____

Procedure 2 (Lichen Symbionts)

1. Search for lichens on your campus. Examine sides of buildings, monuments, and tree trunks. Are lichens found in shady or sunny areas? Note color, texture, spore-forming "fruiting bodies," and shape. Count numbers of types that you find in each of the categories in Figure 12.6. Draw some of the lichen types in the space below:

2. Use a compound microscope to examine a prepared slide of the lichen thallus in cross section. Find layers illustrated in Figure 12.7. Note details in your specimen that differ from the illustration. Draw your specimen in the space below:

Questions for Method B on Lichen Symbionts

1. What, if any, kinds of lichens have you found on campus? What does lichen diversity tell you about air quality where you are?

    _____

    _____

    _____

    _____

    _____

2. From your observations of lichens and from Figure 12.6, which kinds of lichens have the lowest surface area per unit volume of tissue? Which have the most? How would you expect the form of a lichen to affect its nutrient retention ability? Its vulnerability to air pollution?

    _____

    _____

    _____

    _____

    _____

3. Lichen-forming fungi are generally capable of reproducing by the production of **ascospores**: single-celled reproductive units small enough to be carried away on the air. What questions does this fact pose about the obligate/facultative nature of this mutualism?

    _____

    _____

    _____

    _____

    _____

4. The blue-green bacteria (also called cyanobacteria) in some lichens are capable of nitrogen fixation. Why would this be especially advantageous in early stages of ecological succession?

    _____

    _____

    _____

    _____

    _____

5. Does the biological species concept apply to lichens? Should lichens be given species names? On what criteria should two similar-looking lichens be classified as different species?

    _____

    _____

    _____

    _____

    _____

METHOD C: FLOWER/POLLINATOR MUTUALISM
[Outdoor activity, spring]

Research Question
Do flowers on campus attract specialist or generalist pollinators?

Preparation
Most campus landscaping includes flowering trees and shrubs, as well as beds of annuals or perennials. Depending on your latitude, flowers may be blooming in fall or winter months, but in northerly locations spring is the best time to observe pollination. Two or more flowering plants must be available for student observations at the same time.

Some experience in identifying insects, particularly Diptera and Hymenoptera, is useful prior to the exercise, but not essential. Field identification of insect pollinators as butterflies, honeybees, beetles, etc., is adequate for preliminary analysis, but you may wish to collect and preserve voucher specimens for more accurate identification. Denatured ethanol (80% aqueous solution) and isopropyl alcohol are good preservatives for most kinds of insects.

Although observers near flowers are at low risk of bee stings, students should be asked about allergies to bee stings before participating in this laboratory.

Materials (per laboratory team)
Guide to insect identification

Insect net (optional)

Collecting vial with alcohol (optional)

Watch or timer

Procedure
1. Choose a flowering tree, shrub, or flowers of one species in a flower bed that are attracting insects. Set up a chair or something to sit on, and remain still during your observations. Insects will generally ignore you as they fly in to the flowers.
2. For a period of 30 minutes, record numbers and kinds of insects coming to the flower species you have selected. If you do not know the name of the insect, describe it as accurately as you can. As each new insect species arrives, give it a descriptive name and add it to the Preliminary Data Table for Method C. After that, make tally marks for each additional visit by the same species on the line next to the name. If too many insects are moving through your study area to observe individually, confine your observations to a smaller group of flowers. If the same individual insect moves from one flower to another within the patch of flowers you are observing, you do not need to record its presence more than once. At the end of your observation period, you may wish to collect specimens of the most common pollinators for more detailed taxonomic identification.
3. Move to a different site with a different flower species in bloom. Observe a patch of flowers approximately equal in size to the first site. Again record numbers and kinds of insects coming to the flowers for a period of 30 minutes. You may wish to collect specimens if new species are observed in this site.
4. From the data table for both sites, select the three most common pollinators overall. Fill in the Chi-square Table for Method C with numbers of these three species in each of the two sites. If a species was present at one site but not the other, then enter a 0 for the frequency at that site. Follow directions in Appendix 4 to complete a Chi-square contingency test on your data. The null hypothesis for this test

will be that pollinators visit the two floral species in equivalent frequencies. This means that the proportions of bees, butterflies, beetles, etc., would be the same for both flower types.

If you get a significant Chi-square value from the test, a fair conclusion is that the insects are exhibiting preferences, and that the flowers are attracting subsets of the total array of pollinator species flying at this time.

5. Use your field observations, any collections you have made, and your frequency data to answer the Questions for Method C.

Preliminary Data Table (Method C)

INSECT POLLINATOR (Name or briefly describe each species observed.)	FLORAL SPECIES A Name_____	FLORAL SPECIES B Name_____
1.		
2.		
3.		
4.		
5.		
6.		
7.		
8.		
9.		
10.		

Chi-square Contingency Table (Method C)

	FLOWER SPECIES A	FLOWER SPECIES B	Row Totals
Insect Species 1 <u>observed:</u>			
(expected):	( )	( )	
Insect Species 2 <u>observed:</u>			
(expected):	( )	( )	
Insect Species 3 <u>observed:</u>			
(expected):	( )	( )	
Column Totals:			
			⇑ **GRAND TOTAL**

1) For this analysis, you will need records for three insect species most common in the data set as a whole. Write the name of the first species in the box labeled "Species 1," the name of the second in the box labeled "Species 2," and the third in the "Species 3" box.

2) Inside the boxes marked "Flower Species A" and "Flower Species B," write plant names or descriptions of the two kinds of flowers you observed.

3) For Species 1, record the number of times you observed this species on Flower Species A inside the cell at the top left corner of the graph, *above the parentheses*. Going across the top row, enter the number of times you saw this insect on Flower Species B as well. Then enter observations for Species 2 in the second row, and Species 3 in the third row.

4) After entering your observations, add the two numbers in the top row and record the sum in the column labeled "Row Totals" at the right. Repeat for the other species, summing the total number of observations for each species as a row total.

5) Then add all the observations in the first column to determine the total number of insects you saw visiting Flower Species A. Enter the column total at the bottom of the Flower Species A numbers. Calculate column totals for Flower Species B in the same way.

6) Sum up all the row totals and enter the result in the cell labeled "Grand Total" at the bottom-right corner of the table. To check your math, sum the two column totals, and you should get the same result.

7) You should now have the table filled out as below. Three insect species are observed in this example.

	Flower Species A	Flower Species B	Row Totals
Insect Species 1	A_1	B_1	Row Total 1
Insect Species 2	A_2	B_2	Row Total 2
Insect Species 3	A_3	B_3	Row Total 3
Column Totals:	Column Total A	Column Total B	**GRAND TOTAL**

Expected values will be calculated from the **null hypothesis** that insects visit flowers randomly. Using the laws of probability (see Chapter 6), we can calculate how many insects we would expect to observe on each type of flower, based on species abundance and total numbers of visits. If a pattern in the data does not conform to these expectations, we can reject the null hypothesis and follow the alternative reasoning that insects select flowers in nonrandom patterns.

For each cell in the table, you will need to calculate expected values using a simple formula: (Expected value) = (Column Total)(Row Total)/(Grand Total). For example, the expected number of observations for Insect Species 1 visiting Flower Species A would be calculated as follows:

$$\text{Expected number (A1)} = \frac{(\text{Column Total A})(\text{Row Total 1})}{\text{GRAND TOTAL}}$$

8) Perform this calculation, and *enter the expected number for Insect 1 on Flower A within the parentheses* in the top-left cell of the data table.
9) For each of the other cells, multiply the cell's row and column totals, divided by the grand total, to get an expected number. Enter the expected value within the parentheses at the bottom of each cell in the contingency table.
10) For a Chi-square test, you will need to calculate the number of independently varying cells in the table. This is called **degrees of freedom**, abbreviated as **d.f.** For a contingency test, degrees of freedom (d.f.) is calculated using the formula below. Calculate d.f. for your contingency table, and write the result in the box below.

The table has spaces for three insect species and two flower species, so d.f. should be $(3 - 1)(2 - 1) = 2$.

d.f. = (# insect species – 1)(# flower speciess – 1) =

11) Now calculate Chi-square by completing the table that follows. As you fill in the table, you will be calculating a Chi-square value for your data using the following formula:

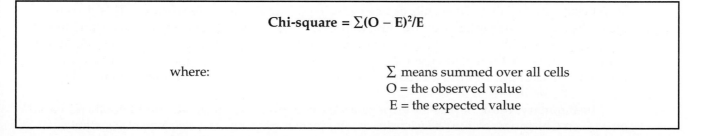

Chi-square $= \sum(O - E)^2/E$

where:

$\sum$ means summed over all cells
O = the observed value
E = the expected value

Chi-square Calculation Table

CELL LABEL Flower letter, Insect number	OBSERVED NUMBER (O)	EXPECTED NUMBER (E)	(O – E)	(O – E)^2	(O – E)^2/E
A1					
A2					
A3					
B1					
B2					
B3					
				Chi-square value =	

12) Enter a cell label for each cell you completed in your earlier table. For example, label the data for Flower B and Insect 2 as "B2." Note that each cell in the earlier table gets its own row in this one.

13) From your entries in the first table, copy all observed and expected values in the columns labeled "O" and "E."

14) Calculate the amount of deviation by subtracting the expected value from the observed value in each row. Enter the difference, with sign, in the column labeled "(O – E)." As a test of your math, check to make sure this (O – E) column sums to zero.

15) Square each deviation. Note that positive and negative signs disappear when you square these values. Enter the squared value in the column labeled "(O – E)^2."

16) Divide each squared deviation by the original expected value. Enter the result in the last column, which is labeled "(O – E)^2/E."

17) Finally, add up all the (O – E)^2/E values, and record the sum at the bottom-right corner of the calculation page. This is the **Chi-square** value for your data set. Because it is calculated from deviations of observed values from expected values, the greater the departure from null hypothesis expectations, the higher the value of Chi-square. If we observed no differences between observed and expected values, the value of Chi-square would be calculated as 0, meaning that insects are selecting flowers exactly as your null hypothesis predicted. Even if insects were selecting flowers at random, some degree of pattern could be expected due to sampling error, so we could anticipate a small Chi-square value most of the time, just due to chance.

As deviations from random expectations grow larger, Chi-square grows too. At some breaking point, called the **critical value** of Chi-square, we can no longer accept the null hypothesis as an explanation for data patterns generating these large deviations. At this point, we say an alternative, nonrandom explanation is warranted. By convention, scientists agree that the critical value of Chi-square has been reached if the odds of generating deviations of this magnitude by random sampling error alone are less than 5%. This 5% probability, written as **p = 0.05** by statisticians, means that our results can be considered **"statistically significant."** This does not mean the alternative explanation we pose is always right, but it does mean we have eliminated the null hypothesis with 95% confidence. An even higher value of Chi-square, corresponding to a p value of 0.01, is the threshold customarily designated as a "highly significant" departure from random expectations. In this case, our conclusion is the same, but our confidence in rejecting the null hypothesis has risen to 99%.

In summary, think of p as the error rate we can expect if we reject the null hypothesis based on a given Chi-square value. *A high value of Chi-square corresponds to a high level of statistical significance, which in turn corresponds to a low p value.*

18) Equating a Chi-square value (along with the degrees of freedom in your experimental design) with its corresponding p value is easily accomplished using a Chi-square table (Appendix 5). A simplified Chi-square table is included below, for your convenience. First, find the row equivalent to the **degrees of freedom** you calculated for this data set. Remember this is (#Flowers − 1)(# Insects − 1) so your degrees of freedom should be 2 if you observed 3 insect species and 2 flower species. Next, compare your Chi-square value with the two Chi-square numbers in the table.

Under the heading "p = 0.05" and in the row corresponding to your degrees of freedom, you will find a critical value of Chi-square. *If your calculation of Chi-square yielded a number larger than the critical value, then your results can be considered significantly nonrandom.* In the context of your experiment, this supports the hypothesis that your insect species are partitioning themselves among flowers in a nonrandom fashion.

Under the heading "p = 0.01," you will find a Chi-square value indicating a higher level of significance. If your calculation of Chi-square yielded a number larger than this, you can have even greater confidence that your data show a nonrandom pattern. The conclusion is the same as for p = 0.05, but your evidence is stronger in this case.

A Chi-square value lower than the number in the p = 0.05 column could have come from observations of insects randomly selecting flowers. This may or may not be the case, but low Chi-square values mean you do not have sufficient evidence to reject the null hypothesis of random flower visitation.

Simplified Chi-square Table of Critical Values*

DEGREES OF FREEDOM	CHI SQUARE GREATER THAN THIS VALUE INDICATES SIGNIFICANCE p = 0.05	CHI SQUARE GREATER THAN THIS VALUE INDICATES HIGHER SIGNIFICANCE p = 0.01
2	5.99	9.21

*Critical values from Rohlf, F. J. and R. R. Sokal. 1995. *Statistical Tables*, 3rd ed. W. H. Freeman, San Francisco.

Questions (Method C)

1. Did you confirm any differences in pollinator preference between the two plant species? Do you think these plants have evolved mechanisms to limit mutualistic interactions to one or a few pollinator species, or are they acting as generalists?

   _____

   _____

   _____

   _____

2. What did you notice about flower color, scent, shape, or orientation that might affect attractiveness to different kinds of pollinators? Does flower structure explain any patterns in insect preference you observed?

   _____

   _____

   _____

   _____

3. How were the flowers you observed shaped to maximize pollen transfer? Were the insects forced to brush against the male parts of the flower (called anthers) in order to enter the flower or in order to access the nectar? How do you think pollen might be transferred to the receptive surface (called the stigma) of the female floral parts?

   _____

   _____

   _____

   _____

4. An insect's resource-seeking behavior is called a **foraging strategy**. What factors other than flower characteristics might influence the foraging strategies of the insects you observed? Is it possible that wind or sun exposure, proximity to shelter, humidity, or other site characteristics are in part responsible for differences in pollination frequency?

   _____

   _____

   _____

   _____

5. In natural plant communities, different species often open their flowers at different times of day or at different times of year. Why might this be advantageous to all members of the community?

   _____

   _____

   _____

   _____

FOR FURTHER INVESTIGATION

1. Try the pea and *Rhizobium* experiment with different varieties of peas. Are some pea varieties better able to accommodate the symbiont than others?
2. Observe and classify lichens on your campus. When you compare the lichens growing on trees on campus with lichens on trees in a less populated area nearby, do you see evidence of air quality differences between the sites?
3. Compare results of your pollinator foraging study at different times during the spring. Do some pollinators shift their preferences as different kinds of flowers become available?

FOR FURTHER READING

Borror, D. J. and R. E. White. 1998. *A Field Guide to Insects*. Houghton Mifflin.

Margulis, Lynn. 1999. *Symbiotic Planet: A New View of Evolution*. Basic Books, New York.

Ramsay, Marylee and John Richard Schrock. 1995. The yucca plant and the yucca moth. *The Kansas School Naturalist*. Emporia State University, Emporia KS. 41(2). http://www.emporia.edu/ksn/v41n2-june1995/KSNVOL41-2.htm

Richardson, D. H. S. 1992. Pollution monitoring with lichens. *Naturalists' Handbook No. 19*, Richmond Publishing Co. Ltd.

Chapter 13

Biodiversity

Figure 13.1 A coral reef off Key Largo in Florida. Tropical coral reefs are among the earth's most diverse systems.

INTRODUCTION

Biologists have always been awed and challenged by the diversity of life on earth. At the conclusion of "The Origin of Species," Charles Darwin contemplates with astonishment a riverbank covered in a tangled growth of vines, shrubs, fungi, birds, mammals, worms, insects, and scores of other living things. After hundreds of pages of carefully marshaled evidence supporting his concept of speciation as a natural and ongoing process, the methodical Victorian naturalist finds the outcome of biological evolution too amazing for words. **Biodiversity**, which Darwin described as "endless forms most beautiful and most wonderful," remains an object of scientific curiosity and passion. A contemporary biology student, snorkeling over a coral reef for the first time (Figure 13.1), is likely to share Darwin's wonderment at the "grandeur" of life. Coral reefs are incredibly species-rich communities. Surrounded by hard and soft corals, echinoderms, annelids, molluscs, arthropods, and fishes of all sizes, shapes, patterns, and colors, it is hard to fathom how so many different species can coexist in one place.

How many species are there? This question is easier to ask than to answer, because we have discovered and described only a fraction of the earth's biota. Most large terrestrial organisms, such as birds and mammals, are so well inventoried that discovery of a new species in these taxonomic groups is a newsworthy event. At the other extreme are soil bacteria, often impossible to culture in standard media, and so poorly studied that there are probably a number of undescribed species thriving beneath your campus grounds. In his influential book *The Diversity of Life*, ecologist E. O. Wilson (1999) reports an estimate of 1.4 million described species, based on interviews with taxonomists specializing in a wide variety of organisms, and comprehensive reviews of databases and museum records. As for the numbers of living species not yet known to science, estimates range from 5 million to 100 million. As ecologists discover new taxonomic categories and new microhabitats, they are constantly revising their estimates as previously unknown groups of organisms are discovered (e.g., Ellwood and Foster, 2004). Wilson's argument for more attention to taxonomic questions in biology is compelling, since intelligent management decisions for global species protection begin with some idea of the number of species we have to protect.

How can we improve our appraisal of biodiversity yet to be discovered? One practical approach is based on repeated sampling of a type of organism in a particular place, using the growing database to develop a **species accumulation curve**. To show how this works, let's visit La Selva, Costa Rica's national rainforest preserve. Here entomologists John T. Longino and Robert Colwell have been collecting ants from the leaf litter of the forest floor in an extended survey of the insect fauna of the park. One of their methods for trapping specimens is the Berlese apparatus (Figure 13.2). Leaf litter collected from the forest floor is returned to a lab and placed in a funnel lined with screening. A lightbulb placed over the top of the funnel heats up the leaf litter, and all the tiny invertebrates from that sample go down through the screening and into the funnel. The bottom of the funnel leads to a jar of preservative, so the species in this sample can be counted and identified (Figure 13.3).

Figure 13.2 Berlese funnel apparatus for collecting arthropods from leaf litter.

Figure 13.3   Arthropods sampled with a Berlese apparatus.

Longino and Colwell's Berlese results for La Selva are shown in Figure 13.4. The x-axis shows the number of samples analyzed, and the y-axis shows cumulative numbers of ant species found in all the samples up to that point. The curve rises steeply at first, because new species are discovered in nearly every sample at the beginning of the study. As more and more samples are examined, it becomes harder and harder to find species not already counted, so the slope of the curve gets less and less steep as the sampling effort continues. From a quick examination of the figure, you could predict that biologists could find more ant species if the study were continued, but you could also place an upper limit on the expected number, based on the decreasing slope of the curve. Species accumulation curves have been developed for many kinds of organisms in many kinds of habitats. The shape of the curve varies somewhat, depending on habitat patchiness and the relative frequency of rare species, but declining rates of return on sampling effort are common to all. At some arbitrary stopping point, say less than one new species per 1000 samples, we can conclude for all practical purposes that the fauna in this locale have been adequately described.

Check your progress:

Estimates of the total numbers of species on earth vary by orders of magnitude. How can we find out which estimates are correct?

Answer: More study of poorly investigated taxonomic groups and species-accumulation analysis in species-rich systems could lead to better global estimates.

Although we are not sure how many species exist in the world, we do know that ecosystems vary significantly in biodiversity. In general, numbers of species per unit area are highest in tropical systems, declining as you travel from the equator toward the north or south pole. A tropical forest may have as many as 300 tree species in a randomly selected hectare of land (Figure 13.5), but in a similar area of boreal forest in interior Alaska, you are likely to find only one or two species of spruce trees (Figure 13.6). Why does biodiversity per unit area decline with increasing latitude? One explanation is that **primary productivity**, which is the rate of biomass accumulation per unit area via photosynthesis, is highest in the tropics where the growing season lasts longer and high temperatures speed up biochemical reactions. If we assume the total **biomass** of an ecosystem can only be divided into a finite number of species, then higher productivity gives an ecosystem the capacity to support more kinds of organisms. This explanation presumes ecosystems are subject to **bottom-up control** by the plants at the base of the food chain. Better growing conditions, in this way of thinking, lead to greater biodiversity.

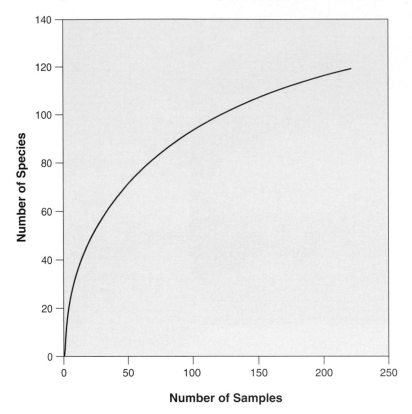

Figure 13.4   Accumulation curve for ants collected with Berlese trapping in La Selva Biological Station in Costa Rica. (After John T. Longino, http://www.evergreen.edu/ants/alascollns/ 2001.expeditions/reports/formicidae/home.html#fig1)

An alternative explanation recognizes the longer geologic history of tropical ecosystems. Since arctic and north-temperate communities were disrupted in the relatively recent past by glaciation, evolution has had a longer undisturbed period to produce species in the tropics. A third possibility is that stable climate in the tropics means organisms migrate less, so tropical populations remain more isolated within valleys or on mountains. Since speciation is thought to occur more quickly when populations are geographically separated, we might expect higher rates of speciation in the tropics as a result of greater climatic stability. All of these explanations may be true in part. As is so often the case in ecology, the trend we observe may have more than one contributing cause.

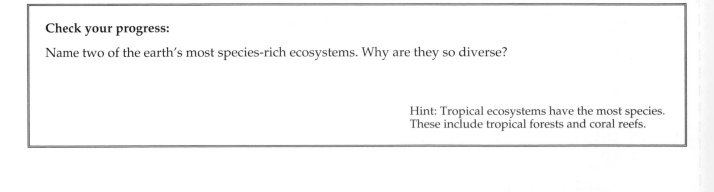

Check your progress:

Name two of the earth's most species-rich ecosystems. Why are they so diverse?

Hint: Tropical ecosystems have the most species.
These include tropical forests and coral reefs.

Figure 13.5   A tropical forest exhibits high biodiversity, with hundreds of species of trees in a single community.

Figure 13.6   Boreal forests of interior Alaska have relatively low biodiversity, with a community dominated by two species of spruce trees.

We also know that biodiversity is unequally distributed among taxonomic groups. As impressed as we may be by the number of living mammals (4000 described, according to E. O. Wilson), there are more than 10 times as many mollusks as mammals (over 50,000 species). That is certainly a lot of snails and clams, but there are twice that number of butterflies and moths (Order Lepidoptera 112,000), and six times that many beetles (Order Coleoptera 290,000). Broad-leaved flowering plants make up 2/3 of the plant kingdom, and more than half of all animal species are insects. Why have some taxonomic groups been so successful at generating new species? Given our understanding of competitive exclusion (Chapter 8), we understand that each species in a community must exploit a unique ecological niche. Small body size would therefore seem conducive to rapid speciation. It is reasonable to assume creatures as small as mites or beetles could find more ways to divide up an environment into distinct microhabitats or unique feeding niches than could a moose-sized animal occupying the same habitat.

Another factor made clear from the fossil record is that numbers of species in an emerging taxonomic group can expand rapidly when the normally slow evolution of organismal forms crosses a threshold into a truly novel body plan. The first winged insects, for example, were suddenly presented opportunities for hunting, grazing, escaping predators, and finding mates that were unprecedented in the history of arthropod life. In the absence of any serious competition save one another, these novel organisms spread across continents, adapting to newfound modes of existence like homesteaders in a land rush. Natural selection to local environments and unprecedented access to far-flung resources generated genetic differences among geographically distinct populations and pushed the process of speciation into overdrive. The result, which we see reflected in patterns of biodiversity in the present day, is a diverse array of related species, each applying the group's common evolutionary advantage as individual variations on the common morphological theme. Evolutionary biologists call this phenomenon **adaptive radiation.** In addition to the insects, the rapid development of flowering plants in the Cretaceous period (over 100 million years ago) presents a noteworthy example. A "breakthrough" to reproduction through flowers and fruits made plants much more successful in terms of species numbers than the older established lines of ferns and conifers. Emerging insect families, adapting to new niches as pollinators, seed predators, and herbivores, rode the tide of floral diversity to become the most species-rich animal group on planet Earth.

Check your progress:

Why are there so many kinds of beetles?

Answer: Beetles are small, and as a consequence of adaptive radiation, exploit a variety of niches.

In addition to species richness, we need to recognize that biodiversity exists in a hierarchy of variation. Genetic diversity within the species is an important consideration in understanding the history and predicting the future of a population (Chapter 6). Above the species level, diversity of higher taxa such as orders or families is also of paramount importance to conservation efforts. Biologists first want to save taxonomic groups represented by one or a few species, because these "end of the line" species have a significant number of unique genetic traits. For example, the mammalian order Proboscidea includes many fossil species of mammoths and mastodons, but only two surviving species: the Asian elephant and the African elephant. In contrast, the Order Rodentia has approximately 2000 living species, so extinction of a rare species of vole or squirrel would not have the same impact on global animal diversity as losing a species of elephant.

Below the species level, biodiversity exists in subspecies and varieties. Florida panthers (*Puma concolor coryi*) were once broadly distributed through the American Southeast, but are now reduced to a few individuals in the Everglades. Although they are considered the same species as more abundant cougars (also known as mountain lions) in Western states, sufficient genetic differences exist among populations to merit separate protection for Florida panthers. At the bottom of the hierarchy of diversity is genetic variation within a population. Maintaining critical levels of genetic variation within populations is important too, because species reduced below a critical threshold lack the genetic variation needed to avoid inbreeding depression. This term refers to genetically related failures in vitality and reproduction due to mating of genetically similar individuals. This becomes a problem in many species when the breeding population drops below 100 or so.

Documenting and managing biodiversity is especially important in the twenty-first century because species extinction represents an increasing problem across the globe. Expanding human populations and an exponentially growing demand for primary resources such as coal, water, cropland, and timber take habitats away from more native species every year. In the oceans, pollution and changes in average sea temperature are stressing coral reefs and all their associated life forms. The vanishing rainforest is well publicized, but biologists are currently voicing serious concerns about the extent of habitat loss within the United States as well. For example, a particularly destructive form of mining in the Appalachians removes entire tops of mountains to expose seams of coal (Figure 13.7). The overburden of rock layers above the seam is dumped into surrounding creek valleys, so mountaintop removal affects water quality downstream as well as the obvious impact on high-altitude habitat for species not found at lower elevations. Although mining companies usually plant grass on the flattened peaks after the coal is removed, the resulting "reclaimed" land cannot support the same biological community that was destroyed.

Check your progress:

Given the rising extinction rates around the world, and the limited funding available for conservation, in what types of habitats should we focus our preservation efforts?

Hint: Consider species richness, unique taxa, and the economic status of the nations involved.

Making sound decisions about biodiversity management and protection begins with accurate data, so measuring diversity is important. Species lists are a good start, but relative numbers of individuals of each species in a community should also be considered. The number of species in a sample is called species **richness**. The degree to which total organism number is distributed equally among these species is called species **evenness**. Both richness and evenness need to be incorporated in a biodiversity index.

Figure 13.7   Mountaintop removal threatens biodiversity in the Appalachians.

To illustrate with an example, suppose ecology students at two hypothetical colleges held a contest to see which had the most biologically diverse bird community (or avifauna) on their campus. College A students spent a Saturday morning walking the campus, counting numbers of each species they encountered. Students at College B engaged in a similar sampling routine. Assume that both colleges had the same species richness (five species) and the same sample size of 100 birds, but that numbers of each kind of bird were distributed differently. (For a visual representation of the two campus data sets, see Figure 13.8.)

Number of Birds of Listed Species Identified on Campus

TYPE OF BIRD	CAMPUS A	CAMPUS B
Pigeon	96	20
Robin	1	20
Starling	1	20
Purple grackle	1	20
House sparrow	1	20

Campus A - Dominated by One Species

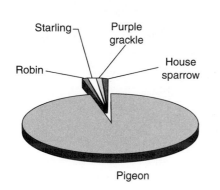

Campus B - Even Species Distribution

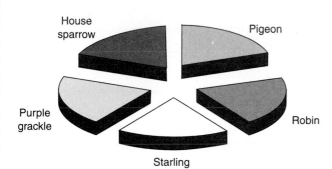

Figure 13.8   Species distribution contributes to biodiversity. Campus A, dominated by one bird species, is less biologically diverse than Campus B, even though both sites have 5 species.

Which school has the more diverse avifauna? Clearly, four of the birds on Campus A's species list make little contribution to the bird community. Campus B's sample, illustrating maximum evenness, demonstrates greater biodiversity. On Campus B, interactions between different species are much more frequent because the community has no single dominant type. Another way of thinking about this as an observer is to ask, what is the probability of encountering the same species twice in a row? On Campus A, the probability is very high, but on Campus B it is much lower.

This probability of change in encounters was used along with species richness to design a way to measure biodiversity that takes both numbers of species and their proportions into account. Called a **Shannon diversity index**, this methodology was originally derived for calculating variations in electronic signals. This index is also known as the Shannon-Wiener Index and as the Shannon-Weaver Index, because its author Claude Shannon collaborated both with Norbert Wiener and with Warren Weaver (Perkin, 1982). In any case, the index measures the likelihood of repetition in adjoining samples. The diversity index is calculated as follows:

Shannon Diversity Index:

$$H = \sum_{i=1}^{S} -(P_i \cdot ln\, P_i)$$

where:

H = the Shannon diversity index
P_i = fraction of entire population made up of species i
S = numbers of species encountered
$\sum$ indicates the sum from species 1 to species S

To calculate the index, first divide the number of individuals of species #1 you found in your sample by the total number of individuals of all species. This is **P₁**, which should be expressed as a decimal value between 0 and 1. Then multiply this fraction times its own natural logarithm. This gives you the quantity **(P₁ · *ln* P₁)**. Since the natural log of a fraction yields a negative number, a minus sign is placed in front of the parentheses in the equation to convert the negative product back to a positive number. Next, plug in species # 2 numbers to calculate – **(P₂ · *ln* P₂)**. Repeat for all species through the last on your species list, which is species number **S**. Finally, sum the – **(Pᵢ · *ln* Pᵢ)** products for all species to get the value of the index, **H**.

The following table demonstrates calculations of the diversity indices for the bird data on Campus A and Campus B.

CAMPUS A BIRDS	N_i	P_i	$ln\ P_i$	$-(P_i \cdot ln\ P_i)$
Pigeon	96	.96	−.041	.039
Robin	1	.01	−4.61	.046
Starling	1	.01	−4.61	.046
Purple grackle	1	.01	−4.61	.046
House sparrow	1	.01	−4.61	.046
TOTAL	**100**			**H = 0.223**

CAMPUS B BIRDS	N_i	P_i	$ln\ P_i$	$-(P_i \cdot ln\ P_i)$
Pigeon	20	.20	−1.61	0.322
Robin	20	.20	−1.61	0.322
Starling	20	.20	−1.61	0.322
Purple grackle	20	.20	−1.61	0.322
House sparrow	20	.20	−1.61	0.322
TOTAL	**100**			**H = 1.610**

High values of H represent more diverse communities. A community of only one species would have an H value of 0, since P_i would equal 1.0, and it would be multiplied by ln (1.0) = 0. Campus A's H value is small, because its community is dominated by one species. If all species are equal in numbers, the equation yields a maximum H value equal to the natural logarithm of the number of species in the sample. For example, Campus B has five species. The H value of 1.61 = ln (5), so Campus B is as diverse as a five-species community can possibly be.

Check your progress:

What added information about biodiversity does the Shannon index convey that could not be derived from a simple species count?

Hint: The distribution of species abundance contributes to biological diversity.

METHOD A: MEASURING BIODIVERSITY OF CAMPUS BIRDS

[Outdoor activity, any time of year]

Research Question

How diverse is the community of bird species on campus?

Preparation

Birds can be censused almost any time of year. Early morning is best, especially in breeding season. A walk across a campus landscaped with trees and shrubs should allow observation of sufficient numbers of species to calculate a diversity index. If available to you, local parks, cemeteries, and residential areas are generally good habitats for songbirds as well. In northern states, migratory songbirds are present only in summer, but the birds that stay all winter are actually easier to see when the leaves are off the deciduous trees. If a bird feeder has been established for observation of feeding niches (Chapter 8), you can collect data by counting birds visiting the feeder during a specified length of time.

Good field guides are available for identification of local birds, but a knowledgeable mentor is the best way to master bird identification. Local Audubon societies are a great source of expertise. You may want to participate as a class in the annual winter bird count that many local organizations hold, and use these data for biodiversity analysis.

This method can be used to accumulate species counts for a species accumulation curve if multiple laboratory sections standardize their observation time. Treat each class observation as a sample, and plot the cumulative number of species found vs. number of samples analyzed. In small institutions, you can accumulate data from multiple years for the same purpose, although changing habitat conditions could become a factor over a period of years.

Materials (per laboratory team)

Binoculars (One for every person is ideal, but binoculars can be shared.)

Field guide to local birds

Data from previous surveys, if available

Procedure

1. Establish a time (e.g., one hour) for your census of campus birds. Designate at least one observer and one recorder for your group. As you walk a survey route or watch a bird feeding station, all members of the class can look for birds and relay information to the recorder. The recorder should write the name of each bird on the Data Sheet for Methods A, B, or C. Each time this species is encountered, record numbers of individuals in the second column of the Data Sheet. If you see a flock of birds, have everyone in the group estimate numbers, and then choose the middle (median) estimate as your group record.
2. When your observation time is complete, sum the number of individuals of each species, and include these totals in the right-hand column of the data sheet.
3. Transfer these totals to the Shannon Calculation Page for Methods A, B, and C. Calculate H, the diversity index for this sample, following the example from the Introduction. Record your estimated value of H.
4. If multiple samples are available, either from your instructor for past classes or from multiple sections of your class, arrange the data in chronological order and fill out the Species Accumulation Data Sheet for Methods A, B, and C.
5. Plot numbers of species as a function of number of samples to create a species accumulation curve for these data sets on the graph paper provided at the end of the chapter.
6. Answer Questions for Methods A, B, and C to interpret your results.

METHOD B: MEASURING INVERTEBRATE BIODIVERSITY

[Outdoor or laboratory, two class meetings required]

Research Question

How diverse is the detritus invertebrate community?

Preparation

Small invertebrates are surprisingly common in decaying leaf litter, but organic landscaping such as shredded bark mulch works very well too. Look for last year's landscaping mulch or leaf litter around the bases of trees or shrubs. Partially decayed organic material in contact with damp soil is best. Samples should be collected using a standardized methodology. A 1-pound coffee can or large empty tin can will work fairly well for taking a standard core sample.

A Berlese apparatus can be made with a ring stand and a large funnel. Drywall joint tape is readily available in hardware and home supply stores, but cheesecloth can be substituted if necessary. A goosenecked lamp with a metal shade and a 60-watt incandescent bulb works quite well for the heat source. If your lab benches do not have an elevated middle section, use a box or overturned bucket to elevate the lamp above the funnel. *Although the lamp bulb should be positioned 10–15 cm above the mulch, maintain enough space to avoid fire hazard, making sure the bulb is not leaning against the funnel or touching the organic material.*

A wide-mouth glass specimen jar, containing 85% denatured ethanol, can be positioned under the funnel to collect specimens. Driving specimens out of the mulch requires several hours, so this is a two-period procedure. If someone can come by after several hours and turn off the lamps and put tops on the jars, counting the specimens can be completed at a later time.

If multiple samples are scored by different lab groups, the class can complete steps 10 and 11 in the Procedure to develop a species accumulation curve. The class will have to share species lists from 8–10 groups to make this meaningful. Previous years' data can be pooled for this purpose as well.

Most introductory entomology texts have illustrations of noninsect arthropods suitable for identification in this exercise. If species identification is not possible, higher-level taxonomy (such as identification of the family) is adequate, as long as all analyses used in the species accumulation data are conducted in a consistent way. If students find an organism they cannot identify, it can still be used as a species in the analysis.

Materials (per laboratory team)

Large sealable plastic bag

Empty tin can

Hand trowel or spade

Ring stand

Large polypropylene funnel, with top diameter 15–20 cm

Drywall joint tape: two pieces, 15 cm long

Wide-mouthed specimen jar

Goosenecked desk lamp

Dissecting microscope

Guide to invertebrate identification

Data from previous surveys, if available

PROCEDURE

1. Locate an area where decaying mulch or leaf litter is in contact with soil. Push the open end of the can through the litter layer down to the soil. Tip the can sideways, and use your hand trowel to scoop all the loosened material into the can. Empty the contents into a large sealable plastic bag and return to the laboratory.
2. Set up the ring stand so that the funnel can be positioned just above the specimen jar as shown in Figure 13.2.
3. Cut two pieces of drywall joint tape about 15 cm long. Cross them, sticky side down, as shown in Figure 13.9. Push the crossed tape into the bottom of the funnel, with the adhesive side down, to create a screened mesh over the bottom of the funnel, as shown in the figure. This will keep leaf litter from falling into your specimen jar.

Crossed drywall tape   Top view of funnel

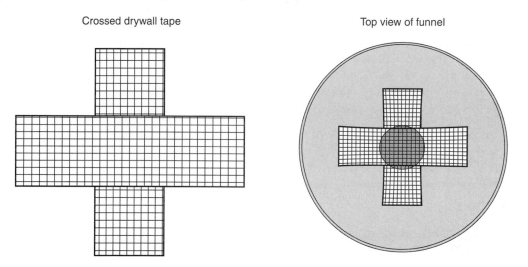

Figure 13.9   Procedure for lining funnel for Berlese apparatus. Cross two pieces of drywall joint tape about 15 cm long, sticky sides down, as shown at left. Then stick the crossed tape onto the bottom of the funnel to cover the hole, as shown at right.

4. Put a sample of leaf litter or mulch into the top of the funnel. Pour alcohol into the sample jar to a depth of 2 cm, and position it under the neck of the funnel.
5. Position the lamp a few inches above the funnel so that the bulb can warm and dry the sample. *Avoid direct contact between the lightbulb and the leaf litter.* Leave several hours or overnight. You should be able to see tiny organisms that have fallen in the jar as the Berlese trap begins to separate invertebrates from the decaying plant material.
6. Pour out your sample into a Petri dish or specimen dish, and examine under a dissecting microscope.
7. Identify or briefly describe species of organisms in your sample. Carefully count the numbers of individuals of each different species you see. Record each species on a separate line in the Data Sheet for Methods A, B, and C.
8. Sum the number of individuals of each species, and include these totals in the right-hand column of the data sheet.
9. Transfer these totals to the Shannon Calculation Page for Methods A, B, and C. Calculate H, the diversity index for this sample, following the example from the Introduction. Record your estimated value of H.
10. (Instructor's option) If multiple samples are available, either from your instructor for past classes or from multiple sections of your class, assign a recording student's name to each group, arrange group data in alphabetical order according to the student names, and use the samples in this order to fill out the Species Accumulation Data Sheet for Methods A, B, and C.
11. (Instructor's option) Plot numbers of species as a function of number of samples to create a species accumulation curve for these data sets on the graph paper provided at the end of the chapter.
12. Answer Questions for Methods A, B, and C to interpret your results.

METHOD C: MEASURING BIODIVERSITY IN SWEEP NET SAMPLES

[Outdoor or laboratory activity, late spring, summer, or fall]

Research Question

How diverse is the insect community in a field habitat?

Preparation

This method requires that students go to a weedy or grassy area not closely mown. Afternoon times are better than mornings for sweep samples, if nights are cool enough to restrict insect activity. The advantage over Method B is that collections can be made in minutes rather than hours.

Killing jars can be made inexpensively from any kind of large specimen jar or recycled jelly jar with a tight-sealing lid. Pour a layer of plaster of Paris, about 3 cm deep, in the bottom of the jar. (Alternatively, cut and layer paper toweling to fill the bottom fourth of the jar.) After the plaster of Paris has hardened, saturate it with ethyl acetate. This is a good killing agent for insects, but not as harmful to humans as other chemical alternatives. Ethyl acetate is the primary ingredient in many kinds of fingernail polish remover, which can be purchased in drugstores and works as well as reagent grade.

If multiple samples are collected by different lab groups, the class can complete steps 7 and 8 in the Procedure to develop a species accumulation curve. The class will have to share species lists from 8–10 groups to make this meaningful. Previous years' data can be pooled for this purpose as well.

Materials (per laboratory team)

Sweep net

Insect killing jar (one per sample)

Sealable plastic bags

Dissecting microscope

Insect identification guide

Procedure

1. Charge your killing jar by pouring 5–10 ml of ethyl acetate onto the absorbent material at the bottom of the jar. All the ethyl acetate should be absorbed; you should not have any pooled in the jar. Screw the lid on tightly and take it with your sweep net to a sample area.

2. In a field of high grass or weeds, collect insects with an insect net by "sweeping." This means to swing the net through the grass with a side to side stroke in front of you, so that insects are knocked off the grass and into the net. Take 30 swings as you walk through the field so that you sweep a new area of grass with each swing of the net. If you do not catch a variety of insects in 30 swings, increase the number of swings per sample, and standardize all further sampling at this number.

3. After you have collected a sample, shake all the insects down into the bottom of the net. Hold the netting closed with one hand to make a "pocket" containing the insects, and place the open mouth of the killing jar into the net adjacent to the pocket. Invert the "pocket" into your killing jar. Quickly slide the lid over the insects and screw it on tight. Insects should be held in the killing jar for at least 15 minutes before you examine them.

4. Examine your insect sample under a dissecting microscope. Identify or briefly describe each species of insect or other invertebrate in the sample. Carefully count the numbers of individuals of each species. Record each species on a separate line in the Data Sheet for Methods A, B, and C. Count the number of each species in your sample, and record those counts as tally marks in the second column of the data sheet.

5. Sum the number of individuals of each species, and include these totals in the right-hand column of the data sheet.
6. Transfer these totals to the Shannon Calculation Page for Methods A, B, and C. Calculate H, the diversity index for this sample, following the example from the Introduction. If you do not need all the rows in the table, leave the bottom rows blank. Record your estimated value of H in the indicated box at the bottom.
7. (Instructor's option) If multiple samples are available, either from past classes or from multiple sections of your class, assign a recording student's name to each group, arrange group data in alphabetical order according to the student names, and use the samples in this order to fill out the Species Accumulation Data Sheet for Methods A, B, and C.
8. (Instructor's option) Plot numbers of species as a function of number of samples to create a species accumulation curve for these data sets on the graph paper provided at the end of the chapter.
9. Answer Questions for Methods A, B, and C to interpret your results.

Data Table for Methods A, B, and C

SPECIES IDENTIFIED	TALLY MARKS (Numbers of individuals)	TOTAL # OF INDIVIDUALS
1.		
2.		
3.		
4.		
5.		
6.		
7.		
8.		
9.		
10.		
11.		
12.		
13.		
14.		
15.		
16.		
17.		
18.		
19.		
20.		
TOTAL # → Individuals in the sample		

Shannon Calculation Page for Methods A, B, and C

SPECIES FOUND IN SAMPLE	N_i	P_i	$ln\, P_i$	$-(P_i \cdot ln\, P_i)$
1.				
2.				
3.				
4.				
5.				
6.				
7.				
8.				
9.				
10.				
11.				
12.				
13.				
14.				
15.				
16.				
17.				
18.				
19.				
20.				
TOTAL			H =	

Species Accumulation Data for Methods A, B, and C

SAMPLE #	CUMULATIVE SPECIES NUMBER	SAMPLE #	CUMULATIVE SPECIES NUMBER
1.		11.	
2.		12.	
3.		13.	
4.		14.	
5.		15.	
6.		16.	
7.		17.	
8.		18.	
9.		19.	
10.		20.	

Species Accumulation Curve for Methods A, B, and C

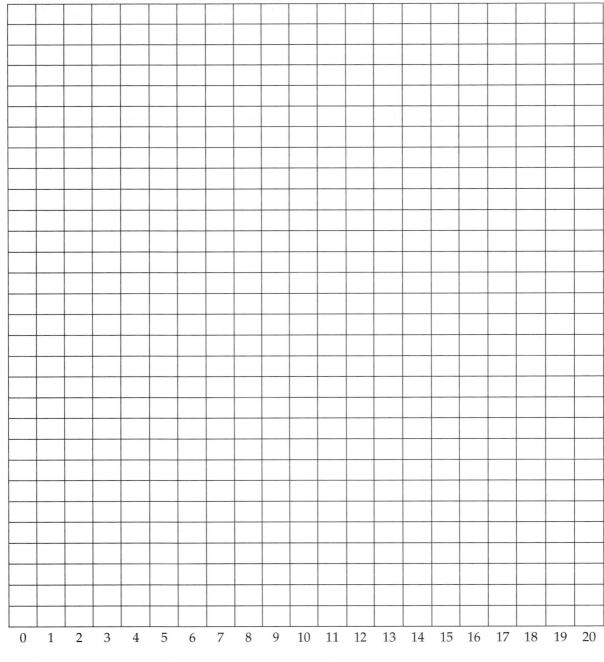

CUMULATIVE SPECIES COUNT
(Number of species in all samples up to this point)

0   1   2   3   4   5   6   7   8   9   10   11   12   13   14   15   16   17   18   19   20

SAMPLE NUMBER

Questions (Methods A, B, and C)

1. Describe the animal life you observed during your sampling process. Did you see species you had never seen before?

   _____

   _____

   _____

   _____

2. For any given sample, the highest possible value of H is equal to the natural logarithm of the number of species in the sample. For your sample, how does H compare with its theoretical upper limit? Explain what this means.

   _____

   _____

   _____

   _____

3. If the Shannon index were used to calculate a value of H for mammals in a multiuse national forest, would this be a good indicator of wildlife management practices? What management practices might increase H? Decrease H?

   _____

   _____

   _____

   _____

4. Which would you expect to yield the highest H value among the methods explained in this chapter: Method A or Method C? Why?

   _____

   _____

   _____

   _____

5. If you were able to draw a species accumulation curve, what conclusions can you draw about species richness at your site? Do you think your collection of samples includes more than half of all species present, or not? Explain.

   _____

   _____

   _____

   _____

FOR FURTHER INVESTIGATION

1. National Audubon Society members have been conducting bird surveys called "Christmas bird counts" for over 100 years. Data for your state and region can be accessed at http://cbc.audubon.org/cbccurrent/current_table.html. Use the census data from a survey conducted near your campus to develop a diversity index. You may also compare counts from a series of locations or a series of years to develop a species accumulation curve for birds in your state.
2. Longino and Colwell suggest the best mathematical description of their species accumulation curve is a logarithmic function of the following form:

Species Accumulation Curve:

$$S_T = \frac{ln\,(1 + Z\,A\,T)}{Z}$$

where:

S_T = cumulative numbers of species discovered up to and including sample T
T = numbers of samples taken
A is a constant (= 6.818 for ants at La Selva)
Z is a constant (= 0.013 for ants at La Selva)

Using any of the methods in this chapter, generate a series of samples and produce a species accumulation curve. If you use A and Z values similar to those found empirically by Longino and Colwell, does the theoretical curve match your data? Can you find A and Z values more appropriate for your organism to fit your data to the theoretical curve?

3. Repeat one of these methods of species sampling in a natural area off campus, and compare results. How much difference can you document in the biodiversity of the two sites?
4. Devise a method for calculating a diversity index for plants. Compare animal diversity and plant diversity for a series of sites.

FOR FURTHER READING

Ellwood, Martin D. F. and William A. Foster. 2004. Doubling the estimate of invertebrate biomass in a rainforest canopy. *Nature* 429:549.

Longino, J. T. and R. K. Colwell. 1997. Biodiversity assessment using structured inventory: Capturing the ant fauna of a tropical rain forest. *Ecological Applications* **7**, 1263–77.

Perkin, J. L. 1982. Shannon-Weaver or Shannon-Weiner? *Journal of Water and Pollution Contribution* 54: 1049–1050.

University of Arizona College of Agriculture and Life Sciences and The University of Arizona Library. 2005. *The Tree of Life Web Project*. http://tolweb.org/tree/phylogeny.html

University of California Berkeley Museum of Paleontology. 2005. http://www.ucmp.berkeley.edu/historyoflife/histoflife.html

Wilson, Edward O. 1999. *The Diversity of Life*. Norton, N.Y.

Chapter **14**

Succession

Figure 14.1 Beach evening primrose claims a sand dune at Padre Island National Seashore on the Texas Gulf coast.

INTRODUCTION

Padre Island National Seashore on the Gulf coast of Texas is a continuously changing landscape. Sea breezes carry powdery sand from the beach, piling up dunes 3–5 meters high in long ridges, parallel to the shoreline and extending for most of the 100-km length of the barrier island. When storm winds blow over the island, sand is carried off the upwind side of a dune and accumulates on the downwind side. The entire dune can shift its position in a remarkably short time. The author once went to sleep in a tent behind a dune on Padre Island, and awoke at the bottom of a deep crater of sand. The massive dune, which had seemed a reliable shelter from the strong onshore wind, had moved 10 meters overnight and almost completely buried the campsite by morning.

In spite of its unstable infrastructure, this habitat supports a vigorous plant community. Beach evening primrose (Figure 14.1), sea oats, and beach croton can tolerate dry surface conditions and salt spray from the nearby surf. Ecologist Patricia Moreno-Casasola (1986) discovered another survival secret of these pioneer species. Dune plants are actually buried by blowing sand from time to time, but grow their way out with upward-slanting lateral shoots that climb rapidly to the surface. By continually "treading sand" these hardy pioneers manage to survive in a landscape too unpredictable for other coastal flora.

As the dune specialists become established, their spreading roots and mat-like vegetation stabilize the surface of the blowing sand. Organic material accumulates, holding water for longer periods between rains. Over time, less specialized grasses and annual plants take advantage of improved growing conditions on the older dunes. These newcomers are superior competitors in the altered environment, so the pioneers are inevitably replaced by the plant community they helped create. Willows, saltgrass, cordgrass, and bushy bluestem dominate the protected side of the dunes of Padre Island. Buttonbush grows in the marshy sloughs, and cacti move into the more elevated sites. Given enough time and sufficient protection, woody shrubs and trees would take over. The southeast coast of Texas supports dense woodland communities of palmetto and live oak trees. On the island, however, the frequency of storms and wind-eroded "blowouts" in the dunes limit the establishment of woody plants. Each time the stabilizing layer of vegetation is stripped back to the bare sand, it creates more habitat for the pioneer species, and so botanical colonization of the dunes never ends.

Ecologists refer to the natural replacement of one biological community with another as **ecological succession**. Early field botanists, following the theoretical work of Frederic Clements (Chapter 9), thought of succession as analogous to the growth and development of an organism. In this view, a predictable series of communities, called **seres**, leads ultimately to a **climax** community capable of maintaining itself in a state of equilibrium. The climax community was understood to persist without significant change until some natural disaster disturbed the equilibrium and restarted the process of succession. Contemporary ecologists recognize that the outcome of succession can be contingent on local conditions, and sometimes occurs in cycles rather than a linear course of development. Even so, following ecosystem disturbance, many natural communities do go through a predictable series of changes, returning eventually to a regionally identified climax.

Check your progress:

Why do sea oats and beach evening primrose persist on Padre Island, even though they do not compete well in a mature community stabilized by their own earlier growth?

Answer: Early-succession species persist in frequently disturbed environments.

As an example, bare rock exposed by a landslide in a moist temperate environment goes through a number of seral stages on its way to becoming a forest. This centuries-long process is called **primary succession**. First among the rock colonizers are lichens. (See Chapter 12 for details on lichen natural history.) As the fungal elements of a lichen extract minerals from the underlying rock face, and the algal components accumulate organic products of photosynthesis, the lichen initiates a slow process of soil formation. Cracks in the rock trap organic matter, sheltering the next stage of succession (Figure 14.2). In very damp habitats, mosses may replace lichens. On dryer sites, annuals and small herbaceous perennial plants begin to appear wherever trapped debris holds sufficient moisture between rains. Their roots continue to absorb nutrients leaching from the substrate, and they contribute organic material at an accelerated rate. Still, it may be centuries before enough soil accumulates on the rock to support grasses or woody perennials. With deeper soil comes taller vegetation: shrubs and trees ultimately replace lower-growing forms as light rather than soil or water becomes the limiting resource for the maturing community.

Figure 14.2 Lichens overgrow a rock face and plants take root in crevices.

1983

2004

Figure 14.3 Vegetation on Mount Saint Helens begins to regenerate a forest destroyed by the 1980 volcanic eruption. Upper and lower photos were taken at the same site, illustrating recovery over 21 years.

Disturbance does not always strip the ecosystem down to bare rock. Fires, windstorms, and human agricultural activity typically leave a layer of soil behind. Even the devastating explosion of Mount Saint Helens in 1980 left ash-enriched soils open to plant colonization (Figure 14.3). Nitrogen-fixing lupines were among the first plants to grow back in the blast zone, and young evergreens are now reforesting the mountain's slopes. A progression of plant forms beginning on disturbed soil is called **secondary succession**. Because this type of community development is not retarded by the slow process of soil building, secondary succession can occur in a matter of decades rather than centuries. An abandoned farm presents a classic example of secondary succession (Figure 14.4). As every gardener knows, annual weeds are very well adapted to cultivated ground. Small, easily dispersed seed, rapid maturation, and high reproductive rates allow short-lived annuals to cover a site in the first year it is left fallow. After one or two seasons of weedy annuals, perennials such as thistle, blackberry, broom sage, and goldenrod move in. Within five years, these are replaced by taller shrubby plants such as sumac and elderberry. Within ten years, woody plants begin to dominate the community. On limestone-based soils in the American midwest, red cedar (*Juniperus virginiana*) is usually the first tree to appear; on more acidic soils it is more likely that pines will replace the shrub stage. Eventually, slower growing deciduous trees replace evergreens because of their greater shade tolerance. In the transition from evergreens to deciduous woods, the prevalence of broad-leaved trees in the understory clearly demonstrates why shade-tolerant deciduous trees dominate the climax in Eastern and Midwestern forests (Figure 14.5).

Figure 14.4 Old-field succession returns an abandoned farmstead to woodland.

Figure 14.5 A New Hampshire forest in transition from pines (taller trees with dark foliage) to deciduous hardwoods (younger trees in light-colored fall foliage).

Although successional stages are described for convenience as communities of plants, it is important to remember that the fauna exhibits successional change as well. An old field community that supports foxes, rabbits, and bobwhite quail in its early stages of succession will provide habitat for bobcats, grouse, and pileated woodpeckers as it approaches a mixed-hardwood climax. Birds play an important role in succession as seed dispersers. For example, cedar waxwings (Figure 14.6) travel in flocks that feed on small fruits such as wild cherry or hackberry. The flock tends to gobble up all the fruits on one tree, and then fly off in search of another. Since fruits are swallowed whole and the seeds inside are resistant to digestion, the seeds of these trees pass through the bird's digestive tract unharmed. A wild cherry stone swallowed by a waxwing may be deposited, along with some high-nitrogen organic fertilizer, many miles away. Without the help of birds, these trees would invade old fields much more slowly, and the nature of the transitional community would not be the same.

Figure 14.6   Birds play an important role in succession by introducing seeds to open sites. Cedar waxwings swallow small fruits whole, and may fly great distances before digesting the fruit and excreting the seed.

Aquatic succession begins with open water of a lake or pond. Year by year, rooted plants encroach on the edges of the pond and sediments are washed in from the surrounding hills (Figure 14.7). Algae generate organic material which settles to the bottom, so the aquatic system is enriched with nutrients and becomes biologically more and more productive. Over time, the pond fills in to becomes a marsh, with water-tolerant shrubs, rushes, and cattails growing where fish once swam. Eventually, leaves and sediment fill in the marsh and replace the wetlands. Whether the system began as a lake or an old field, the same climax forest is the ultimate result.

Succession driven by photosynthetic accumulation of organic material is called **autotrophic succession**. Over time, inorganic nutrients from water or bedrock are converted by plants to organic matter, so biomass increases as this type of succession proceeds. A different type of community development, called **degradative succession**, begins with a concentration of biomass. Consider the fate of a fallen log. As the log decomposes, nutrients are released and the energy trapped in its biomass fuels the growth of a series of small-scale communities which follow one another in a predictable sequence. First among the colonizers are fungi, because they have the rare ability to digest cellulose in wood. Spore-bearing "fruiting bodies" of the fungus may appear at the surface of the log (Figure 14.8), but the significant digestive work is being performed by the extensive network of thread-like fungal hyphae growing through the wood. Insects such as the large Bess beetle (Figure 14.9) consume the rotting wood, along with its crop of fungi. Their tunnels allow other organisms into the microhabitat, and they in turn attract the attention of woodpeckers and small mammals. As inorganic nutrients are released by decomposition, other plants thrive on the rich concentration of plant food. Foresters in the Pacific Northwest call rotting trees "nurse logs" because tree seedlings so often get started on the nutrient-rich ridges of decaying wood. Years later, you can find grown trees standing in a line where the "nurse log" gave them life (Figure 14.10). Although exhibiting predictable community changes, there is no "climax stage" in degradative succession, since the source of nutrients driving community development is ultimately consumed, and the system returns to its original state.

Figure 14.7 Rooted vegetation and accumulating sediment drive aquatic succession in a freshwater pond.

Figure 14.8   Fungi play an important role in degradative succession. Nutrients in a fallen log are released for use by other organisms on the forest floor.

Figure 14.9   Wood-boring insects such as the Bess beetle (*Passalus cornutus*) speed the process of degradative succession. Larvae and adults (see photo) both tunnel through rotting wood, speeding its decomposition.

Figure 14.10   A "nurse log" supports new tree growth.

Even decomposition of animal carcasses occurs in identifiable stages as one community of decomposers is replaced by the next. Forensic entomologists can identify the length of time a body is exposed to the elements by identifying the kinds of insects collected from the crime scene. Dead whales falling to the ocean floor create patches of invertebrate life, which gradually change in species composition over a period of years as the whale's elements are returned to the sea. A surprisingly extensive literature exists on degradative succession on cow dung, describing communities of flies and beetles that come and go in predictable seral stages as their nutrient source dries, grows fungi, and decomposes. Although decomposition is part of the larger picture of nutrient cycling in ecosystems, the changes in communities of decomposers on a log or a cow pat give ecologists a chance to study the process of succession on an experimentally manageable scale.

Whether observing a deer carcass or a mountainside, ecologists aim to move beyond descriptions of succession to understand the causes of community change. Interactions among organisms are of several types. When one species changes the environment in such a way that a different species can move in, we use the term **facilitation**. Dune plants stabilizing sand for colonization by other species provide a good example. In some kinds of community change, greater **tolerance** of competition allows a species to replace less tolerant types. The hardwood trees growing under a canopy of pines will ultimately replace the evergreen forest because they have greater tolerance of shade. Finally, **inhibition** influences succession when one species prevents establishment of another or even of itself. For instance, alfalfa plants produce chemicals that interfere with the germination of their own seedlings. Clearly, the failure of a plant population to replace itself would promote community change.

Check your progress:

How might facilitation influence the course of degradative succession in a fallen log?

Hint: How could the decomposing work of an early species, such as a fungus, prepare the way for a later-successional species, such as a woodpecker?

An ecological concept closely related to our understanding of succession is the theory of r and K selection. This theory identifies contrasting trends in the life history adaptations of organisms. Stated simply, **r-selected** species are adapted to early stages of ecological succession, and **K-selected** species are adapted to later stages. These terms come from the constants in the logistic growth equation, with r representing growth rate and K representing carrying capacity (see Chapter 4). The rationale is that r-selected life histories promote rapid population growth in the colonization stage of community formation, while K-selected traits promote competitive ability at the climax, when most populations are near carrying capacity.

Early succession habitats are temporary, so r-selected species respond with rapid establishment and accelerated development to reproductive maturity. Since the pioneer community usually transitions to another seral stage before the second generation is born, these colonizers must produce a large number of offspring and scatter them far and wide. Young r-selected animals strike out across country when population pressure builds, and r-selected plant seeds have wings, burs, or edible fruits to help them disperse as far as possible. A fortunate few offspring find themselves in a newly opened habitat somewhere else, and survive long enough to reproduce again. For this reason, r-selected organisms are sometimes called "fugitive species," always escaping competitive exclusion in one place by dispersing just in time to found another short-lived population in the next available site.

In contrast, K-selected species are adapted to the more stable and competitive communities of later successional stages. They tend to disperse short distances, mature slowly, and endure the effects of competition well. They produce fewer offspring, but invest more resources in each one. In animals, parental investment may take the form of feeding young for a longer period or helping them learn effective foraging behaviors. In plants, parental investment takes the form of large seed size. A seed containing more oils or carbohydrates is more expensive in terms of maternal energy to produce, but better provisioned to achieve maturity in a shady environment. As a result of parental investment, juvenile mortality is much lower in K-selected than in r-selected species. Since K-selected offspring are likely to encounter the same climax condition as their parents, dispersal is not as important. A few large, poorly dispersed seeds in the climax community result in more viable offspring than a large number of small, broadly scattered seeds.

Check your progress:

Complete this chart, comparing typical life-history traits of r-selected vs. K-selected species.

Life history trait	r-selected species	K-selected species
Body size		
Dispersal ability		
Numbers of offspring per parent		
Parental investment in each offspring		
Age at first reproduction		
Life span		

Answer: r-selected traits include smaller size, good dispersal, many offspring, low parental investment, short time to first reproduction, and short life span. K-selected traits include larger size, poorer dispersal, fewer offspring, higher investment, longer time to first reproduction, and a longer life span.

When comparing life histories, it is important to remember that predators generally have slower maturation, fewer offspring, greater parental investment, and larger body size than their prey, regardless of their community classification. Subjecting rabbits and foxes to an r-selected vs. K-selected comparison would therefore not be very meaningful. However, comparing life history traits of rabbits to deer or foxes to panthers would help explain why foxes and rabbits are found in earlier successional stages, while deer and panthers are found in later ones. Although not all species fit neatly onto the r-selected vs. K-selected life history spectrum, this theory can help us interpret trends in life histories of species within comparable ecological niches.

METHOD A: SUCCESSION IN HAY INFUSIONS

[Laboratory activity]

Research Question

How do physical and biological characteristics of a microbial community change during degradative succession in a hay infusion?

Preparation

This exercise requires monitoring a miniature ecosystem over a long period. Students should work in rotation, checking on the culture once per week for 3–6 weeks.

Directions for hay infusion cultures are described in Method C, Chapter 9. The freshwater hay infusions established as part of that exercise can be followed for a longer time to observe community changes during the process of degradation.

Because hay infusions are dominated by bacterial degradation of polysaccharides in hay for the first week or two, oxygen depletion by bacteria in the early stages severely limits some kinds of protozoans. Bacteria can therefore be seen as agents of inhibition, since they remove oxygen needed by many protists. Bacteria are also agents of facilitation, since the inorganic nutrients released by their enzymatic degradation of the hay support subsequent algal growth. Aeration of a control tube separates these two variables by replacing the oxygen removed by bacteria. Succession in aerated tubes should take a different course as a result.

An oxygen meter with a small probe is needed to monitor dissolved oxygen concentrations in the cultures. If the probe will not fit in the large test tubes as described in Chapter 9, transfer the cultures to larger containers so that this important variable can be monitored.

Materials (per laboratory team)

Two hay infusions (see Chapter 9, Method C) with no salt added

Aquarium air pump with tubing and adjustable air flow

Rubber band

Aluminum foil (small sheet)

Compound microscope

Glass slides, cover slips

Pasteur pipettes for sampling cultures

Dissolved oxygen meter with small probe, sensitive to concentrations 0–14 mg/L (one per laboratory).

Procedure

1.   Prepare two hay infusions as described in Chapter 9, Method C, using 10 g hay or grass clippings and 30 ml aquarium or pond water in a 50 ml tube. Mark the fluid level in the tube with a marker or wax pencil so that the water level can be maintained over time.
2.   Set the air pump at its lowest flow rate, and run a tube from the pump into one of your test tubes. Insert tubing from the air pump into the bottom of one of the hay infusion tubes, and secure it with a rubber band, as shown in Figure 14.11. Bubbles should come through the hay infusion, but not so violently that they cast fluid out of the top. Cover the culture loosely with aluminum foil.

3. On a rotating schedule with laboratory partners, do the following each week: a) turn off air pump, b) measure oxygen in aerated and non-aerated tubes, c) take a drop from the middle depth of each tube and survey it for protists, d) add deionized water as needed to maintain culture depth, and e) restart air pump. Record oxygen concentration of aerated and non-aerated tubes, along with any differences you see in the community of protozoans. Bacteria are difficult to observe with light microscopy, but anaerobic decomposition creates a cloudy suspension and a strong unpleasant smell. Take note of these indicators of bacterial growth as well. Although bacteria in hay infusions are not usually pathogenic, it is best to wash hands after handling cultures.

4. After 3–6 weeks, draw a graph on the Results for Method A pages below. Write a description of changes you observe in the aerated and non-aerated cultures. Try naming communities you saw replacing one another during degradative succession in your hay infusions, and explain differences you observed between them.

5. Answer Questions for Method A to interpret your results.

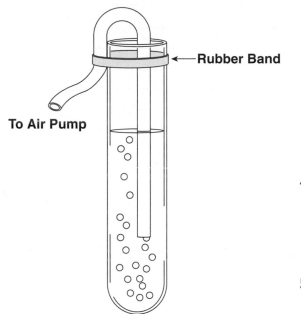

To Air Pump

←—**Rubber Band**

Figure 14.11   Aeration system for hay infusion.

Results for Method A:
Dissolved Oxygen Concentration vs. Time

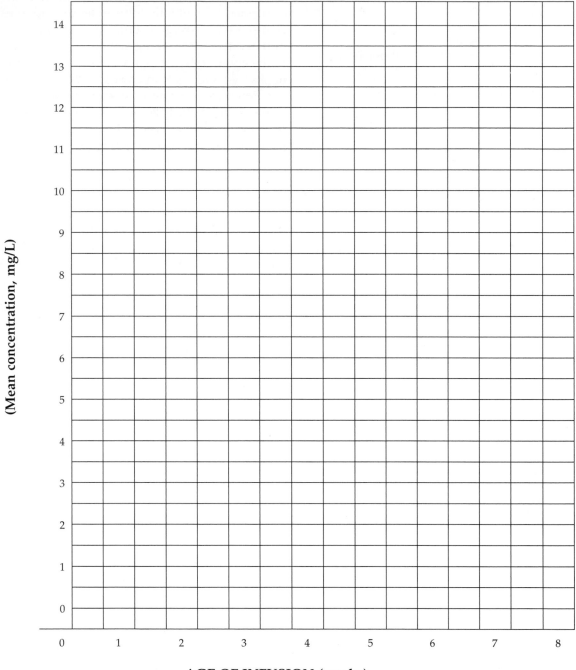

DISSOLVED OXYGEN (Mean concentration, mg/L)

AGE OF INFUSION (weeks)

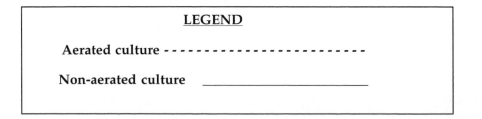

LEGEND

Aerated culture -

Non-aerated culture _____

Questions (Method A)

1. Describe the series of microbial communities you observed during degradative succession in the non-aerated hay infusion, beginning with the bacterial growth stage in the first week. Does the concept of seral stages apply in this system? Explain.

_____

_____

_____

_____

_____

2. How do community changes in the hay infusion correlate with changes in dissolved oxygen? Does this environmental variable explain some of the changes you observed in community structure?

_____

_____

_____

_____

3. How did the aerated culture differ from the non-aerated culture? Can you explain these observations using the same logic you applied in your answer to question 2?

_____

_____

_____

_____

4. Bubbling air through a hay infusion not only adds oxygen, but also constantly mixes the culture. How might this contribute to the course of successional change in the aerated tube?

_____

_____

_____

_____

5. When algae begin to appear, succession shifts from degradative to autotrophic. If both are going on concurrently, bacteria and protists are using oxygen simultaneously released by algae. How could you separately measure rates of photosynthesis and respiration with a simple apparatus like this one?

_____

_____

_____

_____

METHOD B: r-SELECTION VS. K-SELECTION IN TREES
[Laboratory activity]

Research Question
Do trees exhibit patterns of growth and reproduction consistent with r- and K-selection theory?

Preparation
This is a good hands-on activity for a laboratory session, but it requires significant startup effort by the instructor to build a seed and wood collection. The basis of the lab is to compare growth rates (measured by mean width of growth rings in wood) with seed size (measured by weight or by volume) in at least 10 species of trees. This exercise is thus most relevant in regions where successional change from softwood to hardwood forests can be observed. It is important to include a variety of trees with seeds of different sizes. Fruits or seed pods should be opened so that students can weigh or measure individual seeds. If native trees are planted on your campus, or are available to you at an off-campus site, you may be able to collect seeds yourself in the summer or fall preceding the lab date. Alternatively, many kinds of tree seeds can be purchased for study.

Tree ring data can be obtained in several ways. The instructor can use an increment borer to sample trees on campus or in nearby sites. Older residential areas are an excellent place to collect cores for tree ring analysis. With careful handling, tree cores can be used repeatedly, and have the advantage of taking up little storage space.

Alternatively, sample blocks of wood can be purchased for comparison of tree species from a number of suppliers. Because these are primarily used to demonstrate economically important wood types, they tend to include more slow-growing hardwoods than faster-growing softwood species. As trees are trimmed or cut down in your campus or neighborhood, you might ask the removal crew to saw out and save a thin cross section of the trunk or larger limbs for study. You will probably need to smooth the wood surface with sandpaper to make growth rings more visible, but once prepared, these samples can be used for many years.

Materials (per laboratory team)
Wood samples or cores from 10 or more species for tree ring measurements

Seed collection from the same species of trees

Hand lens or dissecting microscope

Ruler, marked in millimeters

Electronic balance, accurate to 0.001 g

(Alternatively, calipers accurate to 0.1 mm)

Field guide to trees, with shade-tolerance information

Procedure

1. From a core sample or wood block from an identified tree species, find a cross section with clearly visible growth rings. (See Figure 5.5 for an illustration of growth rings.) Midway between the center of the tree and the bark, count 10 rings. Measure the distance from the inside of the first ring to the outside of the tenth ring. This is the amount of wood added to the radius of the tree in 10 years. Record the result, in mm per decade, in the Data Table for Method B. Repeat for all the other tree species.
2. Measure the size of representative seeds from each of the tree species. If an accurate electronic balance is available, measure the weight of each seed to the nearest 0.001 g, and record your results in the data table. If your instructor directs you to measure volume rather than weight, use calipers to measure seed length, width, and depth to the nearest 0.1 mm. Multiply these three measurements together to calculate seed volume in cubic mm. Enter your results in the data table beside your growth rate data.
3. On the Method B Graph, plot growth rate (in mm per 10 years) on the x-axis and seed size (in either mg or in cubic mm) on the y-axis. Make a point on the graph for each species, and circle the points so they are easily visible. If you discern a linear relationship between growth rate and seed size, use a ruler to draw the best straight line you can through the points to represent this relationship. Approximately equal numbers of points should fall on either side of the line.
4. If you have access to computer software with statistical software, perform a correlation analysis on these data. (See Appendix 3 – Correlation and Regression.)
5. Answer Questions for Method B to interpret your results.

Data Table for Method B:
Growth Rates and Seed Sizes of Local Tree Species

TREE SPECIES	GROWTH RATE (mm per decade)	SEED SIZE (g or mm³)
1		
2		
3		
4		
5		
6		
7		
8		
9		
10		
11		
12		
13		
14		
15		

Results for Method B: Growth Rate vs. Seed Size

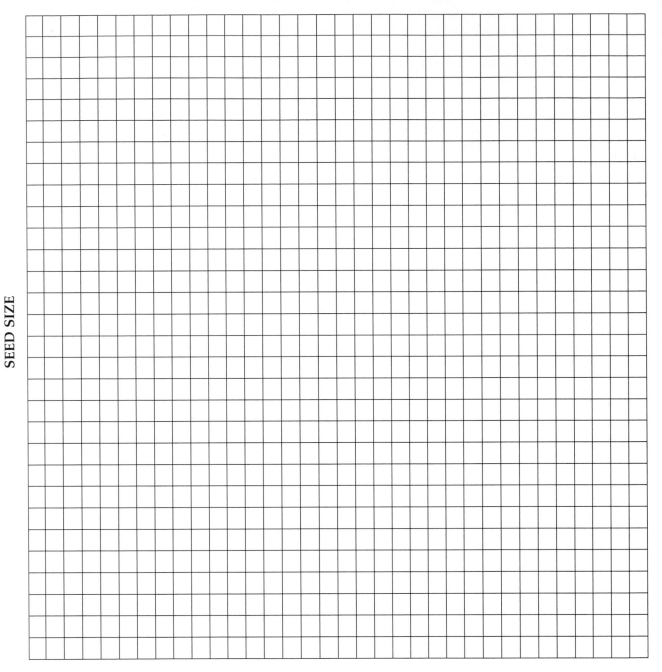

SEED SIZE

GROWTH RATE (mm per 10 years)

CORRELATION ANALYSIS RESULTS

Correlation coefficient r =   _____
Statistical significance =   _____

Interpretation:

Questions for Method B

1. According to the theory of r- and K-selection, which should have larger seeds, trees from early successional stages or from climax communities? Explain.

_____

_____

_____

_____

2. According to theory, which should be faster growing, trees from early successional stages or from climax communities? Explain.

_____

_____

_____

_____

3. Based on your answers to 1 and 2, what kind of relationship would you expect to see represented by the points on the graph for Method B? Did your observations generally conform to your expectations?

_____

_____

_____

_____

4. Look up natural history information for trees that, based on your graph, are most r-selected. Are these trees found in sunny, open sites as expected? Do you see winged seeds or other dispersal mechanisms in these species? Look up trees you would interpret to be most K-selected. Are these species found in mature forests?

_____

_____

_____

_____

5. Plant growth is in part genetically controlled and thus subject to life history adaptation, but environmental factors also influence its rate. Could environmental factors such as water availability or shading have influenced your measurements in this study? How could you design a sampling method to control these extraneous variables?

_____

_____

_____

_____

FOR FURTHER INVESTIGATION

1. Limestone and brick exposed to weather are subject to the same colonization by lichens that we see in primary succession on exposed rock. Measure rates of succession by comparing sizes of lichen patches or area of lichen coverage on buildings or monuments of different ages. Can you estimate a rate of primary succession on these sites?

2. In the hay infusion experiments, try measuring other nutrients such as ammonia nitrogen during the course of succession. Do changes in the concentrations of these nutrients explain some of the changes in the biota you observe?

3. What other attributes of trees, other than seed size or growth rates, could you use to place them in an r-selected or K-selected category? Design a way to measure one of these variables and test its correlation with growth rate or seed size.

FOR FURTHER READING

Howe, Henry F. and Maria N. Miriti. 2004. When Seed Dispersal Matters. *Bioscience* 54(7): 651–660.

Lichter, John. 1998. Primary Succession and Forest Development on Coastal Lake Michigan Sand Dunes. *Ecological Monographs* 68(4): 487–511.

Moreno-Casasola, Patricia. 1986. Sand Movement as a Factor in the Distribution of Plant Communities in a Coastal Dune System. *Plant Ecology* 65(2): 67–76.

National Tree Seed Laboratory, National Forest Service. 2005. http://www.ntsl.fs.fed.us/ntsl_fsstc.html

Pianka, Eric R. 1970. On *r* and *K* Selection. *American Naturalist* 104: 592–597.

Reznick, David, Michael J. Bryand, and Farrah Bashey. 2002. R and K-selection Revisited: The Role of Population Regulation in Life-history Evolution. *Ecology* 83(6):1509.

Chapter 15

Soils

Figure 15.1 Tree roots fight a losing battle against soil erosion along the banks of the White River in Indiana.

INTRODUCTION

Soil is a natural resource easily ignored until it is gone. Water flowing across the surface of the ground after a heavy rain mobilizes soil particles and dissolved nutrients, especially where soils have been disturbed by agriculture, mining, or construction, and carries them into rivers and streams. Along stream banks, soil erosion constantly moves sediments out of terrestrial and into aquatic ecosystems. Trees growing along river banks retard soil loss, but extensive human engineering of waterways for improved agricultural drainage exacts a steep environmental cost. When flow rates are artificially increased by digging deeper and straighter stream channels, tree roots can no longer hold the embankments, and erosion takes another load of soil into the river every time it rains (Figure 15.1). The unfortunately common practice of stripping protective vegetation from the **riparian zone** along streams exacerbates this problem, as does harvesting too many trees from hillsides within the watershed. In alluvial bottomlands, moving water eats away at soils that were deposited by historically slower-moving river systems over a vast expanse of geologic time. Many hectares of productive farmland can be lost in a few short years of poor management. The silt and nutrients washed into streams have negative consequences for the aquatic system as well. Rivers are muddied, silt buries bottom-dwelling invertebrates and fish eggs, and overnourished algae bloom only to decompose, stealing life-sustaining oxygen from the water. Even after a river enters the ocean, marine organisms flee or die from freshwater effluent laden with eroded soils, agricultural chemicals, and untreated wastewater (Figure 15.2). At the mouth of the Mississippi River, a "dead zone" bigger than Lake Ontario develops every summer in the Gulf of Mexico, and it continues to expand as we flush more and more of our ultimate source of agricultural wealth into the sea.

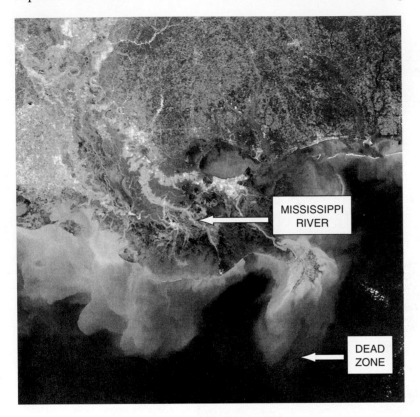

Figure 15.2 Sediments cloud ocean waters in a "dead zone" of hypoxia at the mouth of the Mississippi River.

In the Great Plains, plant communities dominated by native grasses built up and protected deep, rich prairie soils until much of the region was converted to row-crop agriculture in the 1800s. Replacing native perennials with annual crop plants such as wheat left the soil exposed to the weather much of the year. When crops failed during an extended drought in the 1930s, the bare soil dried to a powder and began to move on the wind. Dust storms choked people and animals, covered up what was left of the vegetation, and displaced rural populations throughout a "dust bowl" that extended from Texas to Minnesota (Figure 15.3). Impressed with the gravity of this disaster, the U.S. government developed an extensive system of advisory support, financial incentives, and agricultural restrictions to control soil erosion. Contour plowing, cover crops, and windbreaks became common conservation practices. As we came to understand the negative effects of erosion on aquatic systems, the federal government took steps to reduce sediment runoff into streams. Lumbering practices in national forests were altered to protect watersheds, and construction engineers were required to manage sediment runoff more effectively. Although conservation practices have improved, the total area of land impacted by farming, forestry, and development continues to expand. Soil erosion remains a global concern, especially in the developing world, where every shovelful of soil on every hectare of farmland is needed to support a growing human population.

Check your progress:

Describe five human activities that accelerate the rate of soil erosion.

Answer: Exposing soil to wind in dry regions, channelizing streams, disturbing riparian vegetation, harvesting too many trees from steep slopes, and disturbing soil during construction

As Conservationist Aldo Leopold pointed out over half a century ago, sound land management begins with an understanding of the role of soils in sustaining natural communities. Soils are the interface between the **biotic** and **abiotic** components of a terrestrial ecosystem. Here bedrock is weathered to release minerals taken up and incorporated into organic molecules by plants. Here dead leaves decompose, converting what was biotic back to abiotic material. Here bacteria fix nitrogen from the atmosphere, sheltered within the roots of their host legumes. Here water is sequestered between particles of silt and clay, minerals are leached from humus, and soil organisms stir and transform the ingredients of life.

These chemical and biological interactions produce a characteristic **soil profile**. Road cuts or embankments are good places to see how the soil is structured in your area (Figure 15.4). You can also use a soil sampler to extract a core of soil for analysis in an area of interest (Figure 15.5). Because different processes occur at different depths, soil develops in layers, called **soil horizons** (Figure 15.6). Lying on top of the soil is a layer of litter (dead leaves and other plant material) and "**duff**" (plant material degraded by soil organisms). Since this layer is composed primarily of organic material, it is called the O horizon. Below this is a layer of topsoil, called the A horizon. Mineral soil is enriched with decomposing organic material in this layer. Humic acids, resulting from the breakdown of plant material, give the A horizon a dark color. Plant roots grow through this layer, and earthworms burrow through it, consuming leaf mold and speeding up the process of decomposition by bacteria and fungi.

Figure 15.3 Unusually dry summers and poor soil conservation practices led to devastating wind erosion during the dust bowl years of the 1930s.

Figure 15.4   A scientist examines a soil profile in Ontario.

Beneath the topsoil is a layer of subsoil called the B horizon. This layer has little organic matter, and tends to resemble the underlying rock in its chemical makeup. Fine particles in this layer bind positively charged ions washed down from above, including calcium [Ca^{++}], ammonium [NH_4^+], and potassium [K^+]. Iron and aluminum ions leached out of the upper layers are also recaptured here, their oxidized forms creating horizontal bands of red or yellow within the B horizon. The zone of maximal leaching at the bottom of the A horizon is sometimes given a separate classification, the E horizon. The E stands for **eluviation**, which refers to materials dissolving in water and moving down through the soil. **Illuviation**, the opposite process, refers to deposition of dissolved substances as they are recaptured in the subsoil.

Under the subsoil lies the C horizon, where bedrock weathers and breaks up into small particles over time. This rocky layer is called "parent material" because it gives rise to the mineral components of the soil. In glaciated regions or coastal plains, parent material may be made up of loose geologic deposits rather than bedrock. The type of rock under the soil makes a large difference in its resultant characteristics. Limestone-based soils are rich in calcium and other minerals, and the calcium carbonate in limestone buffers pH, making these soils neutral or slightly alkaline. Sandstone-based soils lack buffering capacity, so they tend to be more acidic. Granite, a common form of igneous rock, also lacks buffering capacity, but tends to weather very slowly, so soils on granite tend to be shallow, with a limited B horizon.

Figure 15.5   A sampler is pushed into the ground to extract a soil core for analysis.

Check your progress:

Name four major soil horizons and briefly describe processes occurring in each.

Answer: O Horizon: litter accumulation
A horizon: decomposition of organic matter and leaching of nutrients
B Horizon: recapture of dissolved ions as they pass through mineral soil
C horizon: weathering of parent material

Soil Profile

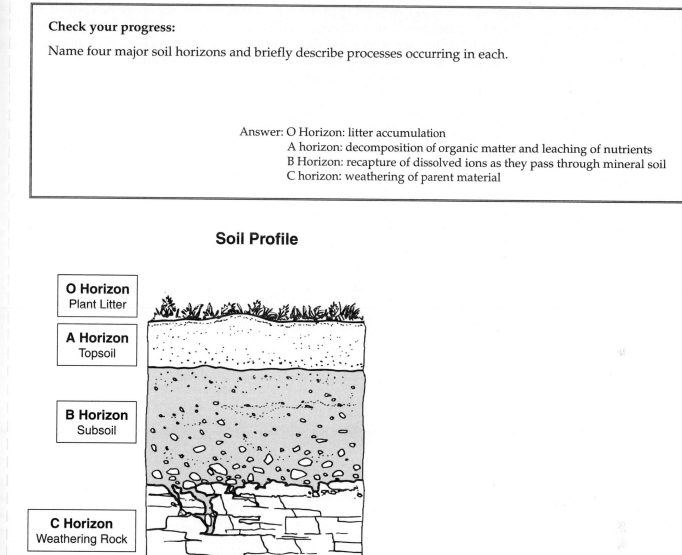

O Horizon
Plant Litter

A Horizon
Topsoil

B Horizon
Subsoil

C Horizon
Weathering Rock

Figure 15.6   A soil profile is made up of distinctive layers, or horizons. Decomposing plant material (O Horizon) adds humus to a layer of topsoil rich in nutrients (A Horizon). Below this is a layer of subsoil (B Horizon), whose clay particles recapture ions leaching out of the topsoil. Weathering parent rock (C Horizon) adds minerals and particles to soil as it gradually breaks down.

Soil particles come in many sizes, which are classified for convenience as clay, silt, or sand. Clay particles are the smallest, defined as smaller than 0.002 mm. Silt particles are of intermediate size, between 0.002 and 0.05 mm. Sand particles are largest, from .05 to 2 mm. Anything larger than 2 mm in diameter is classified as gravel. Soil containing a mixture of sand, silt, and clay particles is called loam. The size of the soil's particles determines its capacity to hold water, which is ecologically very important. Three critical measures of water in the soil are saturation capacity, field capacity, and wilting point. **Saturation capacity** refers to the amount of water a soil contains when all the spaces between soil particles are filled with water. Sand has the highest saturation capacity, because its relatively large particles cannot be packed as tightly together as other soil types, so there is more room for water in between the grains of sand. Silt has an intermediate saturation capacity, and clay has very little, since its tiny particles fit closely together, without much space left for water.

Field capacity measures the amount of water remaining in the soil after the force of gravity has pulled all it can down through the soil column. Because water adheres to soil and to itself, drained soil still contains water clinging to soil particles against the pull of gravity. Field capacity is a more ecologically important measure of soil water than saturation capacity, because soils do not remain saturated unless they lie below the water table. After water drains down through the soil after a rain, field capacity measures how much water is actually left to support plant life. To measure field capacity, wet soil is allowed to drain, and then weighed. The sample is then dried in an oven and weighed again. The difference between the wet weight and dry weight represents the grams of water that were in the drained soil. This water weight is divided by the dry weight and multiplied by 100% to calculate field capacity. A good way to think of field capacity is to consider how much water, in comparison to its own dry weight, can soil take up and hold against the pull of gravity? Particle size significantly affects field capacity of the soil (Figure 15.7). The smaller the particles in soil, the greater their surface to volume ratio. (See Chapter 2 for an explanation of surface/volume ratio.) A gram of clay therefore has much more surface area among its tiny particles than a gram of sand has among its fewer, larger particles. Clay has the highest field capacity, losing very little water to drainage after saturation. Silt has an intermediate field capacity and sand has a low field capacity. This means that clay soils dry out very slowly after rains, but sandy soils dry out much sooner.

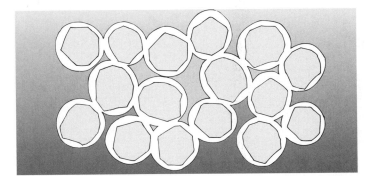

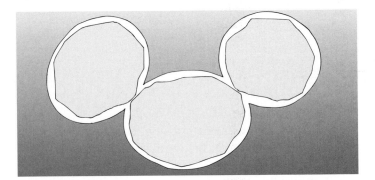

Not all water held in the soil against the pull of gravity is available to plants. Although the higher solute concentration within root tissues creates osmotic potential, which draws water from the surrounding soil, there is a limit to this water movement. The force needed to move water against its tendency to remain in place is measured in pressure units, and is called **water potential**. At a water potential equivalent to −15 atmospheres of suction, water is attracted equally to soil and roots, so no more water movement occurs. At this point, we say the water content of the soil has dropped to a **wilting point**, because plants start to droop when they can no longer extract soil water. Sand and silt contain almost no water at the wilting point, but tiny clay particles hold much of their water too tightly to be taken up by plant roots.

Figure 15.7   Small soil particles (top) have higher field capacity than large soil particles (bottom) because the smaller particles have more surface area per unit volume of soil. Water clinging to particles is illustrated in white.

The amount of water actually available to plants is the difference between the field capacity and the wilting point. This difference is not great in sandy soils because they have a low field capacity. It is also not ideal in clay soils, because they have a high wilting point. Silty soils make the most water available to plants because they have both a relatively high field capacity and a relatively low wilting point (Figure 15.8).

Check your progress:

Why do silty soils make more water available to plants than either sandy or clay soils?

Answer: Sand does not hold as much water against the pull of gravity, and clay holds water so tightly that roots cannot extract all the water held among its particles.

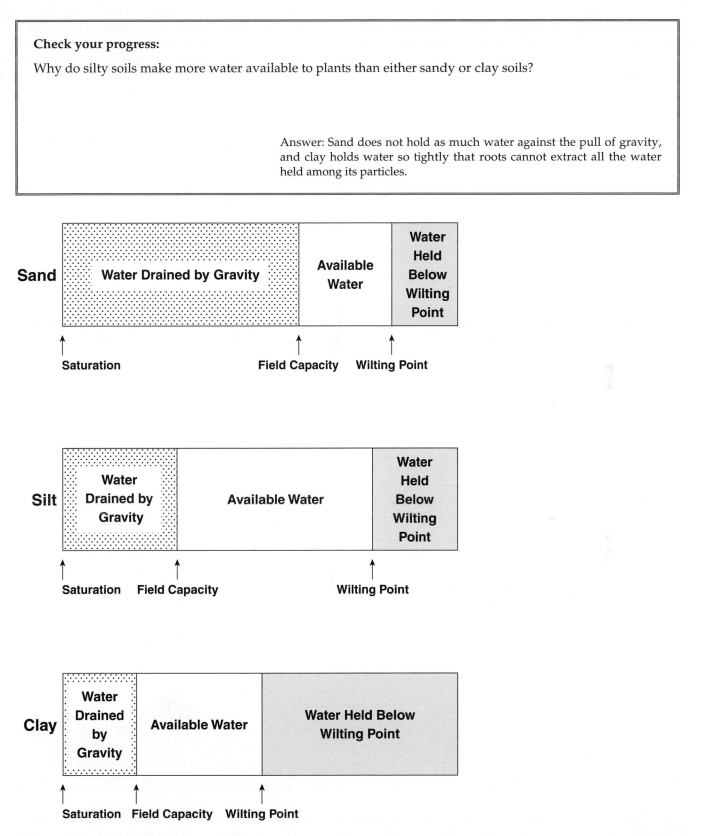

Figure 15.8 Comparison of available water among three soil types. Silty soils make more water available to plants because of their high field capacity and low wilting point.

Nutrient cycling, a property of all sustainable ecosystems, occurs primarily in the soil. Nitrogen serves as a good case study. It is an especially important nutrient because of its presence in the amino acid residues of proteins, and often is the limiting factor for plant growth. Since the atmosphere is 78% nitrogen gas, this element is available everywhere on the planet. However, the gaseous form of nitrogen (N_2) is very stable. Its two nitrogen atoms are not easily split apart, which must be done as a first step in synthesizing nitrogenous organic compounds. This critical step in the nitrogen cycle, called **nitrogen fixation**, occurs occasionally when nitrogen is oxidized by lightning strikes, but more often through metabolic activity of a few specially adapted microorganisms. Blue-green bacteria like *Anabaena* can capture nitrogen in a wet environment, and *Rhizobium* bacteria do so in root nodules of their host legumes (see Chapter 12 for details). Since fixation uses a low-energy substrate (N_2) to make a high-energy product (NH_4^+), this chemical transformation uses a lot of energy. According to V. P. Gutschick (1981), it takes approximately 10 grams of sugar to fuel fixation of one gram of nitrogen. To put this in perspective in English units of measure, bacteria would need to consume more than two tons of sugar to convert all 420 pounds of nitrogen gas contained in the air within a 30×30 foot classroom into organic nitrogen. It is not surprising, therefore, that nitrogen fixation is frequently dependent on the chemical energy captured by photosynthesis, either inside the cell as we see in *Anabaena*, or through the cooperation of a host plant, as we see in *Rhizobium*.

Once nitrogen is fixed as ammonium, the soil bacteria *Nitrosomonas* and *Nitrobacter* can extract energy for their metabolic needs by oxidizing ammonium ions. Ions of NH_4^+ are converted first to nitrite (NO_2^-), and then to nitrate (NO_3^-). Oxidized nitrogen can be taken up by plants and converted to amino acids with some photosynthetic energy expended for the additional biomolecular construction costs. Amino acids are used by plants and by animals farther up the food chain to build proteins. Organic nitrogen excreted by animals and released by decay of organic matter is converted back to ammonium in a series of energy-releasing reactions called **ammonification**. Enzymes within most cells are capable of this step, so it tends to happen as a consequence of protein digestion. Cycling of nitrogen from ammonium through oxidized nitrogen to organic nitrogen and back to ammonium occurs readily within organisms and in the soil because the energy involved in these steps is comparatively modest. However, there are bacteria that disrupt this cycle by converting nitrates in soil all the way back to nitrogen gas. These organisms, called denitrifying bacteria, are **facultative anaerobes**. In the absence of oxygen, they capture chemical energy by reducing oxidized forms of nitrogen, which escapes from the soil as N_2 gas. Denitrifying bacteria gain back a great deal of the energy that was expended to fix nitrogen in the first place, but the community loses soil nitrogen as a result. Since denitrification occurs in poorly aerated soils, such as marshlands and flooded plains, waterlogged soils tend to be low in available nitrogen. It is no wonder that carnivorous plants such as pitcher plant and Venus' fly trap, which catch insects as a supplementary source of nitrogen, find their niche in boggy habitats.

Check your progress:

Diagram steps in the nitrogen cycle, naming organisms facilitating major steps.

Hint: You should show nitrogen gas converted first to ammonium, then to nitrite, then to nitrate and finally to organic nitrogen. Ammonification and denitrification should be included to complete the nitrogen cycle.

Humans alter the nitrogen cycle by artificially fixing nitrogen and applying tons of soluble nitrogen in the form of ammonium, nitrites, or nitrates to agricultural fields. Because these nitrogen fertilizers are highly soluble, they tend to flow with rainwater down to the water table and across the ground to nearby streams. Since nitrates and nitrites are toxic to humans, and since excess nitrogen overstimulates algae, leading to oxygen depletion as discussed above, managing nitrogen in agro-ecosystems is of paramount importance. It is significant that when the Ecological Society of America planned a series of reports designed to educate the public about important ecological issues, they chose human intervention in the nitrogen cycle as their very first topic (Vitousek *et al.*, 1997).

Ecosystems less impacted by human intervention tend not to overnourish stream water, because decomposition of leaf litter and other organic debris releases nitrogen and other essential nutrients gradually, to be taken up by plants and used over and over again. Fungi and bacteria are the primary decomposers in soils, but their relative importance depends a great deal on the temperature. Which do you think thrives best in cold soils? Think about organic decay inside your refrigerator. A refrigerator set at 50° F will stop almost all bacterial growth, but fungi can grow, albeit slowly, on the neglected carton of cottage cheese left too long on the bottom shelf. Similarly, northern or high-altitude soils are generally too cool to support much bacterial growth, so fungi are the primary decomposers in these ecosystems. Fungi release acid to digest their food, which also retards bacterial growth. Because soils are actually frozen much of the year, decomposition may not keep up with the rate of litter accumulation. Humic acids from these slowly decomposing materials tend to add more acidity. Temperate soils therefore tend to be acidic, and relatively rich in organic matter.

Tropical soils, by contrast, support both fungi and bacteria that quickly break down organic material. Faster plant growth in warmer climates takes up nutrients almost as quickly as they become available. Although the lush vegetation of the rainforest would seem to indicate rich supporting soils, most of the nutrients in warm-climate ecosystems are retained in the biotic parts of the system. A rainforest tree crashing down in a storm decomposes with remarkable speed, contributing the nutrients in its massive body to surrounding trees within just a few weeks of its demise. In rainforest regions, heavy rains can quickly flush free nutrients from the soil. As a result, clearing away tropical forest to create new farm land quite often exposes soils holding few remaining nutrients. One solution to this problem is to cultivate perennial plants in the tropics rather than annuals, for better nutrient retention and soil protection. The plowed fields that yield so much food in temperate climates may not represent ecologically sound agriculture in large regions of the tropics.

Check your progress:

Compare the chemical and biological properties of tropical and temperate soils.

Answer: Organic matter is rapidly decomposed in tropical soils. As a result, these soils tend to retain fewer nutrients than northern soils.

Figure 15.9   The nightcrawler *Lumbricus terrestris*, lives in permanent vertical burrows, which significantly enhance air and water movement through the soil.

As important as soils are in determining patterns of life on our planet, the reverse is also true. The biological community living in a place exerts lasting effects on the soil. Soil pH in a maple forest, where fallen leaves decompose very quickly, will be measurably higher than the pH of soil in a pine forest in the same climate and on the same parent material. Animals, such as earthworms and burrowing ants (Figure 15.9), play key roles in aeration, mixing, and pulling organic materials down into the soil. As soil scientists learn more about soil function, we realize that the total soil ecosystem should be considered in management decisions. Pesticides that kill insects above ground also kill soil organisms, with consequences that should not be ignored. Nowhere is it more clear that all the parts of an ecosystem are tied together than in this hidden, but vitally important, ecological realm.

METHOD A: PHYSICAL PROPERTIES OF SOILS

[Laboratory activity]

Research Question
How does particle size affect field capacity of soil?

Preparation
This exercise requires comparison of a soil sample from your campus with samples of clay and of sand (method adapted from Bouyoucos, 1951.) *If using a soil sampler on campus (Figure 15.5) check with your facilities manager to make sure you are clear of buried pipe, drainage lines, electrical conduit, and data cable.* Push a soil sampler into the ground to a depth of several inches, twist a half-turn, and pull out the soil core. Each laboratory group will need at least 100 g of soil for the two procedures, so repeat core sampling as needed. Alternatively, construction sites on campuses often expose soil banks. If this is the case, a hand spade can be used to collect a sample.

As controls, the soil sample is compared against samples of clay and sand. Moist clay can be ordered from ceramic supply sources. *If using powdered clay or dry sand, avoid inhaling dust, as silicates in sand and clay can cause respiratory problems.* Wetting the material and then air-drying prior to student use is good practice. Also, silica-free sand is available through building supply companies, sold for the purpose of filling sandboxes in playgrounds.

A 0.1 normal solution of sodium hydroxide will be needed to prepare soil samples for the separation of particles. Add 4 g NaOH pellets to 1 liter of water for a laboratory class to share. Although this is not a very strong solution, exercise caution and rinse with water if skin contacts the solution.

A Bouyoucos hydrometer that reads *grams of soil per liter* is essential for this experiment. Bouyoucos hydrometers, also called soil hydrometers, are available from *Hogentogler & Co., Canadawide Scientific,* and from *Analytica.* These glass instruments measure the amount of suspended sediment by floating higher in more dense solutions and lower in less dense solutions. Since heavier particles settle out of a soil suspension first, carefully timed hydrometer readings indicate how much sediment remains after sand and then after silt has fallen out of the suspension. Since water changes density as it warms or cools, hydrometers are calibrated to be used at a particular temperature, generally 21° C. Fill a liter flask for each lab group with tap water ahead of time, and allow the water to equilibrate to room temperature.

For the second procedure, students will need access to a laboratory sink to drain flowerpots and a drying oven or warm, well-ventilated location to dry samples. A drying oven set at 105° C will dry samples overnight, but air drying for a week between laboratory sessions yields fairly good results.

Materials (per laboratory team)
For Procedure 1

Large mortar and pestle

0.1 normal NaOH solution

1-liter flask with water at room temperature

Thermometer, to check water temperature

Soil sample (50 g)

Small bucket or plastic bin

Wash bottle, filled with tap water

Spatula or plastic spoon

Long stirring rod

Thermometer, to check water temperature

1-liter graduated cylinder

Magnetic stirring platform

Stirring bar, shorter than diameter of graduated cylinder

Bouyoucos hydrometer

Access to balance, accurate to 0.1 g

Clock or lab timer

For Procedure 2

Three 3" plastic flower pots with drainage holes in the bottom

3 aluminum weighing trays, or 3 squares (6" each) of aluminum foil

Paper toweling

Samples of campus soil, clay, and sand (50 g each)

Access to balance, accurate to 0.1 gram

Access to drying oven

Procedure 1: Estimating sand, silt, and clay components of campus soil

1. Weigh out 50 grams of air-dried soil. Place sample in mortar cup, add enough NaOH solution to cover the soil, and grind with pestle for 5 minutes to separate all soil particles.
2. Your hydrometer is calibrated for use in water of a particular temperature. (Commonly, this is 21° C.) Make sure the water in your liter flask is of the correct temperature. If not, add warm or cool water to adjust it, and stir well before beginning your test.
3. Place a stirring bar in the bottom of the graduated cylinder, add 500 ml water, and turn on the stirrer. The speed should be adequate to create a vortex in the water, but not so high that the stirring bar skips around in the cylinder.
4. With the magnetic stirrer running, use the spatula to scoop soil into the cylinder. Then use the wash bottle to rinse soil off the spatula into the cylinder. Finally, rinse the mortar cup into the cylinder. Make sure the stirring bar keeps spinning as you add soil. If it stops, use a long stirring rod to keep the sediments afloat.
5. Fill the graduated cylinder up to the 1 liter mark with water. Stir thoroughly. When the soil "milkshake" is completely blended, place the hydrometer carefully into the suspension (Figure 15.10). When it stops bobbing up and down, turn off the magnetic stirrer and mark the time. Note how the hydrometer begins to sink as sand settles out and the density of the suspension decreases. After 2 minutes and 40 seconds, read the level of the suspension on the neck of the hydrometer. Mark this on the Data Sheet for Method A-1 as the "silt plus clay" reading.

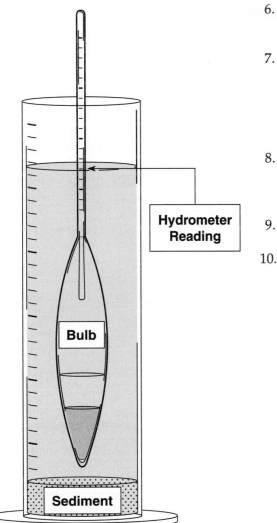

6. You now have two hours to wait for your next reading. This would be an ideal time to begin the tests of field capacity (Procedure 2).

7. After two hours, take another hydrometer reading. At this point, the silt will have settled out, so this reading tells you how much clay remains suspended in the water. (Longer settling time is required in some hydrometer methods, but two hours gives a good estimate.) Record your "Clay" reading in the data table.

8. Observe the pattern of sediments in the bottom of the graduated cylinder. Can you see layers of sediment, with larger particles at the bottom and smaller ones at the top? Record your observations at the bottom of the results page.

9. Calculate portions of sand, silt, and clay as directed on the Calculation Page for Method A-1.

10. Interpret your results by answering Questions for Method A.

Figure 15.10 Soil hydrometer method. As sediments fall out of suspension, the density of the mixture in the graduated cylinder becomes less dense, so the hydrometer floats lower in the water. Read the scale on the hydrometer stem where it breaks the surface of the suspension. Units are in grams of soil per liter.

Procedure 2: Measuring field capacity of soil samples

1. Line three flowerpots with a layer of paper toweling in the bottom. Place about 50 grams of sand or sandy soil in one pot, 50 grams of clay soil in a second, and 50 grams of your campus soil sample in a third.
2. Saturate the three samples completely with water. Set them in a sink or in a bucket to allow the pots to drain for 15 minutes. When the pots stop dripping, the soil is at field capacity.
3. If you do not have aluminum weighing trays, make a tray of aluminum foil by folding a 12 × 12 sheet in quarters, then folding up the edges to hold soil. Label each weighing tray, one for sand, one for clay, and one for your soil sample. Weigh each aluminum tray and record these weights in the Data Table for Method A-2.
4. Using a spoon or spatula, scoop as much of the drained soil as possible into the appropriate weighing tray. Weigh the tray with wet soil, subtract the weight of the tray, and record "wet weight" for this sample in the data table. Repeat for the other samples.
5. Place aluminum weighing trays with soil in a drying oven overnight, set at 105° C and with a ventilation fan running. Alternatively, let your samples air-dry for a week.
6. Weigh the tray with dried soil, subtract the weight of the tray, and record "dry weight" for this sample in the data table. Repeat for the other samples.
7. Complete the Calculations for Method A-2, and compare the field capacities of your three samples.
8. Interpret your results by answering Questions for Method A.

Data Table for Method A-1:
Hydrometer Readings

TIME ELAPSED SINCE SOIL BEGAN SETTLING	BOUYOUCOS HYDROMETER READING (grams/liter)	MATERIAL REMAINING IN SUSPENSION
0 minutes	50	sand + silt + clay
2 minutes, 40 seconds		silt + clay
2 hours		clay

Calculations for Method A-1:
Percent Sand, Silt, and Clay

PARTICLE TYPE	GRAMS OF PARTICLES IN 50-GRAM SAMPLE	PERCENT COMPOSITION = grams/50 g × 100%
SAND	(sand + silt + clay) − (silt + clay) =	%
SILT	(silt + clay) − (clay) =	%
CLAY	(clay) =	%

Description of Sediment Layers in Cylinder

_____

_____

_____

_____

_____

_____

Data Table for Method A-2:
Results for Three Soil Samples

DATA FOR SAND	WT. OF ALUMINUM WEIGHING TRAY	WT. OF TRAY WITH SAND	WT. OF SAND
WET WEIGHT			
DRY WEIGHT	————		

DATA FOR CAMPUS SOIL	WT. OF ALUMINUM WEIGHING TRAY	WT. OF TRAY WITH SOIL	WT. OF SOIL
WET WEIGHT			
DRY WEIGHT	————		

DATA FOR CLAY	WT. OF ALUMINUM WEIGHING TRAY	WT. OF TRAY WITH CLAY	WT. OF CLAY
WET WEIGHT			
DRY WEIGHT	————		

Calculations for Method A-2
Field Capacity

$$\text{Field capacity} = \frac{(\text{wet weight of soil}) - (\text{dry weight of soil})}{(\text{dry weight of soil})} \times 100\%$$

	SAND SAMPLE	CAMPUS SOIL SAMPLE	CLAY SAMPLE
WET SOIL WEIGHT			
DRY SOIL WEIGHT			
FIELD CAPACITY	%	%	%

Questions for Method A

1. Based on your hydrometer readings, how would you describe the soil on your campus? Sandy, silty, mostly clay, or loamy? Given what you know about the geology and topography of your location, is this the soil composition you would expect?

_____

_____

_____

_____

2. From your observations of sediments accumulating in the bottom of the cylinder, what kinds of water movement in a river would keep sand suspended long enough to be transported downstream? Silt? Clay? Can you explain why silt, sand, and clay tend to be deposited in different locations along a river system?

_____

_____

_____

_____

3. Based on the composition of your soil, what hypothesis would you make about its field capacity, in comparison to sand at one extreme and clay at the other? Did your actual measurement of field capacity conform to this expectation? Explain.

_____

_____

_____

_____

4. How might the frequency of irrigation on your lawns in summer be related to the field capacity of the soil?

_____

_____

_____

_____

5. How might you design an experiment to measure the wilting point of your campus soil? Can you guess, from the hydrometer data, whether the wilting point would be comparatively high or comparatively low? Explain.

_____

_____

_____

_____

METHOD B: SOIL pH IN TWO MICROHABITATS
[Outdoor activity]

Research Question
Does the type of plant cover affect soil chemistry?

Preparation
This exercise involves field testing pH, comparing microhabitats on your campus. A portable pH meter with soil probe is the easiest way to take multiple measurements. Soil pH measurement depends on some free water in the soil, so plan this exercise at a time of year when the soil is not completely dry. Depending on the trees and shrubs on your campus, you may have several kinds of microhabitats to choose from. Evergreen shrubs such as juniper or *Taxus* and evergreen trees such as spruce or pine affect pH if their needles are allowed to decompose where they fall. If leaf litter is completely removed by your grounds-keeping staff, effects of plants on the soil will be harder to discern. Be aware also that fertilizers affect pH by adding salts to the soil. It is wise to check with facilities management to find out if any fertilizer applications are planned, and to allow several rains to remove free salts after fertilizer application before doing this exercise.

Materials (per laboratory team)
Portable pH meter with soil probe

Soil thermometer (if meter does not also record temperature)

Procedure
1. In consultation with your instructor, choose a tree or shrub that you suspect may affect soil pH. If this species is represented by several specimens on your campus, decide on a sampling plan that will distribute sample sites around several plants.
2. A lawn area nearby will serve as your control plot. Alternate readings between your shrub/tree sites and your lawn sites.
3. To take a reading, push the soil probe into the ground, taking care to sample at the same depth each time. Give the meter a minute to generate a reliable reading, and write down the pH in the appropriate column of the Data Table for Method B. As you are taking each reading make note of the depth of leaf litter and any other details you see associated with this site. Add these observations to the right-hand column in the data table. Record the temperature, either by reading directly from your meter or with a separate measurement using a soil thermometer, and record that number as well. Repeat your measurements for 10 tree/shrub sites and 10 lawn sites.
4. Because pH is a logarithmic scale, these data are not likely to be normally distributed. To test differences in pH between tree/shrub and lawn sites, it is therefore best to use a **non-parametric** statistical test. In Calculations for Method B, follow instructions for a Mann-Whitney U test.
5. Interpret your data by answering Questions for Method B.

Data Table for Method B:
Temperature and pH Readings

LAWN SITES (CONTROL GROUP)		
SITE #	pH	TEMP.
1-a		
2-a		
3-a		
4-a		
5-a		
6-a		
7-a		
8-a		
9-a		
10-a		

TREE/SHRUB SITES (TREATMENT GROUP)		
SITE #	pH	TEMP.
1-b		
2-b		
3-b		
4-b		
5-b		
6-b		
7-b		
8-b		
9-b		
10-b		

Calculations for Method B:
The Mann-Whitney U-Test

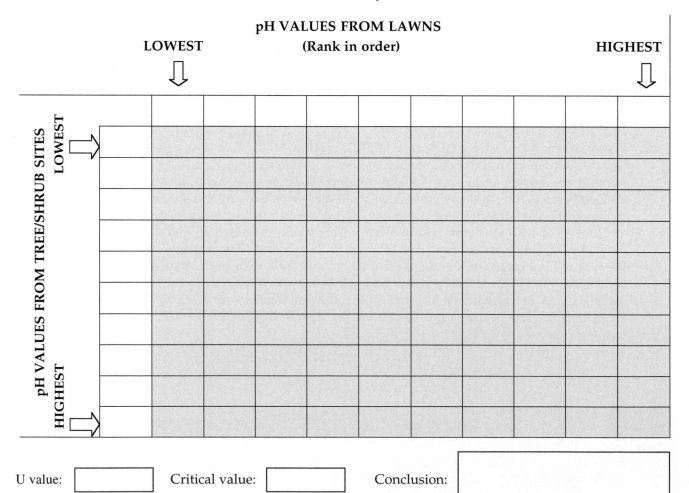

U value: [   ]   Critical value: [   ]   Conclusion: [   ]

1. In the Mann-Whitney U-Test table, first enter the 10 pH values for lawns in the unshaded cells across the top, ranking them in order from lowest on the left to highest on the right.
2. In the unshaded cells down the left-hand side, enter the 10 pH values for tree/shrub sites, ranking them from lowest at the top to highest at the bottom.
3. For each of the shaded cells in the body of the table, compare the lawn pH value at the column heading with the tree/shrub pH value at the row heading for that cell. If the lawn pH above is higher than the tree/shrub pH at the left, write a + sign in that cell. If the lawn pH is lower, write a – sign. If both are equal, write ½.
4. Count plus signs for the entire table, scoring each ½ marked cell as half a plus sign. Now count minus signs for the entire table, scoring each ½ as half a minus sign. Compare the number of + and the number of – signs counted. Whichever number is greater is the Mann-Whitney U value. (See example below.)
5. The null hypothesis for this test is that the two sites do not differ in pH. The critical value for this test, with 10 columns and 10 rows, is 77. (See Appendix 4 – The Mann-Whitney U test.) Higher U values mean greater significance. If the Mann-Whitney U value is 77 or greater, you have demonstrated a difference in pH between the two microhabitats at a significance level of p = 0.05. If the U value is less than 77, the pH values do not differ significantly between the sites.

Sample Table

pH VALUES FROM LAWNS

		6.3	6.4	6.7	6.9	7.2	7.2	7.4	7.5	7.6	7.8
pH VALUES FROM TREE/SHRUB SITES	5.9	+	+	+	+	+	+	+	+	+	+
	6.1	+	+	+	+	+	+	+	+	+	+
	6.1	+	+	+	+	+	+	+	+	+	+
	6.3	½	+	+	+	+	+	+	+	+	+
	6.5	–	–	+	+	+	+	+	+	+	+
	6.6	–	–	+	+	+	+	+	+	+	+
	6.8	–	–	–	+	+	+	+	+	+	+
	6.9	–	–	–	½	+	+	+	+	+	+
	7.0	–	–	–	–	+	+	+	+	+	+
	7.2	–	–	–	–	½	½	+	+	+	+

The number of + signs, and hence the U value, is 80. This exceeds the critical value of 77. In this hypothetical example, tree/shrub sites are significantly more acidic than shrub sites.

Questions for Method B

1. The Mann-Whitney U method tests the null hypothesis that differences in pH among the sites are due to sampling error. Did you reject or accept this null hypothesis? Explain your conclusion.

   _____

   _____

   _____

   _____

2. Did you notice any differences in temperature between the sites under trees or shrubs vs. the lawn sites? Aside from the immediate effects of temperature on pH assessment, how might a cooler temperature affect soil chemistry? (Hint: Think about soil organisms.)

   _____

   _____

   _____

   _____

3. Did you notice variation in leaf litter depth among your tree/shrub sites? Do your pH and temperature readings suggest that litter depth might be an important factor? Explain.

   _____

   _____

   _____

   _____

4. Suppose in this exercise your results showed a mean difference in pH of the two kinds of sites, but the Mann-Whitney test yielded a U value of 70. Can you conclude with certainty that pH is the same in these two habitats? Can you conclude that there is a real difference? How might you resolve any unanswered questions?

   _____

   _____

   _____

   _____

5. Acid rain, caused by air pollutants returning to earth in the form of sulfuric or nitric acid, lowers the pH of the soil. Why does rain of pH 4.5 cause more damage in some ecosystems and less in others?

   _____

   _____

   _____

   _____

METHOD C: NUTRIENT CAPTURE BY SOILS
[Laboratory activity]

Research Question
Does soil texture affect elution and recapture of nitrate ions?

Preparation
This exercise is based on movement of nitrate ions through soil. This is a critical issue for agriculture, since nitrates are commonly applied as fertilizer and loss through the soil column means higher crop production costs. It is also a water quality issue, since nitrates and nitrites that percolate through soil end up in groundwater and surface waters to the detriment of human and ecosystem health. Because nitrates are commonly evaluated in water quality tests, kits for nitrate assay are available at a range of prices and sophistication. Choose a kit that measures nitrates in the 5–50 ppm range. You may want to use the same kit for water quality analysis in an outdoor activity (see Chapter 16).

Clay, sand, and campus soil can be obtained as described in Method 1. *If powdered soils are used, take care to avoid inhaling dust.* Wetting the soils and letting them air dry is a good precaution. An optional, and interesting, addition to the exercise is to compare a piece of intact sod to the loose soil samples. A hand trowel can be used to cut a small circle of sod from intact turf. Trim the sod with a sharp knife to fit the Buchner funnel. Weigh the sod, and remove soil from the bottom until it weighs 50 grams. Place the sod on top of a layer of filter paper in the funnel, and compare elution rates through sod with elution rates through loose soils. You may wish to do one demonstration with sod for the entire laboratory.

Potassium nitrate is a strong oxidizing agent, and should be handled with care in dry form. Make 500 ml of 0.05 Normal solution per laboratory group. Alternatively, potassium nitrate can be purchased as a 0.1 N solution and diluted by half for this exercise.

Materials (per laboratory team)
Kit for testing nitrates in water samples (5–50 ppm sensitivity)

3 Buchner funnels (ceramic or polypropylene), 5–7 cm diameter

3 sheets filter paper to fit Buchner funnels

3 flasks (250 ml each) for collecting eluted solution

Separatory funnel (100 ml or larger), glass or polypropylene

Ring stand

350 ml potassium nitrate solution (0.05 N)

50 g sample of sand

50 g sample of clay soil

50 g sample of campus soil

Marking pen or wax pencil

Procedure

1. Place a Buchner funnel in each of the three flasks, and put a circle of filter paper in the bottom of each funnel.
2. Weigh 50 g of sand and place this sample on top of the filter paper in the first funnel. Similarly, prepare a 50 g sample of clay and a 50 g sample of campus soil, placing them in the other two funnels. Mark the three flasks "sand," "clay," and "campus," indicating the type of soil in the funnel in each flask.

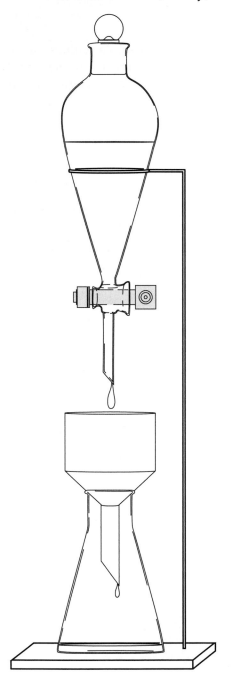

3. Fill the separatory funnel with 100 ml of 0.05 N potassium nitrate (KNO_3). Position it in the ring stand so that the tip of the funnel is just a little higher than the top of your Buchner funnels as they sit in their flasks. (See Figure 15.11.)
4. Adjust the stopcock on the separatory funnel so that it drips steadily onto the soil surface. As it flows through the soil, you should collect eluted material in the flask. After the 100 ml of solution has all run into the first soil sample, set this sample aside to continue eluting through the soil.
5. Position the second Buchner funnel and its flask under the separatory funnel. Refill the separatory funnel with 100 ml more of potassium nitrate solution, and drip this through your second soil sample. Repeat for the third sample.
6. After all the Buchner funnels have stopped dripping, use a graduated cylinder to measure how much of the solution passed through the soil. Enter your results in the Data Table for Method C.
7. Use a water quality test kit to assess how much nitrate nitrogen is present in the water that ran through each soil sample. Perform a control analysis on the unused potassium nitrate solution, so you know what level of nitrate entered the soil. Record your results in the data table.
8. Analyze your results by answering Questions for Method C.

Data Table for Method C:
Effects of Aerobic and Anaerobic Incubation on Soil Nutrients

SOIL TYPE	VOLUME OF SOLUTION ELUTED	NITRATE CONCENTRATION (PPM)
SAND		
CLAY		
CAMPUS SOIL		
CONTROL (No soil exposure)	Not applicable.	

Figure 15.11 Buchner funnel apparatus for testing nitrate elution through soils.

Questions for Method C

1. Which type of soil retained the greatest volume of solution? Explain this result, based on your understanding of particle size in each soil type.

_____

_____

_____

_____

2. Which kind of soil recaptured nitrate as it passed through the soil column? Explain why you think this result was obtained.

_____

_____

_____

_____

3. Why do soil conservation agencies strongly recommend getting a soil test before applying fertilizers on farm fields?

_____

_____

_____

_____

4. Although septic sewer lines go to a wastewater treatment plant in most communities, water entering storm drains is usually diverted into nearby streams or rivers. In areas of moderate to high rainfall, groundwater also feeds surface streams. If too much nitrogen fertilizer were applied to lawns on your campus, what are the consequences for organisms and people living downstream from you?

_____

_____

_____

_____

5. Discuss advantages and disadvantages of "organic" fertilizer, applied in the form of animal manure, as opposed to potassium nitrate as a means to enhance soil nitrogen levels. Consider both the farmer's perspective and the perspective of the greater community.

_____

_____

_____

_____

FOR FURTHER INVESTIGATION

1. Collect soils from a large transect across a river valley near your campus. Analyze soil particle composition using the Bouyoucos hydrometer method, and plot silt content as a function of elevation. You may also wish to compare plant communities on different soils analyzed in your survey.
2. Compare pH under different kinds of trees and shrubs on your campus. Do some species have a greater impact on soil chemistry than others?
3. Try incubating soil in a jar, covered by 10 cm of water. Test soil nitrates and nitrites in the soil before and after incubation. As a control, do the same experiment with soil sterilized by autoclaving or by heating in a 350-degree oven for 2 hours. Do live organisms make a difference in the process of denitrification?

FOR FURTHER READING

Bouyoucos, G. H. 1951. A Recalibration of the Hydrometer for Making Mechanical Analysis of Soils. *Agron. J.* 43: 434–438.

Gutschick, V. P. 1981. Evolved Strategies in Nitrogen Acquisition in Plants. *American Naturalist* 118: 607–637.

Leopold, Aldo. 1949. *A Sand County Almanac.* Oxford University Press, N.Y.

Sugden, Andrew, Richard Stone, and Caroline Ash. 2004. Soils, the Final Frontier. Introduction to special edition on soils. *Science* 304:1613.

Vitousek Peter M., John Aber, Robert W. Howarth, Gene E. Likens, Pamela A. Matson, David W. Schindler, William H. Schlesinger, and G. David Tilman. 1997. Human Alteration of the Global Nitrogen Cycle: Causes and Consequences. *Issues in Ecology I. Ecological Society of America.* http://www.esa.org/science/Issues/FileEnglish/issue1.pdf

Chapter 16

Aquatic Environments

Figure 16.1 Threatened red-legged frogs (*Rana aurora draytonii*) depend on shrinking aquatic habitats, which have been reduced in quality and area in California.

INTRODUCTION

California red-legged frogs (*Rana aurora draytonii*) were once familiar aquatic animals in the Western United States (Figure 16.1). Mark Twain wrote about these frogs, and about the California gold miners who gambled on their leaping ability in "The Celebrated Jumping Frog of Calaveras County." Larger than any other native frog in the region, these amphibians were intensively harvested for food in the 1800s. Overharvesting, along with widespread diversion of surface water for irrigation, pollution of streams, and introduction of aggressive competitors such as the bullfrog, caused a steady decline in red-legged frog populations over the past century, according to the U.S. Fish and Wildlife Service (2001). Now listed as a threatened species, red-legged frogs are emblematic of our general concerns about aquatic ecosystem health.

Amphibians all over the world are declining, not for a single reason, but because their fragile wetland habitats, and the pure water they require, are impacted by so many different human activities. Synthetic hormones released as pollutants in wastewater, nutrients and sediments from agricultural runoff, draining of marshlands, exotic diseases, increased ultraviolet penetration of atmospheric ozone, and gradual changes in climate have all been suggested as possible contributing factors (Davidson *et al.*, 2001). Although rare species like the red-legged frog do eventually merit protection under the Endangered Species Act, recovery efforts are daunting after the species has been reduced to a few fragmented populations. The Endangered Species Act has been a valuable conservation tool, but its application usually involves damage control rather than prevention. Basing conservation policy solely on a species' endangered status is like waiting to see the dentist until you have only one tooth left in your mouth. Ecologists recognize that effective biodiversity protection for wetland and aquatic species depends on ecosystem-level understanding of our environment, and thoughtful management of entire watersheds. Mark Twain's miners may have bet on one frog at a time, but our task is to improve the odds for the aquatic community as a whole.

What are the most important environmental factors affecting life in aquatic systems? Dissolved oxygen is certainly critical, and this variable is always linked with temperature. Cold water is able to carry more oxygen than warm water. At one atmosphere of pressure, a liter of water near the freezing point can dissolve about 14 milligrams of oxygen. The same water heated to 30° C can dissolve only about 7 milligrams of oxygen. In other words, water in a temperate zone creek loses nearly half its ability to carry dissolved oxygen between January and July. Metabolic activity by bacteria or by aquatic organisms reduces oxygen levels below this maximum limit, so accelerated decomposition of organic matter in warm weather can make this difference even more extreme. Add to this the effects of water movement, with stream flow slower and more sluggish in summer, and faster with more aeration in winter, and the difference is greater still.

Figure 16.2 Trees reduce water temperature in shaded upper reaches of a stream.

Within a watershed, a stream tends to be more oxygenated in its upper reaches, where flow rates are greater and overarching trees shade the water (Figure 16.2). As a stream slows down and gets wider in its lower reaches, temperature increases and oxygen levels decline. A slower flow rate downstream provides less mixing, and this reduces oxygen levels as well. These changes in the physical environment affect life forms in the stream. Active fish with high oxygen demand are limited to cool streams with rapidly flowing water (Figure 16.3). Fish inhabiting warm stagnant waters tend to lower their metabolic demand for oxygen by

Figure 16.3   Salmon migrate upstream to spawn in cool, oxygen-rich waters.

moving slowly and spending much of their lives sitting still and waiting for prey to come to them (Figure 16.4). Human activities that increase water temperature, such as cutting shade trees away from stream banks or discharging heated water from power plants into rivers, significantly affect resident aquatic life.

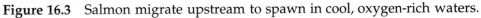

Check your progress:

Aside from the obvious advantage of leaving more time for tadpole development, why might so many species of frogs in temperate zones shed their eggs very early in springtime?

Hint: Frog eggs need high levels of dissolved oxygen for development.

Figure 16.4 Brown bullhead catfish tolerates warm waters with low oxygen.

In a lake, warm water floats on top of cold water, so **stratification** of the water column occurs as the sun warms the surface (Figure 16.5). The warm upper layer, called the **epilimnion**, is the most biologically active. Since light penetrates only a few meters in most freshwater lakes and ponds, photosynthesis is limited to the epilimnion. The cooler, darker zone at the bottom is called the **hypolimnion**. Algae and other plankton "rain" down from the epilimnion above, to be decomposed by bacteria living in the mud at the bottom. There is little mixing of water between layers, so oxygen used up by these decomposers is not immediately replaced. Over the summer, the hypolimnion becomes richer in dissolved nutrients, but poorer in dissolved oxygen. Growth of algae near the surface slows down in midsummer, because their recycled nutrients are trapped in the hypolimnion.

Lake Stratification

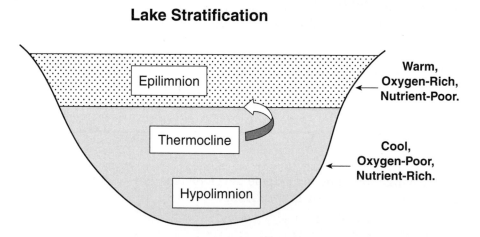

Figure 16.5 Temperature stratification in a freshwater lake.

On the interface between the warm epilimnion and the cold hypolimnion is a dividing line called the **thermocline**. This boundary layer can be found near the surface in early summer. If you have ever been swimming in a northern lake in June, you know the "toe test" may indicate warmth near the surface, but the near-freezing hypolimnion becomes all too apparent when you jump in. Through the summer, the thermocline drops lower and lower as solar heating expands the epilimnion. When fall comes, stratification is reversed.

Water exposed to cold air gets colder and heavier, so surface water sinks to the bottom. This mixes the upper and lower layers of the lake, a phenomenon known as "**fall turnover**." The thermocline disappears as epilimnion and hypolimnion are blended together. Since water resists temperature changes, it can take a long time to cool the entire lake to $4°$ C, which is the temperature at which water is heaviest. Below that temperature, water molecules begin organizing themselves into the lattice of crystalline ice, becoming less dense than liquid water. By the time ice forms on the top of a lake, the entire water column has been cooled to $4°$ C, all the while mixing nutrient-rich waters from below with oxygen-rich waters from above. By the time of the spring thaw, surface waters can again support algae growth as the epilimnion begins to re-form in a new season.

Check your progress:

If a shallow lake and a deep lake both have the same surface area and the same average water temperature of $20°$ C, which will freeze over first when winter comes? Why?

Answer: Deep lakes take longer to freeze over because all the water must be cooled to $4°$ C before any of it turns to ice.

In addition to temperature and dissolved oxygen, light is a critical variable in aquatic systems. Aquatic food chains are built on productivity of microscopic algae, called **phytoplankton**, floating in the water. These algae tend to float near the surface because light intensity declines exponentially with depth. For example, if we measure red light of wavelength 620 nm at 100% intensity at the surface of a perfectly clear lake, its intensity declines to 10% at a depth of 9 m, 1% at 18 m, and 0.1% at 27 m. Not all wavelengths of light penetrate water equally well. Blue light, with wavelengths in the 400–500 nm range, is transmitted much more effectively than red light (600–700 nm). When sunlight composed of all colors enters water, the red wavelengths are preferentially absorbed, so light that has passed through water looks blue. Chlorophyll, the photosynthetic pigment for land plants and many algae, absorbs blue and red portions of the spectrum, so the rapid attenuation of red light near the surface significantly reduces energy available to phytoplankton. In deepwater marine habitats, some algae actually use a different photosynthetic pigment, which absorbs only the blue light available at that depth. Since their cells reflect red light rather than absorb it, they have a red appearance when brought up to the surface. They are appropriately called red algae.

Suspended particles and pigments interfere with light transmission in direct proportion to their concentration. This is the principle behind the spectrophotometer, which you may have used in chemistry laboratory to measure the concentration of a colored solution. Silt and clay soil washing into streams reduces photosynthesis by interfering with light transmission to algae and submerged plants. This reduction in clarity is called **turbidity.** It can be measured in an instrument similar to a spectrophotometer, called a turbidimeter. Standard units of turbidity, called nephalometric turbidity units, or N.T.U., are often included in water quality assessments of streams. A more traditional way to measure water transparency is with a Secchi disc (Figure 16.6). The Secchi disc is a round flat piece of metal or plastic, 20 cm in diameter, painted white and black in alternating quarters and weighted so that it will sink. A ring in the center is tied to a chain or rope which is marked off in meters, and the Secchi disc is lowered over the side of a boat. The observer lets out the line until the Secchi disc disappears from view. Then the observer pulls up the line until the Secchi disc is just visible, and a depth measurement is recorded. Ideally, Secchi disc readings are taken between 10:00 am and 2:00 pm, and observation is off the shady side of the boat for better visibility. The greater the depth a Secchi disc can be seen, the clearer the water.

Check your progress:

What factors determine the amount of light energy available to an algal cell suspended in a lake?

Answer: Colors of light needed for photosynthesis, distance from the surface, and turbidity of the water

---

Secchi Disk

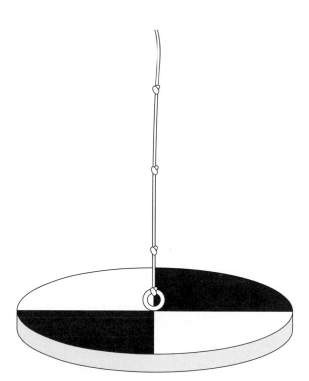

Figure 16.6 A Secchi disc is lowered into a body of water from a boat to measure turbidity of the water.

Finally, dissolved chemicals determine viability of aquatic life. Nutrients such as phosphates, nitrates, and potassium are needed in small amounts for biosynthesis, but can wreak havoc with aquatic communities if added in large amounts (see Chapter 15). A body of water receiving too much fertilizer is said to be **eutrophic**, which literally means "overfed." Algae grow exponentially in "blooms" or floating mats that block out all the light below the surface. Shaded algae die and decompose. Bacterial decomposition uses up dissolved oxygen, with predictable results for other aquatic organisms.

Some sources of pollution are easier to find than others. **Point sources**, such as industrial spills, pulp mill effluent, and insufficiently treated municipal wastewater, enter a body of water at an identifiable location. **Non-point sources**, which include runoff from lawns or farm fields, acid rain, manure from feed lots, silt from road construction, or leachates from mining operations, are more diffuse in their origin and thus harder to identify and control. Toxicity may be acute, resulting in dramatic fish kills, or chronic, causing gradually declining biodiversity in an affected waterway. The growing number of biologically active compounds entering streams from antibiotics in animal feed and from medicines incompletely metabolized by humans has more recently raised concerns about effects on animal fertility, reproduction, and development.

Chemical tests have been designed for many kinds of pollutants in water, but it is difficult to assess how much damage is being inflicted on stream biota in this way. A more direct approach is to test the water directly on organisms, monitoring their viability over time. So called **bioindicator** organisms are chosen for their short life spans and high sensitivity to pollutants. Two organisms used routinely for biomonitoring work are *Daphnia* (also called water fleas—Figure 16.7) and the small fish *Pimephales promelas* (also called fathead minnows—Figure 16.8). The premise of this biological approach is that water clean enough to support bioindicator organisms is probably safe for the aquatic ecosystem as a whole. Bioindicators are studied in two settings. Laboratory experiments expose a subject population to suspected toxins or suspect water sources in carefully controlled experiments. For example, a pesticide might be screened by making up a series of dilutions and exposing fathead minnows in aquaria to these dilutions over a period of time. A concentration of the pesticide just strong enough to kill half of the fish is called the lethal dose for 50%, or **LD_{50} value.** This critical concentration can then be compared with LD_{50} values for other pesticides to determine which has the lowest effect on fish if it runs off of fields into streams.

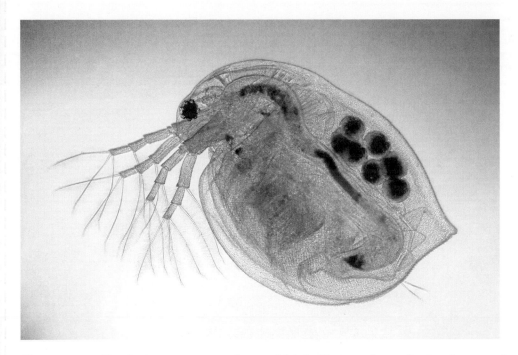

Figure 16.7   Daphnia are commonly used bioindicators for pollution in aquatic environments.

A second biomonitoring approach looks at organisms already living in a stream. A convenient group of bioindicators for on-site assessment are **benthic macroinvertebrates**. Benthic means living on the bottom, and macroinvertebrates are non-vertebrate animals large enough to be seen with the naked eye. This group includes strictly aquatic species, like amphipods and snails, that live all their lives in the water. It also includes the aquatic larvae of many insects, such as dragonflies, that live near aquatic habitats as adults and lay eggs in the water (Figure 16.9). Some of these macroinvertebrates, including the juvenile forms of mayflies, caddis flies, and stoneflies, are quite sensitive to pollution. If any toxins have entered the stream in the past few months, these organisms will be absent from the benthic community. Other macroinvertebrates, including aquatic annelids and the larvae of midges, can tolerate a high pollution load. By comparing numbers of pollution-sensitive vs. pollution-tolerant species, ecologists can develop a water quality index that is more inclusive than a battery of chemical tests. In a sense, the bioindicator species have been monitoring pollution in the stream 24 hours a day, 365 days a year, for every pollutant that can harm organisms, right up to the time of your arrival.

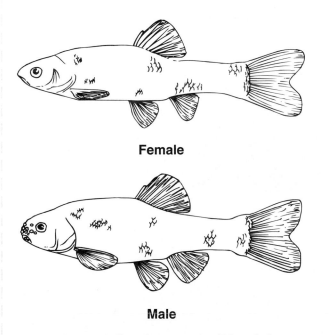

Female

Male

Figure 16.8   Fathead minnows (*Pimephales, promelas*) are commonly used as bioindicators for toxins affecting survival and development of vertebrate animals.

Check your progress:

To monitor non-point sources of pollution, what advantages would you see in a series of chemical tests, such as pH, ammonia nitrogen, and dissolved oxygen assays? What advantages would you see in using bioindicators instead?

Hint: One approach is more comprehensive, the other potentially yields a more specific diagnosis of the problem.

Figure 16.9   A dragonfly deposits eggs on aquatic vegetation.

METHOD A: DISSOLVED OXYGEN AND TEMPERATURE
[Laboratory/outdoor activity]

Research Question
How does temperature affect dissolved oxygen concentration?

Preparation
A series of five water baths at different temperatures between $0°$ C and $40°$ C will be needed for the laboratory portion of this experiment. If you do not have water baths that can be set at a range of temperatures, a fair substitution can be made with inexpensive fish tank heaters placed in 1000-ml beakers. Beakers can also be left overnight in an incubator or refrigerator. Make sure the water stays uncovered at temperature for several hours so that oxygen has a chance to equilibrate before testing.

Kits that include a chemical test for dissolved oxygen are inexpensive and widely available. These are adequate for the experiment if groups cooperate on testing and share data to save time. A portable meter with a dissolved oxygen probe is a better way for an individual to collect all needed data in a short time. For greater accuracy, groups can be instructed to pool their data, and to plot a standard curve through mean values for the class.

Outside water sources depend on your campus environment. Fish ponds, fountains, puddles, ditches, and adjacent streams are all possible sample sites. If outdoor sources do not exist, try measuring dissolved oxygen in aquaria as a model system.

Materials (per laboratory team)
Access to five water baths, of varying temperature

Access to at least one outdoor water source

Dissolved oxygen test kit or portable dissolved oxygen meter

Thermometer for assessing water temperature

Procedure
1. In each of the water baths, measure temperature and dissolved oxygen. Record your results in the Data Table for Method A.
2. Generate a graph on the Results for Method A page, showing the relationship between temperature (x-axis) and dissolved oxygen (y-axis). Draw a smooth curve through your points. This will be your standard curve.
3. Find a water source outdoors (or use an aquarium if necessary). Measure temperature and dissolved oxygen. Note water clarity, organic debris, flow rate, any organisms present, and other factors that may affect dissolved oxygen concentration. Repeat for more than one site if you can.
4. Draw circled points representing your field or aquarium data on the graph with your standard curve. Use your observations and measurements to answer Questions for Method A.

Data Table for Method A

PURE WATER SAMPLE NUMBER	TEMPERATURE °C	DISSOLVED OXYGEN (mg/l)
1		
2		
3		
4		
5		
FIELD SAMPLE(S) (Describe sites below.)	**TEMPERATURE** °C	**DISSOLVED OXYGEN** (mg/l)

Results for Method A:
Dissolved Oxygen as a Function of Water Temperature

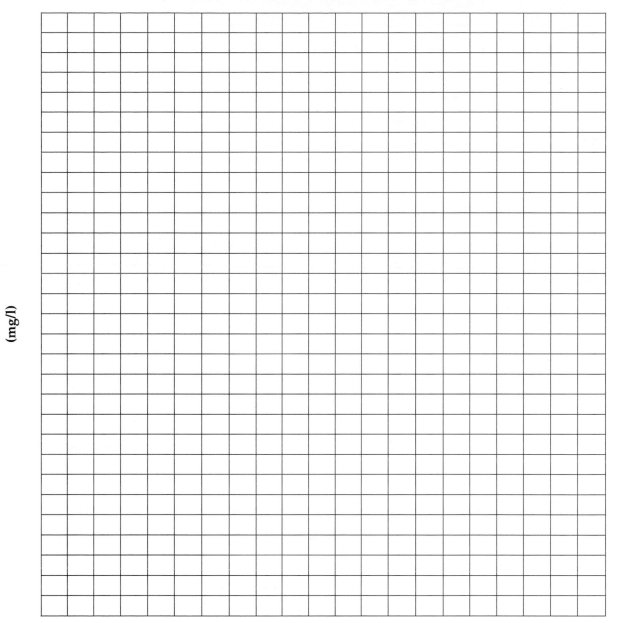

WATER TEMPERATURE (°C)

DISSOLVED OXYGEN (mg/l)

Field Observations:

_____

_____

_____

_____

_____

_____

_____

Questions for Method A

1. Describe the shape of your standard curve. Is this relationship linear or curvilinear? Explain.

   _____

   _____

   _____

   _____

   _____

2. If oxygen is dissolved in cold water, what happens to the oxygen when the water warms up? Have you ever observed this phenomenon?

   _____

   _____

   _____

   _____

   _____

3. In your field (or aquarium) site, how did the data compare with your standard curve? Was the water at this site saturated with oxygen, based on its temperature, or not? Explain other factors that may have affected oxygen levels, based on your observations.

   _____

   _____

   _____

   _____

   _____

4. How might the relationship between temperature and dissolved oxygen explain the adaptive significance of symbiotic algae within the bodies of coral polyps on tropical reefs?

   _____

   _____

   _____

   _____

5. Trapping of infrared radiation by carbon dioxide in the atmosphere has the potential to warm the atmosphere a few degrees over the next century. This does not seem very significant to a terrestrial mammal like yourself, but how might this change affect tropical aquatic habitats?

   _____

   _____

   _____

   _____

METHOD B: SEDIMENT LOAD AND WATER CLARITY

[Laboratory/outdoor activity]

Research Question

How do suspended sediments affect water clarity?

Preparation

A simple turbidity meter can be made from a clear plastic tube, with inside diameter about 5 cm, and about a meter long (Figure 16.10). A 2-inch diameter plastic mailing tube 3 feet long will work, but extruded acrylic is sturdier. Next, you will need a black, one-hole rubber stopper to fit the bottom of your plastic tube. Use white waterproof paint to make a Secchi disc design on the inside surface (small end) of the stopper. (See Figure 16.6 for a standard Secchi pattern.) Insert a short piece of glass tubing into the hole in the outside surface (larger end) of the stopper. Slip a 2–3 foot length of surgical tubing over the glass tubing, making sure you have a tight fit. Place a pinchcock on the surgical tubing to control water flow. A meter stick can be used to measure depth of water in the tube. A funnel for adding water to the top of the tube and a bucket for catching water let out completes the apparatus.

To create a standard, soil is suspended in water at a known g/l ratio. Clay or fine silt stays suspended much better than sandy soil. Silty soils may be obtained near rivers or stream banks. If stream water can be collected from your local watershed, you may wish to use soil from the campus itself. Since each soil is unique, try mixing 1 g soil per liter as a starting point. Put the soil into a 4-liter reagent bottle or plastic milk jug, add a liter of water, and shake well. Then add the rest of the water and shake again to suspend all the particles. Fill the tube to a depth of 20 cm, and take a turbidity reading as described below. If the disc is not visible, add more water. If it is very easily visible, add more soil. When you have a suspension that yields a Secchi disc reading of approximately 20 cm, make a 3-liter stock suspension at this soil/water ratio for each laboratory group.

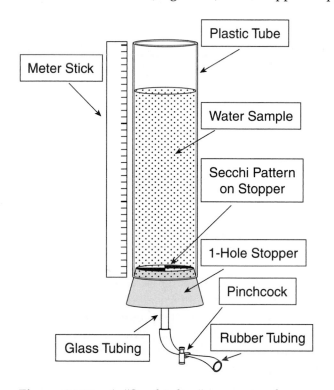

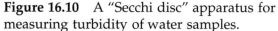

Figure 16.10 A "Secchi disc" apparatus for measuring turbidity of water samples.

For the environmental water sample, collecting water from a campus lake, stream, or drainage ditch would be ideal. If not, you can collect samples or make up an unknown sample to simulate collected stream water. *Use caution when collecting water samples, especially from steep-sided channels and near fast moving water.* When filling a 4-liter bottle for a sample, be sure to avoid stirring up sediments from the bottom. If wading into the stream, stand downstream of the collecting point.

Materials (per laboratory team)

Local topographic map (or access to Web-based maps)

Turbidity apparatus (described above)

Meter stick

5-gallon plastic bucket for waste water

Large plastic funnel

1-liter graduated cylinder

4-liter reagent bottle or gallon plastic milk jug, for mixing

3-liter sample of stock soil suspension

1-liter sample of water from a local stream or lake

Access to tap water

Procedure

1. Obtain from your instructor the concentration of sediments in the stock suspension, expressed in grams of sediment per liter of suspension. Record this figure beside the label "Stock suspension concentration" in the Data Table for Method B.
2. Assemble the turbidity tube as shown in Figure 16.10, with the stopper securely in place and the pinchcock closed. Put the surgical tubing in the waste bucket to catch water drained from the tube, or conduct your trials outside.
3. Pick up the jug containing the stock suspension and shake it up well to suspend the sediments. Place a funnel in the top of the tube and fill it half full of the standard suspension.
4. Remove the funnel. Have one of the laboratory team hold the tube upright and look down through the water column while another operates the pinchcock. Open the pinchcock to let the water sample run slowly out. When the Secchi pattern on the stopper at the bottom of the tube is just visible, stop the water flow and measure the height of the water column from the Secchi disc pattern to the surface. If you overshoot the mark, pour in stock suspension until you cannot see the pattern, and try again. Record your turbidity measurement, in cm, in the data table beside the concentration figure you entered before.
5. In the empty jug, mix 500 ml of the stock solution with 500 ml of tap water. Call this "stock × 1/2." Divide the concentration of the stock solution by 2 and place this number (expressed in mg/l) in the second row of the data table. Shake the new suspension well, and rinse out your turbidity tube. Fill the turbidity tube about 2/3 full with the new mixture. Measure turbidity as described above. Record your turbidity measurement in the data table.
6. Empty your mixing jug into your waste bucket, and rinse it out with tap water. Then add 500 ml of stock solution to 1000 ml of tap water. Shake well. Call this suspension "stock × 1/3." Divide the concentration of the original stock solution by 3 and place this number in the third row of the data table. This time, fill your turbidity tube all the way to the top to begin your measurement. Measure and record turbidity as before.
7. Repeat step 5, but fill the rinsed mixing jug with 500 ml stock solution and 1500 ml tap water. Call this suspension "stock × 1/4." Record concentration and turbidity readings in the fourth row of the data table.
8. Repeat step 5, this time filling the mixing jug with 500 ml stock solution and 2000 ml tap water. Call this suspension "stock × 1/5." Record concentration and turbidity readings in the fifth row of the data table.
9. Complete the graph on the Results for Method B page by plotting sediment concentration (in milligrams of suspended material per liter of water) on the x- axis and turbidity tube reading (in cm) on the y-axis. Draw a smooth line through these points to produce a standard curve.
10. Collect water from at least one site on your campus, or use a water sample supplied by your instructor. Shake up the sample to ensure sediments are suspended, and then measure its turbidity.

11. Plot the turbidity readings as circled points on your graph. Based on the standard curve you have prepared, estimate sediment concentrations in your field sample or samples.
12. Consult a local topographic map, or on-line map of your region to find the streams that drain the sites you sampled. Interpret your results by answering Questions for Method B.

Data Table for Method B

SOIL SUSPENSION SERIES	SEDIMENT CONCENTRATION (mg / liter)	TURBIDITY TUBE READING (cm)
STOCK SUSPENSION		
STOCK × 1/2		
STOCK × 1/3		
STOCK × 1/4		
STOCK × 1/5		
FIELD SAMPLE(S) (Describe sites below.)	**SEDIMENT CONCENTRATION** (est. from standard curve)	**TURBIDITY TUBE READING** (cm)

Field Observations:

_____

_____

_____

_____

_____

_____

_____

Results for Method B:
Turbidity Tube Reading vs. Sediment Concentration

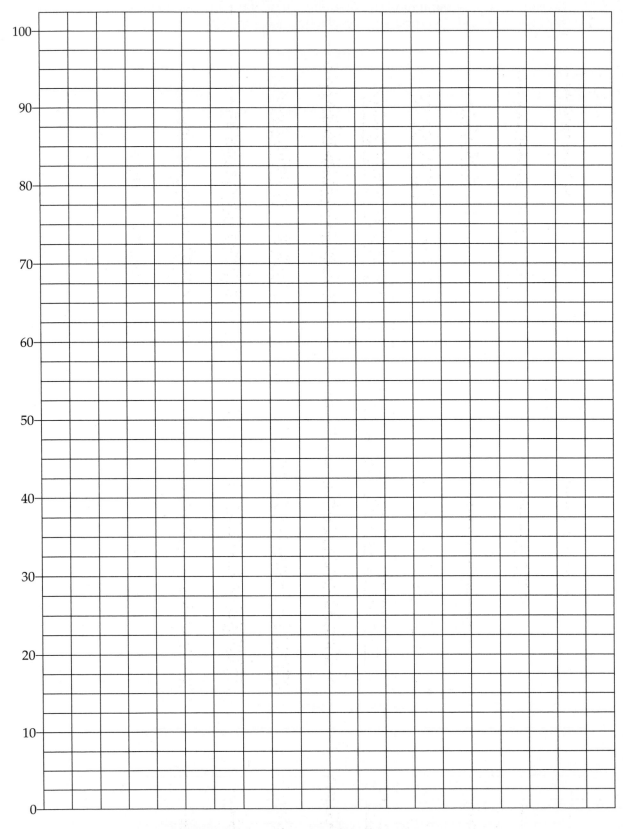

SEDIMENT CONCENTRATION (mg/l)

Questions for Method B

1. Describe the shape of your standard curve. Is this relationship linear or curvilinear? How did your data compare with solubility values mentioned in the Introduction?

_____

_____

_____

_____

_____

2. How did your field sample(s) compare with your standard samples? Do you think turbidity at this site might be high enough to restrict photosynthesis?

_____

_____

_____

_____

_____

3. Consult a topographic map to find the watershed in which your campus is situated. What streams or rivers drain the land you are on right now? What sources of sediment might be entering streams in this watershed?

_____

_____

_____

_____

_____

4. Why is water clarity so important in aquatic ecosystems? What are the consequences of increased turbidity due to sediments washed into streams?

_____

_____

_____

_____

5. Suppose a lake received both suspended particles, such as silt, and dissolved particles, such as humic acids from leaf litter, after a heavy rain. Over time, will these two components of turbidity have similar or different impacts on the lake system? Explain.

_____

_____

_____

_____

METHOD C: LD$_{50}$ DETERMINATION FOR PESTICIDE

[Laboratory/outdoor activity]

Research Question

What concentrations of common pesticides are toxic to *Daphnia*?

Preparation

Daphnia cultures should be ordered close to the time they are to be used, since their culture in aquaria is sometimes difficult. *Daphnia pulex* is a good bioindicator, and somewhat easier to maintain than the larger *D. magma*. Chlorine in tap water is toxic to *Daphnia*, so water for dilutions and controls should be either spring water purchased for this purpose or aged tap water which has been left open to the air to allow chlorine to evaporate before use.

A 1% copper sulfate solution can be used as the pesticide. Under its common name of "bluestone," copper sulfate is commonly added to lakes and is even incorporated in swimming pool chemical systems to control algae. If desired, other home and garden pesticides such as carbamate insecticides or glyphosphate herbicides can be used for comparison by different laboratory groups. For the initial full-strength concentration, mix soluble pesticide according to package directions. *Make sure to follow warning labels on any pesticides used, and instruct the class to handle these chemicals carefully.*

Materials (per laboratory team)

Test tube rack

Nine 15-ml test tubes

100 ml spring water or chlorine-free water

Daphnia culture (at least 60 individuals)

Large-bore dropper for capturing *Daphnia*

1% solution of copper sulfate

10-ml pipette

8 disposable 1-ml pipettes

Pipette bulbs or pipette pumps

Vortex test tube mixer (optional)

Marking pen, for labeling test tubes

Dissecting microscope

Petri dish or depression slide

Procedure

1. Label your test tubes as shown in the top row of the following table. Take a minute to look at the pesticide concentrations that will be added to each tube. Note that each tube in the **dilution series** will hold a pesticide concentration 1/10 as concentrated as the one before. Also, note that we can express dilute concentrations as fractions of a gram per liter, or as parts of pesticide per parts of solution. One gram per liter is one part pesticide per thousand parts of the solution (or 1 **ppt**). Similarly, 0.001 gram per liter is equivalent to one part pesticide per million parts solution (1 **ppm**), and 0.000001 gram per liter is equivalent to one part pesticide per billion parts solution (1 **ppb**).

Tube Label	#1	#2	#3	#4	#5	#6	#7	#8	Control
Concentration (grams per liter)	10	1	0.1	0.01	0.001	0.0001	0.00001	0.000001	0
Concentration parts per thousand, parts per million, parts per billion	10 ppt	1 ppt	100 ppm	10 ppm	1 ppm	100 ppb	10 ppb	1 ppb	0

2. Line up your test tubes in the rack as shown in Figure 16.11. Using a 10-ml pipette, fill Tubes #2 through #8 with 9 ml clean water each. Add 9 ml to the control tube as well.

Serial Dilution of a Pesticide Solution for Bioassy

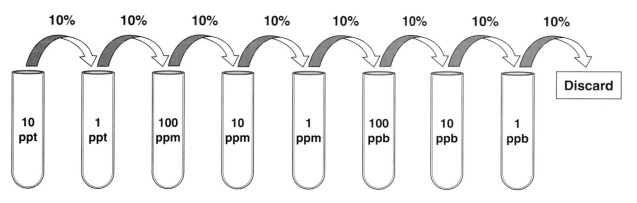

Figure 16.11 Serial dilution of a pesticide (copper sulfate) is accomplished by transferring 10% of each tube in the series into the next. A 1:10 dilution in each tube results in a pesticide concentration 1/10 as strong with each step.

3. Use the large pipette to measure 10 ml of 1% copper sulfate into Tube #1.
4. Using a 1-ml pipette with bulb or pipette pump, transfer 1 ml of the copper sulfate solution from Tube #1 into Tube #2. *Do not mouth-pipette chemical agents.* Mix tube #2 with a vortex or by placing the pipette into the tube and carefully stirring the solution. You have now completed a 1:10 dilution, so the concentration in Tube #2 is 1 ppt.
5. Using a new pipette, transfer 1 ml of solution from Tube #2 into Tube #3. Mix.
6. Repeat Step 5 for each tube in the dilution series, transferring 10% of each solution into the next tube to dilute by factors of 10, as shown in Figure 16.11. After mixing tube 8, take out and discard 1 ml so that all tubes will contain the same volume of 9 ml.
7. Using a large-bore dropper, transfer *Daphnia* into each of the eight solutions. Try to add six individuals to each tube. Note the time that *Daphnia* were exposed to the pesticide.
8. Save at least one individual *Daphnia* for observation under the microscope. Put several drops of water on a Petri dish, and place the *Daphnia* into the water. Observe under a dissecting scope. Make drawings and notes about *Daphnia* morphology and normal behavior.
9. After *Daphnia* have been exposed to copper sulfate for one hour, observe the tubes carefully. Any ill effects you notice after this brief time would be attributed to **acute toxicity**, meaning an immediate effect. Note changes in viability and behavior among the treatments.
10. If possible observe your *Daphnia* over 2–3 days to determine longer term effects of the toxin.
11. Record your results in the Data Table for Method C. Answer Questions for Method C.

Data Table for Method C

Tube Label	#1	#2	#3	#4	#5	#6	#7	#8	Control
Concentration	10 ppt	1 ppt	100 ppm	10 ppm	1 ppm	100 ppb	10 ppb	1 ppb	0
Number of viable *Daphnia* after 1–2 hours									
Number of viable *Daphnia* after 1–2 days									

Drawing of *Daphnia*

Notes on Behavior

_____

_____

_____

_____

_____

_____

_____

_____

_____

_____

_____

_____

_____

Notes on *Daphnia* Response to Copper Sulfate:

_____

_____

_____

_____

_____

_____

_____

Questions for Method C

1. What was the LD_{50} concentration for this pesticide? Does your answer depend on the time that has elapsed since exposure?

_____

_____

_____

_____

_____

2. For routine testing of pesticides, what arguments would favor using *Daphnia* over fathead minnows? What arguments would favor fathead minnows over *Daphnia*?

_____

_____

_____

_____

_____

3. This method determines the toxic threshold only to the nearest order of magnitude. How would you adapt this experiment to find an LD_{50} with greater accuracy?

_____

_____

_____

_____

_____

4. *Daphnia* are representative of small aquatic invertebrates that float in freshwater lakes, eat phytoplankton, and in turn are prey for larger species. Based on your experiment, what would you say to fishing enthusiasts who propose to add copper sulfate to a lake to control aquatic "weeds"?

_____

_____

_____

_____

5. Suppose copper sulfate from an abandoned industrial site were leaking into a stream, and your tests of the effluent on live *Daphnia* show it to be in the toxic range. How might you test the hypothesis that dilution in the stream is reducing concentrations of the pollutant below threshold levels that would harm life in the stream?

_____

_____

_____

_____

FOR FURTHER INVESTIGATION

1. It does not take long to learn to identify families of macroinvertebrates and calculate your own biological index of water quality. State natural resource agencies may be able to help get you started. A few sample Web sites of water monitoring agencies who work with volunteers are included in the suggested readings.

2. Sample stream water for dissolved oxygen, temperature, and turbidity just after a rainstorm, and then repeat your sampling several times a day for two or three days afterward. How does the surge of stormwater through a stream affect water quality?

3. Use topographic maps or on-line geographic data to find the watershed you live in. How many hectares of land are drained by the river or stream nearest you? What land use is prevalent in your watershed? What might be done to improve water quality by improved watershed management?

FOR FURTHER READING

Davidson, Carlos, H. Bradley Shaffer, and Mark R. Jennings. 2001. Declines of the California Red-legged Frog: Climate, uv-B, Habitat and Pesticides Hypotheses. *Ecological Applications* 11(2): 464–479.

Ferrando, M. D., E. Andreu-Moliner, and A. Fernandez-Casalderrey. 1992. Relative Sensitivity of *Daphnia magna* and *Brachionus calyciflorus* to Five Pesticides. *Environ Science and Health Bulletin* 27(5):511–22.

National Science Foundation. 2004. "Water on the Web" http://www.waterontheweb.org/aboutus/index.html

University of Wisconsin Extension, Citizen Stream Monitoring Program. 2005. http://clean-water.uwex.edu/wav/monitoring/biotic/index.htm

U.S. Environmental Protection Agency. 2005. Biological Indicators of Ecosystem Health. http://www.epa.gov/bioindicators/html/benthosclean.html

U.S. Fish and Wildlife Service. 2001. http://www.fws.gov/endangered/features/rl_frog/rlfrog.html

Chapter 17

Energy Flow

Figure 17.1 Sunshine powers life in the Great Smokey Mountains.

INTRODUCTION

Every day, a square meter of land at the equator receives about 3000 kilocalories of energy from the sun. A kilocalorie is the amount of energy needed to raise one kg of water 1° C. This means that a pool of water 3 meters deep could be warmed one degree per day of exposure to the tropical sun, provided none of the energy it received was reflected or radiated back into the atmosphere. Solar input declines with latitude, but even as far north as Anchorage, Alaska, solar energy input averages fully half as much as at the equator. What happens to all this energy? A third of it reflects off clouds or snow fields and bounces back into space. Another third warms the land and sea, and about a fourth is absorbed in the process of evaporation, driving the water cycle. A small fraction of the daily solar input is used by plants for photosynthesis; typically 1% to 2% of light striking a forest is captured as chemical energy in plant cells (Figure 17.1). Though little more than a footnote in the solar energy budget, photosynthesis is immensely important to life on earth. Kilocalories converted from sunlight to chemical bond energy by green plants provide the life force that flows through food chains to power the entire biotic community.

Although chemical constituents are cycled from one ecosystem component to another without being used up (see Chapters 15 and 16), energy flows through the ecosystem on a linear path, according to the laws of thermodynamics. The **first law of thermodynamics**, also known as the energy conservation rule, is that matter and energy are neither created nor destroyed. Although matter can be converted to energy in nuclear reactions, this does not happen in normal ecosystem processes. For the biologist, it is safe to assume energy flowing out of a system will ultimately balance energy flowing in. For example, our hypothetical three-meter pool in the tropics will absorb energy until it becomes as warm as the surrounding air. After that, the sum of energy radiating out of the water and energy used in evaporating the water will equal the radiant energy absorbed by the pool each day. Energy in ecosystems follows the same conservation rule. The chemical bond energy captured by photosynthesis is converted to heat when sugars and other biomolecules are metabolized by plants or the animals that eat them, ultimately warming the environment and radiating energy back into space. To put it simply, life on Earth depends on continuous solar radiation because living systems cannot make new energy, nor can they reuse what they expend.

The **second law of thermodynamics**, also called the **entropy** rule, recognizes that energy comes in many forms, which can be converted from one to another. A green plant transforms light energy to chemical energy. A thundercloud transforms mechanical energy to electrical energy. A muscle cell transforms chemical energy to mechanical energy. Energy conversion is a ubiquitous part of nature, but it is never 100% efficient. A portion of the transformed energy is always lost in the form of heat. Since heat is the least useful form of energy, the second law says that useful energy is always lost in energy conversions. As an example, think about the energy stored by plants in the Carboniferous period of our geologic past. Some of this energy has been lying under the ground, trapped in the chemical bonds of fossil fuels for millions of years. Then humans pumped the oil out of the ground and separated its chemical ingredients, saving the shorter, more volatile hydrocarbon chains to make gasoline. If you drive a gasoline-powered car, that fuel is burned in a series of small explosions within the cylinders of your engine. As chemical energy is converted to mechanical energy, the engine turns the wheels and you move down the road. If your engine and drive train were able to translate every kilocalorie of the energy in gasoline into forward motion, your car would be 100% efficient. The second law says this is impossible. Inevitably, much of the energy in auto fuel is converted to heat, which radiates from the engine, warms your tires as they rub against the road, and blows out your exhaust pipe with all the waste products of combustion. A typical gasoline engine is only 30% efficient, and if it is attached to a heavy vehicle that accelerates quickly, brakes often, and pushes tons of air out of its way at high speeds, the efficiency of the vehicle as a whole is significantly less. Thanks to the second law, as much as 4/5 of the energy you purchase at the gas station is used to heat up the atmosphere, and only 1/5 is used to get you to your destination.

Entropy can also be envisioned as a measure of disorganization, so another way to state the second law is that self-contained systems tend to lose organization over time. For example, if you leave a glass of water on a desk when the classroom is closed up for summer break, you would be walking away from an organized system, with water compartmentalized within the glass. Over time, water in the room will become less organized, with molecules evaporating from the glass and mixing randomly with the air in the room. You could return the system to its former organized state by pumping all the air through a chilled condenser and collecting the water back into the glass, but this would require the input of lots of energy from outside the system. In a similar way, ecosystems generate and maintain the organization of living biomass, but this is possible only because of constant energy input from the sun.

The laws of thermodynamics inform our understanding of ecosystem structure and function. Figure 17.2 shows, in highly schematic form, the flow of energy through an idealized terrestrial food chain. After some light is reflected back into the sky, much of the solar energy entering a plant is used up in **transpiration**, which removes water from the leaves, pulling replacement fluids up through the stem. Plants use some of the remaining solar energy to synthesize high-energy organic compounds like sugar and ATP. For this reason, photosynthetic organisms are collectively called **primary producers**. The total amount of energy captured by photosynthesis across the whole ecosystem per unit time is called **gross primary productivity**. After plants have captured energy, they must then use a share of these photosynthetic products to support their own life processes. Respiration at night, and the energetic cost of maintaining non-green plant parts, must be entered on the debit side of the energy ledger. Whatever is left, called **net primary productivity**, accumulates as energy in plant biomass, which may become new tissue in a living plant, humus in the soil, or food for an animal. After animals have consumed their share of the total plant biomass, any remaining energy that accumulates in organic material is called **net ecosystem productivity**.

In aquatic systems, algae are important photosynthetic organisms, and productivity occurs in the epilimnion (Chapter 16). A conventional way to measure productivity directly at representative depths in a lake is to divide a water sample into two bottles: one of clear glass and the other painted black. Dissolved oxygen is measured at the outset of the experiment, then the sample bottles are sealed and resuspended at the same depth on a line attached to a tethered float. After 24 hours, the bottles are pulled up to the surface and dissolved oxygen is measured again. In the dark bottle, the change in oxygen concentration represents oxygen used in respiration only, since there was no light to power photosynthesis. In the light bottle, the change in oxygen represents net productivity, since both respiration and photosynthesis have occurred in this sample. To calculate productivity, we use the simple equation

Gross Photosynthesis – Respiration = Net Photosynthesis

Since the dark bottle gives us a way to measure respiration, and the light bottle tells us net photosynthesis, we can rearrange the equation to get gross photosynthesis as follows:

Net Photosynthesis + Respiration = Gross Photosynthesis

For example, suppose initial readings were 5 mg/l oxygen, and after incubation we measured 3 mg/l in the dark bottle and 6 mg/l in the light bottle. We would know respiration used 5 – 3 = 2 mg/l. Net photosynthesis would be 6 – 5 = +1 mg/l as measured in the light bottle, and gross photosynthesis must have been 1 net + 2 respiration = 3 mg/l gross photosynthesis in the light bottle.

To convert oxygen generated in photosynthesis to energy, we use the chemical formula

$$6\ CO_2 + 12\ H_2O \rightarrow C_6H_{12}O_6 + 6\ O_2 + 6\ H_2O$$

For every 6 moles of O_2 generated in photosynthesis, 1 mole of sugar is added (at least temporarily) to the biomass of the ecosystem. This translates to 192 grams of oxygen for every 180 grams of sugar. Since the caloric content of sugar is 4.2 kcal/g, each gram of oxygen attributed to gross photosynthesis represents $(180/192)g \times 4.2\ kcal/g = 3.9\ kcal$ of energy captured by photosynthetic algae in the lake.

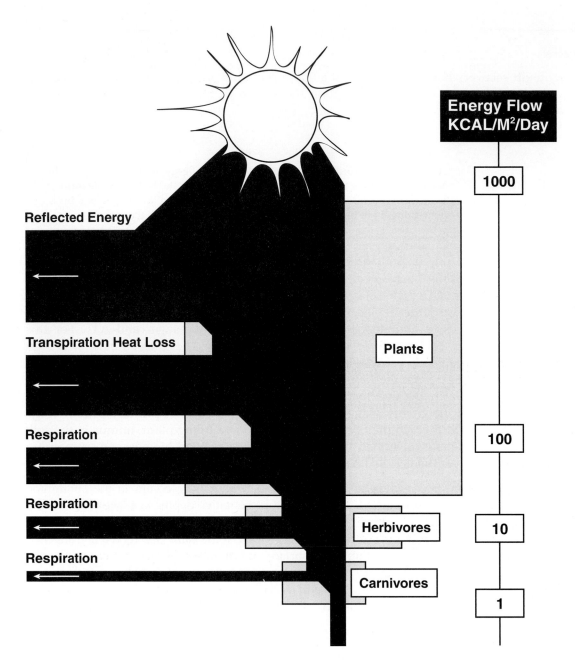

Figure 17.2   Energy flow in ecosystems conforms to the second law of thermodynamics. Thickness of black "pipe" represents energy in kcal/m2/day, flowing down from the sun through three steps in a food chain. Energy lost as heat is shown on the left. The logarithmic scale on the right demonstrates roughly 90% energy loss at each trophic level. Gray boxes represent standing biomass. (Concept based on the work of H.T. Odum, reprinted in Kemp *et al.*, 2004.)

Animals that eat plants are called **herbivores**. Since they represent the first link in the food chain after plants, herbivores are also called **primary consumers**. **Carnivores** are meat-eating animals. Since they consume other consumers, carnivores are also called **secondary consumers**. Larger carnivores eat smaller carnivores, so we can identify additional steps in the food chain as tertiary consumers, quaternary consumers, and so on. Each step in this sequence is called a **trophic level**. Note that available energy declines by about 90% as it passes through a trophic level. The extent of second-law energy loss varies, depending on the type of biological community, but the "ten percent rule" for energy conversion to the next trophic level is a convenient principle to guide our thinking about energy flow.

You can visualize the food chain as an energy pyramid, with total energy available to each trophic level roughly 10% as much as the level just below it (Figure 17.3). By the time we get to the top carnivores, there is not much energy left. This is why large predators at the top of the food chain must range over a large land area to harvest the food they need. To put it another way, the base of their food pyramid must be very large in order to support one pair of hawks or mountain lions on the top tier.

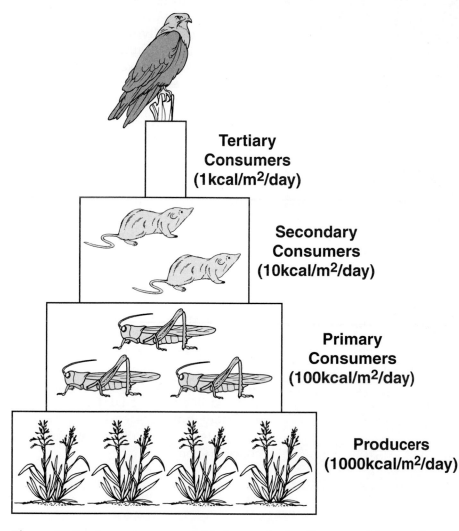

Tertiary
Consumers
(1kcal/m^2/day)

Secondary
Consumers
(10kcal/m^2/day)

Primary
Consumers
(100kcal/m^2/day)

Producers
(1000kcal/m^2/day)

Figure 17.3   An energy pyramid demonstrates the second law of thermodynamics. Each trophic level receives about 1/10 as much energy as the level below it. (Numbers indicate orders of magnitude for energy flow. Boxes and numbers of organisms are not to scale.)

The energy pyramid explains why some kinds of toxic agents become more concentrated as they move through a food chain. This **biological concentration** of toxins was discovered by accident when a water-insoluble pesticide called DDT was widely sprayed on forests, farmlands, and marshes to control insect pests.

The problem, first publicized by Rachel Carson in her classic *Silent Spring*, arises from the fact that DDT is an artificially designed organic compound not metabolized by living cells. Since it is soluble in lipids rather than in water, DDT is not readily excreted by animals. Instead, it accumulates in the fat deposits of an organism's body. As a result, most of the DDT an animal has ever eaten remains in its tissues. If those contaminated animals provide food for a predator, the predator's every meal includes all the DDT accumulated over the prey's lifetime. It is as if all the other biomolecules in food were boiled away by digestion and respiration, leaving a concentrated residue of DDT behind. As a result, concentrations of DDT rise with each step in the trophic pyramid. DDT sprayed on a pasture would be concentrated by a factor of 10 in consumer biomass, by 10×10 in secondary consumers, by $10 \times 10 \times 10$ in tertiary consumers, etc. It is no wonder that top carnivores like the osprey (Figure 17.4) suffered ill effects of DDT poisoning in sufficient numbers early on to alert biologists to the problem of biological concentration.

Figure 17.4   As top carnivores in coastal environments, ospreys were among the first organisms affected by DDT biologically magnified in the aquatic food chain.

Check your progress:

List an example of an organism that occupies each of these trophic levels:
Producers
Consumers
Secondary consumers
Tertiary consumers

Hint: Choose a plant or alga as a producer, an herbivore as a consumer, a small carnivore as a secondary consumer, and a top carnivore as a tertiary consumer.

Ecologist Howard Odum pioneered comprehensive efforts in the late 1950s to measure energy flow and material cycling at an ecosystem level. He chose Silver Springs in Florida as a model system because its spring-fed waters provide stable temperatures and a constant influx of mineral nutrients throughout the year. Accumulation of biomass at each trophic level was carefully monitored, and the energy content of organic materials was determined by burning dried plant or animal samples in controlled conditions to assess their energy content. Odum's results (Figure 17.5) set a standard for research in what is now called **systems ecology.**

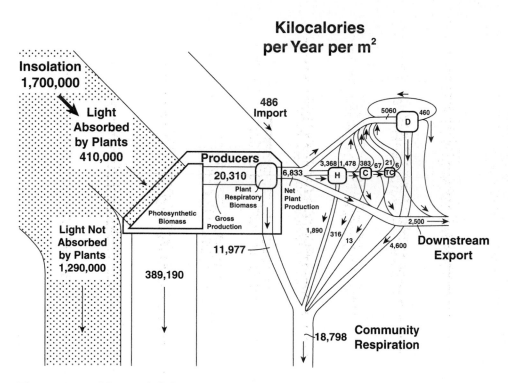

Figure 17.5 Howard Odum's energy flow diagram from the Silver Springs study. Trophic stages for consumers are represented as H = herbivores, C = carnivores, TC = top carnivores, and D = decomposers. As reprinted in Kemp and Boynton, 2004.

One useful concept that emerged from studies of this kind is **ecological efficiency**, or E_t. We define E_t as productivity at trophic level t (in biomass or energy per unit time) divided by the productivity of the next lower level. To express efficiency as an energy ratio, we use the following calculation:

$$E_t = P_t / P_{t-1}$$

E_t = ecological efficiency at trophic level t
P_t = net productivity at trophic level t
P_{t-1} = net productivity at trophic level t–1
Productivity can be measured either in kcal energy or in grams of biomass.

As an example, suppose a pine forest were infested with an outbreak of pine looper caterpillars (Figure 17.6). If we assume these caterpillars have about the same caloric content per gram of tissue as the pine needles they are eating, and if trees produced 200 g needles per square meter each season, while caterpillars feeding on those needles produced 14 g of insect biomass per square meter, the ecological efficiency for the herbivores would be 14/200 = 0.07, or 7%.

Figure 17.6   Northern pine looper caterpillars feed on pine needles.

Check your progress:

Ecological efficiency of carnivorous mammals is usually a little lower than ecological efficiency of reptilian carnivores at the same trophic level. How might this be explained?

Answer: Basal metabolic rate is higher in mammals. They burn more calories at rest, so they do not convert as many of their food calories to biomass.

Not all plant material is consumed by herbivores, and not all animals are consumed by carnivores. Some plant material dies before it is eaten. Some wild animals die of old age. Leaves cover the forest floor in the fall, and mature algae cells drift down to the bottom of a pond. Herbivore digestion is incomplete, so animal manure contains a lot of unused plant biomass. Most of this dead material is consumed by **detritivores** such as bacteria, fungi, and many kinds of invertebrates in the **decomposer** community. If you have observed a hay infusion or fruit flies in the laboratory, you are already familiar with organisms in this ecological category. In Howard Odum's Silver Springs study, detritivores included bacteria and scavengers such as crayfish. They are illustrated in his biomass pyramid as a slender column, taking some of its energy from each of the trophic levels in the food chain (Figure 17.7).

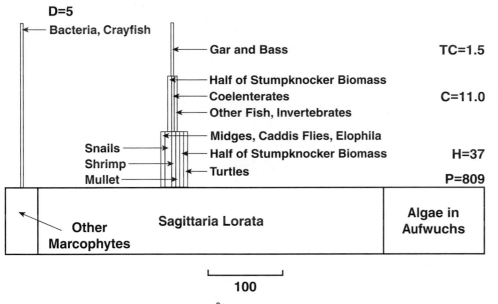

Figure 17.7   Odum's pyramid of biomass from his study of Silver Springs. Trophic stages are represented as P = producers, H = herbivores, C = carnivores, TC = top carnivores, and D = decomposers. (From Howard Odum, 1957, as re-drawn in Kemp and Boynton, 2004.)

The importance of detritivores in an ecosystem depends on the portion of plant material that goes uneaten by herbivores. In aquatic systems, short-lived algae are producers at the base of the food chain. A typical algae cell can double itself every day, but populations do not expand exponentially because most of the cells are consumed by **zooplankton** within hours of their formation. The **standing crop** of algae biomass that we could harvest at any point in time is small in comparison to the high primary productivity of the aquatic system. Trees, in contrast, may produce biomass more slowly than algae in a pond, but they protect a much larger proportion of their cells from immediate herbivory. Tree biomass accumulates in large plant bodies, so a mature forest has a huge standing crop of uneaten plant material. This means that detritivores have a lot more "leftovers" to consume in a forest than in a pond. In a forest, the bacteria, fungi, and invertebrate decomposers in fallen logs and leaf litter make up a much larger ecosystem component than you could find consuming the meager organic residue in a typical freshwater ecosystem.

To understand how a lower rate of production can result in a larger standing biomass, think of productivity as your salary and biomass as the size of your bank account. You probably know "algae people" who make a lot of money, but spend it just as fast, never saving anything. You probably also know "tree people" who put a little of every paycheck into savings so that their net worth accumulates over time. In the same way that a small paycheck can generate a large bank account, a forest gradually amasses a large standing crop by protecting its assets in the form of wood, roots, and leaves over decades of time. For this reason, it is necessary to distinguish between a pyramid of biomass, showing the amount of organic material present at each trophic level, and a pyramid of energy, showing how many kilocalories flow through each level on an annual basis. The second law dictates that the pyramid of energy must become significantly smaller at each step, but the pyramid of biomass can have a larger or smaller ratio of producers to consumers, depending on their relative life spans and energy storing potential. A good example of variation in biomass pyramids is supplied by Carlos Duarte and colleagues in a 2000 study of plankton in the Mediterranean (Figure 17.8). At low nutrient levels, the pyramid of biomass can be inverted because biomass of slow-growing algae never gets a chance to accumulate. By changing the concentration of dissolved nutrients, Duarte *et al.* altered the relative sizes of trophic levels in the biomass pyramid, even though energy always flows in the same pattern of decreasing energy availability through the food chain.

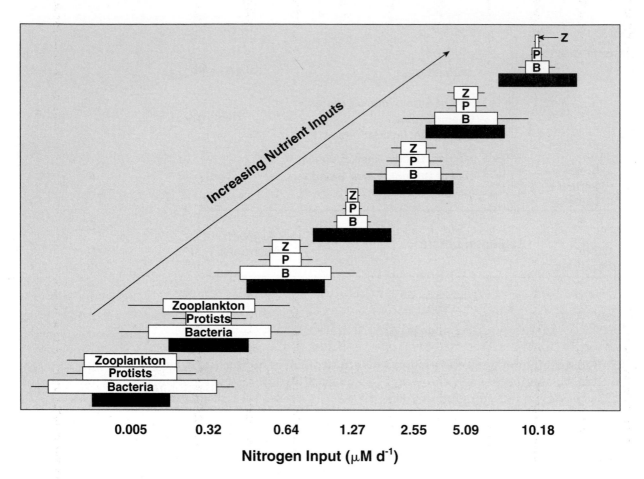

Figure 17.8 Duarte *et al.* (2000) demonstrated that the shape of the biomass pyramid in a planktonic community can be changed by adding inorganic nutrients to the water. Black bars represent producer biomass.

Check your progress:

Given the "ten percent rule," how can producer biomass be less than consumer biomass in an ecosystem?

Answer: Biomass is the result of energy flow over a period of time. If consumer biomass is larger, it is because consumers are living longer than producers in this community, and accumulating biomass over a longer period of time. Their "bank account" is larger, even though their "salary" is smaller.

METHOD A: PRODUCTIVITY OF PLANKTON

[Laboratory activity]

Research Question

What are the magnitudes of gross vs. net productivity in a plankton community?

Preparation

This laboratory method establishes a simple two-species plankton community. The single-celled alga *Chlorella* is chosen to represent freshwater phytoplankton, and *Daphnia pulex* to represent zooplankton. Stock cultures of both are readily available. Order enough *Daphnia* for each laboratory group to use 40 individuals. To culture *Chlorella* in the laboratory, remove labels and rinse plastic 2-liter soda bottles: you will need one bottle for every five laboratory groups. Fill the bottle 3/4 full with 1500 ml of commercial spring water or dechlorinated tap water, and add 5 g 20-20-20 soluble fertilizer to each bottle. Inoculate the bottle with *Chlorella*. As a control, fill another soda bottle 3/4 full of water, add 5 g fertilizer, and maintain it at the same temperature. Shake well, screw on caps, and incubate bottles lying on their sides at room temperature about 50 cm from a 4-tube fluorescent light bank or in a partially shaded part of a greenhouse for about a week. When the water in the *Chlorella* bottle looks green, it is ready to use. Make sure algae suspensions are shaken well before dispensing in smaller containers to students.

For oxygen measurements, obtain water testing kits reading oxygen in ppm, which can be translated to mg/l (Chapter 16). Alternatively, you may wish to use a meter with dissolved oxygen probe for faster analyses. Incubation can be conducted in a greenhouse, in an environmental chamber, or under fluorescent lights. If you put algae cultures in a greenhouse, partially shade them with cheesecloth or other translucent material to avoid the photoinhibition algae experience in direct sunlight. If using an environmental chamber, simulate local summer temperature and day length. (A day/night regime of 16 hours of light and 8 hours of darkness, with a temperature of $25°$ C, is a reasonable simulation of summer in mid-U.S. latitudes.) A fluorescent light fixture about 50 cm from the cultures will provide adequate light for algae growth.

Materials (per laboratory team)

6 wide-mouth specimen bottles with screw caps, 50 ml volume

3 sheets (about 30 cm square) aluminum foil

Dissolved oxygen test kit or meter with dissolved oxygen probe

150 ml *Chlorella* culture

Daphnia culture, with at least 40 individuals

150 ml of control water without algae

3 beakers, 50 ml size

Thermometer

Compound microscope

Slides, cover slips

Large-bore dropper for transferring *Daphnia*

Markers for labeling culture bottles

2 Pasteur pipettes, with bulb

(Optional) photometer or light meter

Procedure

1. Slowly pour a sample of *Chlorella* culture, without mixing in additional air, into a small beaker and perform a test for dissolved oxygen. When performing an oxygen test, always use a thermometer to make sure the tested samples are at the same initial temperature. Record your results in the Data Table for Method A in the pre-incubation column for all four bottles containing *Chlorella*. Then do the same for a sample of control water, recording dissolved oxygen in the pre-incubation column for these two samples.

2. Label four of the 50-ml sample bottles "*Chlorella*—Light," "*Chlorella*—Dark," "*Chlorella* + *Daphnia*—Light" and "*Chlorella* + *Daphnia*—Dark." Fill all four bottles to the brim with *Chlorella* culture. To each of the two bottles labeled "*Daphnia*," add 20 individual *Daphnia* from your stock culture, using a large-bore pipette to avoid harming them. Label the remaining two sample bottles "Control—Light" and "Control—Dark," and fill these two bottles with algae-free control water. Screw all six caps on tightly.

3. Wrap the three bottles labeled "Dark" in aluminum foil. Make sure the foil covers all the glass so that no light can enter the dark bottles.

4. Place the six samples in an environmental chamber, a greenhouse, or under a fluorescent light bank at ambient temperature. If your tops seal well and light is from above, lay the bottles on their sides to allow more light to reach the algae. Record the temperature and note day length if using an environmental chamber. If you have a light meter or photometer, measure light intensity next to your samples, and record this value.

5. Place a drop of the remaining *Chlorella* culture on a glass slide, cover the drop with a cover slip, and observe under high magnification with a compound microscope. Draw a typical *Chlorella* cell on the data page. If you have not already done so in a previous laboratory, place a single *Daphnia* on a slide without a cover slip and draw it as well.

6. After incubation of at least one day, and not more than three days, measure dissolved oxygen again in each bottle. Note the number of days the cultures have been sealed in the sample bottles, and record your results in the data table. The dark bottles containing organisms should have less oxygen than they did in the initial measurement, because without light, these communities have been engaged in respiration but not photosynthesis. In the light bottles, both photosynthesis and respiration have occurred, so their oxygen levels should be higher.

7. Calculate net photosynthesis, community respiration, and gross photosynthesis for each pair of bottles as described in the introduction. If your samples were incubated for more than one day, divide these values by the number of days incubated to get daily rates. Record your results in the combined cells in the three right-hand columns in the data table.

8. Compare your own results with means of the entire class, and interpret your results by answering Questions for Method A.

Data Table for Method A

EXPERIMENTAL CONDITIONS	DISSOLVED O$_2$ BEFORE INCUBATION (mg O$_2$/l)	DISSOLVED O$_2$ AFTER INCUBATION (mg O$_2$/l)	NET PRIMARY PRODUCTION (mgO$_2$/l/day)	COMMUNITY RESPIRATION (mg O$_2$/l/day)	GROSS PRIMARY PRODUCTION (mgO$_2$/l)
Control Light					
Control Dark					
Chlorella Light					
Chlorella Dark					
Chlorella + Daphnia Light					
Chlorella + Daphnia Dark					

Incubation Conditions:

Light intensity:_____   Day length: _____

Temperature: _____   Days incubated: _____

Drawing of algae

Drawing of *Daphnia*

Questions for Method A

1. Define the difference between gross primary productivity and net primary productivity. Explain how you measured each of these values in the pair of bottles containing *Chlorella* culture alone.

   _____

   _____

   _____

   _____

   _____

2. Using figures from the Introduction, calculate the amount of energy algae are capturing in gross photosynthesis in your two sample cultures on a daily basis.

   _____

   _____

   _____

   _____

   _____

3. In the pair of bottles containing *Daphnia*, what was the effect of zooplankton on gross primary productivity? on respiration? on net productivity? Did these figures match your expectations?

   _____

   _____

   _____

   _____

   _____

4. Why was the plain water control necessary? Did either control show any difference in oxygen concentration during incubation? If so, how does this affect interpretations of your other measurements?

   _____

   _____

   _____

   _____

   _____

5. What would the shape of the trophic pyramid look like in your cultures containing both *Chlorella* and *Daphnia*? How would you measure ecological efficiency of *Daphnia*?

   _____

   _____

   _____

   _____

   _____

METHOD B: PYRAMID OF BIOMASS FOR PLANKTON

[Laboratory activity]

Research Question

What is the structure of the biomass pyramid in a sample of plankton?

Preparation

This laboratory exercise analyzes a previously collected sample of planktonic organisms. You will need to purchase only one hand-held plankton sampling net to collect plankton from freshwater or marine habitats for analysis by the class. A tow net with 5" to 8" diameter mouth and an attached collecting bottle works well. Select a plankton net with fine mesh, 20 micrometer size, to collect the smaller phytoplankton species.

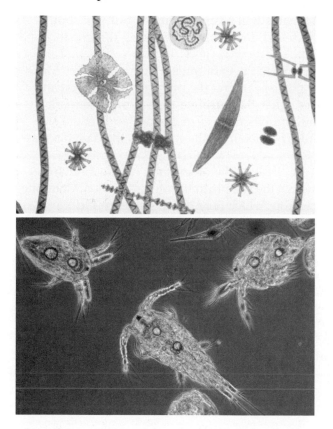

(WildCo Wildlife Supply and Forestry Suppliers carry fine-mesh plankton samplers.) Continue tossing the net until the sample looks turbid, then transfer to a glass sample bottle. A large fishing float attached to the line about 3" from the net is helpful in keeping the net off the bottom. Cast out from a dock or shore and retrieve the net by hand to get a plankton sample. Even in cool weather, you can usually collect algae and zooplankton (Figure 17.9). Although this procedure calls for only enough data to describe a single biomass pyramid, you may wish to take samples at different locations or at different times of year for a more detailed study.

To preserve plankton, make up a solution of Formalin-Acetic Acid (FAA) by mixing equal volumes of 37% formalin with glacial acetic acid. Add 1 ml (20 drops) of FAA to each 50 ml of phytoplankton sample that you collect. Store preserved samples in the dark for longer preservation. A preserved sample should be useable for this laboratory for several years. Thoroughly mix your samples and divide them into 10-ml vials if desired for convenient allocation to laboratory groups.

Figure 17.9   Plankton samples contain a variety of species.

Materials (per laboratory team)

Compound microscope with ocular micrometer (mechanical stage preferable)

Preserved plankton sample (10 ml)

Pasteur pipette, or dropper, with bulb

Glass slide, cover slips

Calculator

Informational materials on phytoplankton and zooplankton

Procedure

1. Shake your sample vial to suspend preserved plankton. Transfer a drop to a glass slide using a pipette or dropper.

2. Place the slide on the mechanical stage of your microscope, and focus on the plankton. Move the slide to a point near the upper-left corner of the cover slip, and scan across the slide until you encounter an organism. Identify the organism, and place it in a trophic level classification. All algae are primary producers. Most protozoans are primary consumers. Larger micro-invertebrates such as *Daphnia* or copepods can be identified as secondary consumers. Tertiary consumers such as fish fry are rarely collected in plankton samples.

3. Find the appropriate section of the Data Table for Method B for this organism's trophic level, and enter its name or description on the first line of that section. Estimate the microorganism's length, width, and height using the ocular micrometer. Since measurements are comparative, you do not need to calibrate the micrometer unless directed to do so by your instructor. Just make sure to measure all organisms on the same scale. Height estimates will be approximate, but you should get a sense whether the organism is cylindrical or flattened, for example, by using the fine focus. Enter these measurements in the data table.

4. Continue on across the slide, classifying and measuring each new organism you encounter. Make sure you enter algae among the primary producers, protists among the primary consumers, and larger invertebrates among the secondary consumers. After measuring a type of organism once, you do not need to measure each additional one you find of similar size. Record an average size, and simply count additional individuals of the same type, making tally marks in the data table as you go.

5. When you finish scanning all the way across the cover slip, use the other knob on the mechanical stage to drop down one field of view, and scan back across the sample in the other direction. Continue until you have counted plankton species in the entire drop, which is 1/20 ml of the sample.

6. After you have counted and measured the organisms in your sample, add up the total number of each type you found. Then multiply the number of individuals × length × width × height to get total volume of this kind of organism in your sample. Record this total, expressed in cubic micrometer units, in the right-hand column of the data table.

7. Add total volume of producers, recording this total at the bottom of the first section in the data table. Repeat for total volume of primary consumers and total volume of secondary consumers. Since plankton are neutrally buoyant, volume is proportional to wet biomass in these organisms. Illustrate your results by drawing a pyramid of biomass on the Results for Method B graph.

8. Interpret your results by answering Questions for Method B.

Data for Method B:
Plankton Counts at Three Trophic Levels

PRIMARY PRODUCERS							
SPECIES NAME OR DESCRIPTION	L	W	HT.	MEAN VOL.	TALLY MARKS	TOTAL NUMBER COUNTED	TOTAL BIOMASS VOLUME

TOTAL VOLUME OF PRODUCERS:

| PRIMARY CONSUMERS | | | | | | | |
SPECIES NAME OR DESCRIPTION	L	W	HT.	MEAN VOL.	TALLY MARKS	TOTAL NUMBER COUNTED	TOTAL BIOMASS VOLUME

TOTAL VOLUME OF PRIMARY CONSUMERS:

SECONDARY CONSUMERS							
SPECIES NAME OR DESCRIPTION	L	W	HT.	MEAN VOL.	TALLY MARKS	TOTAL NUMBER COUNTED	TOTAL BIOMASS VOLUME

TOTAL VOLUME OF SECONDARY CONSUMERS:

Results for Method B:
Pyramid of Biomass for Plankton Sample

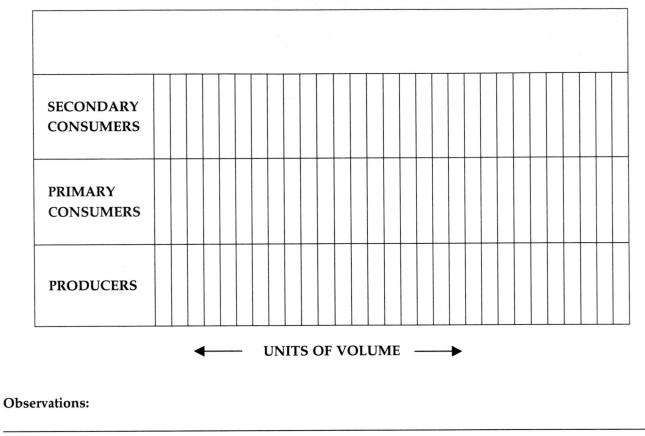

SECONDARY CONSUMERS			
PRIMARY CONSUMERS			
PRODUCERS			

←——   **UNITS OF VOLUME**   ——→

Observations:

_____

_____

_____

_____

_____

_____

_____

_____

_____

_____

_____

_____

_____

_____

_____

Questions for Method B

1. How would you describe the relationships among trophic levels in this study? Is this what you expected to find? Explain.

   _____

   _____

   _____

   _____

   _____

2. Reference to Figure 17.8 demonstrates that biomass pyramids in plankton communities can be inverted; that is, with producer biomass smaller than consumer biomass. How is this possible, given the second law of thermodynamics?

   _____

   _____

   _____

   _____

   _____

3. Is diversity roughly equivalent among the three trophic levels, or are some levels more species-rich than others? Propose one or two hypotheses for this observation.

   _____

   _____

   _____

   _____

   _____

4. Why are decomposers so poorly represented in the plankton community?

   _____

   _____

   _____

   _____

5. Assume a fish farmer is trying to raise fish strictly on the aquatic food chain, without supplementing their food supply. If the farmer has two equal-sized ponds, one stocked with carp feeding exclusively on algae and the other with bass feeding at the fifth tier of their aquatic food chain, how many kilograms of carp could be harvested each year for each kilogram of bass? (Use $E_t = 0.1$ for each trophic level.)

   _____

   _____

   _____

   _____

METHOD C: ESTIMATING ECOLOGICAL EFFICIENCY OF LEAF MINERS

[Laboratory/campus, late summer or early fall]

Research Question

What is the ecological efficiency of leaf miners?

Preparation

Leaf miners are very common herbivores that consume the soft tissue inside leaves. Several insect groups, including flies and beetles, are represented by species in this niche. Oak, Holly, and Chrysanthemum plants are frequent hosts. Hatching from eggs oviposited on or within leaves, the larval leaf miner chews a tunnel that gets wider as the insect grows. The tunnel is easily visible through the transparent epidermis of the leaf (Figure 17.10).

Figure 17.10   A leaf miner leaves evidence of its dietary history in a winding trail through the leaf.

Since these insects eat little or nothing as adults, their entire dietary history is apparent in the winding trail they leave behind them in the leaf. Leaf miners thus present an excellent opportunity to calculate ecological efficiency of an herbivore. Biomass produced by the herbivore can be estimated as volume of the mature larva or pupa. Plant biomass consumed is measured by the volume of the tunnel. The ratio of leaf miner biomass/plant biomass eaten is a measure of ecological efficiency of the herbivore, as explained in the Introduction. Since leaf miners live in an essentially two-dimensional world, the areas as seen from a top view of both the insect and the tunnel can be used to calculate the efficiency ratio.

If you can find affected plants on campus, students can collect their own leaves for analysis. Leaves with leaf miner larvae or pupae still inside are needed for the calculation. Leaves can be pressed, preserving both the leaf and the dried insect, and used for more than one class. An alternate approach is to have students search for pictures similar to Figure 17.10 on the Internet, and do their analysis based on enlarged prints of the photographs.

Materials (per laboratory team)

Dissecting microscope with ocular micrometer

Leaves with leaf miner larvae or pupae inside

Calculator

Ruler, marked in mm

Thread (30 cm length)

Procedure

1. Using a calibrated ocular micrometer, measure length and width of the leaf miner. Calculate the area of the top surface of the leaf miner by multiplying length × width. Enter your area estimate in the Data Table for Method C. (If using enlarged photographs, use a ruler to measure length and width of the larva image.)
2. Measure the length of the leaf miner's tunnel. For small specimens, use the highest power of your dissecting scope. Measure the width of your microscope's field of view by placing a ruler on the stage. Then trace the course of the miner's tunnel, noting "landmarks" at the edge of each visual field and moving the slide to advance one field at a time. Multiply the number of visual fields by the number of mm in a field, and you have the length of the tunnel. For larger specimens, (and for photos) you can do this by laying a thread over the course of the tunnel, and then stretching the thread out to measure with a ruler. Record tunnel length, in mm, in the data table.
3. Measure and record the width of the tunnel at the location of the mature larva. To calculate the area of the tunnel, as seen from the top, you can assume the tunnel is shaped like a long skinny triangle, with the current location of the insect at its base, and the place where the miner's egg hatched at its point. Since a triangle's area is one-half the base times the height, you can multiply the length of the tunnel by one-half the maximum width to get an approximate area.
4. Divide the area of the miner by the area of its eaten food to calculate an ecological efficiency ratio. Remember that both numerator and denominator of this fraction contain a depth measurement equal to the leaf thickness. Since the same depth measurement is in the numerator and the denominator, they cancel, so we will not have to measure the thickness of the leaf.
5. If possible, compare ecological efficiencies for more than one kind of miner or for more than one host species. Photocopy the data page as needed to extend your study to include more specimens.
6. Interpret your data by answering Questions for Method C.

Data Table for Method C

SAMPLE NO.	MINER WIDTH	MINER LENGTH	MINER AREA	TUNNEL MAX. WIDTH	TUNNEL LENGTH	TUNNEL AREA = $(\frac{1}{2}$ W$) \times$ L	ECOLOGICAL EFFICIENCY
1							
2							
3							
4							
5							
6							
7							
8							

Questions for Method C

1. Relate the concept of ecological efficiency, as measured in this exercise, with the trophic pyramid illustrated in Figure 17.3. What would the trophic pyramid look like for leaves and leaf miners?

_____

_____

_____

_____

2. You calculated ecological efficiency in this exercise by comparing biomass of an insect to the biomass it consumed. What kind of measurements would you need to base your efficiency calculation on kcal of energy in the insect vs. kcal of energy in the leaf tissue it ate?

_____

_____

_____

_____

_____

3. How would you expect ecological efficiency of a leaf miner to compare with efficiency of an adult leaf-eating beetle flying from branch to branch and chewing on the same kinds of leaves? Explain.

_____

_____

_____

_____

_____

4. Explain how the second law of thermodynamics applies to ecological efficiency of leaf miners.

_____

_____

_____

_____

_____

5. Do you think a leaf miner qualifies as an herbivore or a parasite? On what basis would you make a distinction between these categories?

_____

_____

_____

_____

FOR FURTHER INVESTIGATION

1. Use the light-bottle, dark-bottle technique of Method A to measure gross primary production and net community productivity for a field-collected sample of plankton. How do results compare with your artificial plankton community?
2. Collect plankton samples from different depths by spacing the net at different distances from a float attached to the line. Do you see differences in phytoplankton densities above and below a thermocline, as explained in Chapter 16?
3. Use the key words "secondary plant compounds" and "tannins" to find out how plants have evolved defenses against leaf-eating species. How might these chemical defenses affect ecological efficiency?

FOR FURTHER READING

Carson, Rachel. 1962. *Silent Spring*. Houghton Mifflin, N.Y.

Duarte, Carlos M., Susana Agustí, Josep M. Gasol, Dolors Vaqué, and Evaristo Vazquez-Dominguez. 2000. Effect of nutrient supply on the biomass structure of planktonic communities: an experimental test on a Mediterranean coastal community. *Marine Ecology Progress Series* 206: 87–95.

Kemp, W. M. and W. R. Boynton. 2004. Productivity, trophic structure, and energy flow in the steady-state ecosystems of Silver Springs, Florida. *Ecological Modelling* 178:43–49.

Martin, Jennifer. 2001. *Marine Biodiversity Monitoring: Protocol for Monitoring Phytoplankton*. Department of Fisheries & Oceans Biological Station, St. Andrews, New Brunswick Canada. http://www.eman-rese.ca/eman/ecotools/protocols/marine/phytoplankton/intro.html

Micscape Magazine. 2000. Pond Life Identification Kit. http://www.microscopy-uk.org.uk/index.html?http://www.microscopy-uk.org.uk/pond/

Needham, J. G. and P. R. Needham. 1989. *A Guide to the Study of Freshwater Biology*. McGraw-Hill, New York.

Chapter 18

Island Biogeography

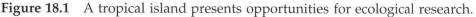

Figure 18.1   A tropical island presents opportunities for ecological research.

INTRODUCTION

Field research is rewarding, but always challenging, and sometimes bewildering. The ecologist attempting to measure one variable typically encounters a hundred more thrown in for good measure by the complex amalgam of chance and causality we call living nature. Population fluctuations, interactions with other species, weather patterns, human interference, and many more unforeseen factors impinge on the ecologist's best-laid experimental plans. When ecologically significant forces are acting all around us, how can we pin down the cause of a significant ecosystem change, such as coral bleaching or amphibian extinction? How can we establish a "control ecosystem" to contrast with our observations of the natural world? Simplicity and replication, the twin pillars of experimental science, are not as easily attained in ecology as in physics or biochemistry. When we discover a natural system that lends itself to controlled investigation, field researchers are understandably excited by the opportunity. This is why islands, with their well-defined boundaries, isolation from interference, and simplified biological communities, have played such an important role in the development of ecological thought (Figure 18.1).

Ecologists Robert MacArthur and E. O. Wilson saw opportunity in islands as natural laboratories for investigating community structure. Drawing on their extensive experience exploring biodiversity on island chains around the globe, they realized islands provided a proving ground for their developing conception of biotic communities as dynamic systems, not just static collections of species.

Through comparisons of islands, MacArthur and Wilson could gather repeatable measurements of ecological processes to find how each influences the resulting flora and fauna. Their seminal 1967 book, called *The Theory of Island Biogeography,* begins with the observation that the number of species on an island is related to its area. On a graph displaying each variable on a logarithmic scale, the relationship can be drawn as a straight line. This **species–area relationship** has since been demonstrated in many kinds of organisms on many kinds of island clusters (for example, see the distribution of bats on Caribbean islands in Figure 18.2). The empirically derived expression for the species–area relationship is as follows:

Species–area relationship:

$$S = C\,A^z$$

Taking the log of both sides: $\qquad\qquad\qquad \log S = \log C + z \log A$

S = the number of species present
C = a constant that varies with the taxonomic group
A = the area of the island
Z = a constant, roughly equal to 0.3 for island habitats

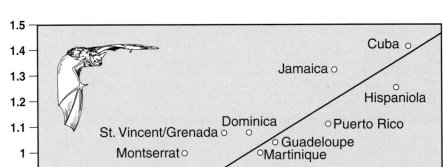

Species-Area Curve for Caribbean Bats

Figure 18.2 caption below plot:

Figure 18.2 Species-area relationship for bats on Caribbean islands. (From Scott Pedersen. 2005. http://biomicro.sdstate.edu/pederses/caribres.html)

Note that taking the logarithm of both sides transforms the equation to linear form. To visualize the biology behind this formula, it is instructive to note that a tenfold increase in the area of an island produces an approximately twofold increase in its species count (Figure 18.3).

Species–Area Relationship

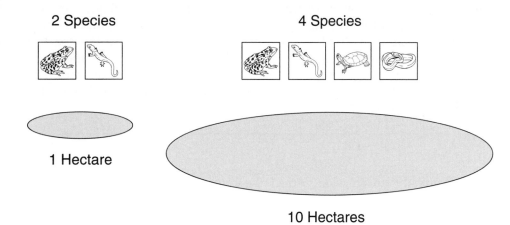

Figure 18.3 An island ten times as large can support twice as many species.

To explain this empirically derived species–area relationship, MacArthur and Wilson theorized that species numbers on islands represent a state of equilibrium—a delicate balance between local extinction on one side of the ecological teeter-totter and immigration across the water on the other. Island size, they reasoned, should have little influence on immigration rates, but a lot to do with extinction. Therefore, as island size increases, extinction rates should slow and the balance should shift toward greater biodiversity (Figure 18.4).

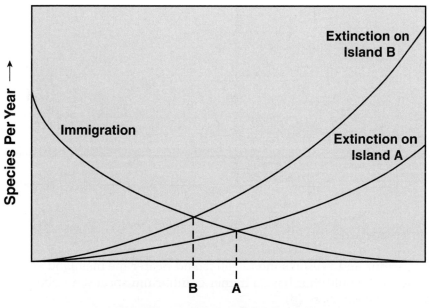

Figure 18.4 The Area Effect. In MacArthur and Wilson's model, a larger island (A) would have a lower rate of extinction, and would thus have a higher equilibrium species number than a smaller island (B).

Distance from the mainland should have the greatest impact on immigration, so all other things being equal, an island farther from sources of immigrants should have fewer species than an island closer to shore (Figure 18.5). Comparative data support this part of the theory as well. Easter island, remotely located in the South Pacific, has only 47 species of native higher plants, according to biogeographer Jared Diamond (2004). The flora of Catalina, an island of roughly equivalent size only 42 km off the coast of California, has more than ten times as many native plant species. The higher flow of immigrant species to Catalina pushes its equilibrium number higher, even though Catalina's rate of extinction is expected to be roughly the same as Easter's. MacArthur and Wilson concluded that size and location both influence the balance between immigration and extinction on islands.

Check your progress:

Why would the number of species already on an island affect the rate at which new immigrants successfully establish themselves?

Hint: Think about niche space.

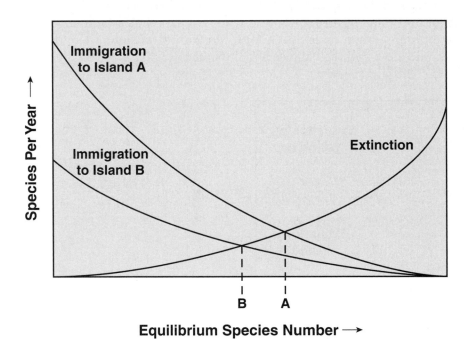

Figure 18.5 The Distance Effect. In MacArthur and Wilson's model, an island nearer the mainland (A) would receive immigrants at a higher rate, and would thus have a higher equilibrium species number than an island of the same size located farther from shore (B).

While still a graduate student working with E. O. Wilson, Daniel Simberloff saw an opportunity to test these theories in the small mangrove islands that dot the shoreline of southern Florida (Figure 18.6). Simberloff and Wilson first censused all the arthropods living in the canopy of several mangrove islands in Florida bay. This was no small task. Mangrove thickets are notoriously difficult to walk through, and arthropods are so diverse that identification to species requires considerable expertise. Through persistence and collaborative taxonomic work, Simberloff and Wilson established solid baseline data on the invertebrate fauna of their miniature islands. These data, according to the theory, represented equilibrium species numbers determined by the islands' sizes and locations.

Figure 18.6 A mangrove island off the coast of Florida provides nesting sites for birds, and a permanent home for many small arthropod species. Simberloff and Wilson (1970) removed invertebrates (but not birds) from four islands and documented their recolonization.

Simberloff and Wilson then manipulated their experimental system. They blanketed and fogged four of the mangrove islands, completely eliminating all their resident invertebrate species. The researchers then carefully documented recolonization of the islands with repeated follow-up visits. After two years, all but the most remote island had returned to their former equilibrium species numbers (Figure 18.7). Interestingly, recolonized islands supported a somewhat altered list of species, even though total species numbers were about the same as the original census figures. The study's conclusion was that biogeographic principles do operate on island communities, regardless of the kinds of species that happen to occur in a particular place.

Check your progress:

Why was the Simberloff–Wilson study of mangrove islands deemed so important by ecologists, given that the species–area curve had already been discovered by comparing species numbers on oceanic islands?

Answer: Comparative studies can show correlations between variables, but a controlled experiment provides more convincing evidence of causal relationships.

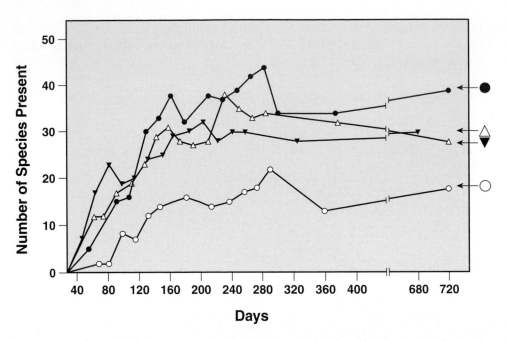

Figure 18.7 Simberloff and Wilson's survey results documenting recolonization of four mangrove islands. Two years after all insects were removed from the islands, biodiversity had recovered to nearly their original species numbers (indicated by arrows at right), with the exception of the most distant island from shore (open circles). (After Simberloff and Wilson, 1970.)

Island biogeography theory has been substantially refined and modified in the 40 years since its introduction, but it remains central to our understanding of community dynamics. Ecologists quickly realized that the same theory explaining immigration and extinction on oceanic islands would also apply to other kinds of isolated habitat. For example, consider a marmot living in the Southwestern United States (Figure 18.8). Marmots are adapted to the kinds of cold, rocky habitats found at northern latitudes or high altitudes. During periods of glaciation thousands of years ago, marmots were able to migrate south across much of North America. When the climate warmed up, marmot populations retreated to the tops of mountains, where they remain isolated today. For marmots, the Southwest consists of "sky islands" of alpine habitat surrounded by a "sea" of inhospitable lowlands (Figure 18.9). Given enough time, communities of mammals adapted to alpine habitats on Southwestern mountain ranges conform to the same kinds of species–area relationships that terrestrial animals exhibit on oceanic islands (Brown 1971). Similarly, a chain of lakes can be considered an archipelago for fish, and a patch of forest is an island for tree-nesting birds.

Because human activities have seriously fragmented habitats around the world, island biogeography theory is especially relevant to conservation biology. To an increasing degree, parks and wilderness preserves are becoming scattered islands of natural habitat in a landscape so altered by human activity that it no longer supports native species. Whether we consider fragmentation of tropical forest in Bolivia (Figure 18.10) or native prairie reduced to bits and pieces in the American corn belt, we find immigration between habitat patches has become increasingly difficult for species facing local extinction on shrinking remnants of their former continental ranges. Guided by application of island biogeography theory, conservationists stress the importance of wildlife corridors between existing parks to enhance rates of population exchange, and larger blocks of preserved land in recognition of the species–area relationship. Debates about preserving a single large or several small preserves have largely focused on the applicability of biogeography theory to particular environments. Island biogeography not only gives us opportunities to understand the history of natural communities, but also helps us find ways to protect biodiversity for the future.

Figure 18.8   A marmot is a medium-sized mammal confined to alpine habitats.

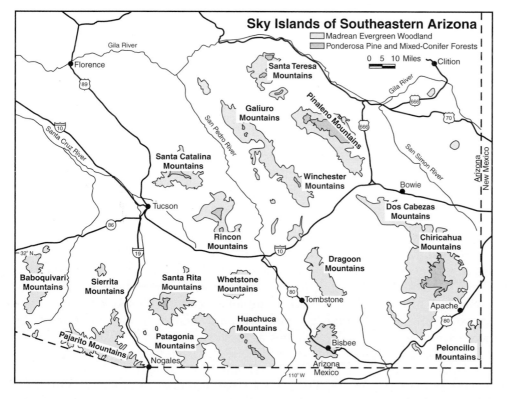

Figure 18.9   "Sky islands" are illustrated as shaded areas on a map of Southeastern Arizona. (From C.J. Bahre, 1998.)

1975   **2000**

Figure 18.10   Habitat fragmentation revealed by satellite images taken in 1975 (left) and 2000 (right) of an area originally covered in tropical dry forest east of Santa Cruz de la Sierra, Bolivia. A formerly continuous forest (dark area) has been reduced to small habitat islands.

Check your progress:

Why are tunnels constructed under highways for wildlife migration an increasingly important strategy for maintaining biodiversity?

> Answer: Highways divide the landscape into habitat islands. If migration across the highway is limited, the species–area relationship reduces biodiversity through elevated extinction in the lands remaining between roads.

Speciation also has a significant influence on the makeup of island communities, but on a different scale of time. Whereas immigration and extinction can transform a community in tens or hundreds of years, evolution at the level of species occurs over thousands or millions of years. When isolated for a very long time from the mainland gene pool, island populations change via **genetic drift** and **natural selection** (see Chapter 6). Both Charles Darwin and Alfred Russel Wallace came away from extensive collecting trips through islands realizing that isolation and speciation go hand in hand. Island species may arrive by happenstance, but many of them experience predictable changes as they adapt to an insular environment. For example, the capability of flight that undoubtedly helped bring birds to islands in the first place tends to be lost in the absence of mainland predators. The endemic Hawaiian goose, called the nene, is a weak flier in comparison to other species of geese (Figure 18.11). Size, too, may be subject to rapid evolutionary change. Large mammals such as the elephant show a tendency toward reduced size when permanently marooned on islands, while small rodents tend to become larger over evolutionary time. (For a detailed and accessible discussion of evolution on islands, see Quammen, 1997.)

Figure 18.11   Nene, the endemic Hawaiian goose, is distinct from mainland species.

Whether or not they develop "island" peculiarities, long-isolated populations develop genetic differences, which may be sufficient to block renewed gene flow if these organisms are later reintroduced to the mainland. By definition, a population no longer able to breed with the ancestral type is a new species. This so-called **allopatric** model of speciation is apparent in the various stages of genetic difference found between island and mainland populations today. Since physical isolation commonly precedes genetic isolation, it is likely that most of the earth's species were once residents of islands, separated by water or by discontinuous habitat from others of their kind, at some critical juncture in their evolutionary past.

Check your progress:

Based on the allopatric speciation model, what role could habitat islands such as mountaintops play in developing biodiversity?

Answer: A population stranded on a mountaintop for a sufficient duration accumulates genetic differences, which prevent interbreeding with ancestral populations, and so becomes a separate species.

METHOD A: BIOGEOGRAPHY SIMULATION GAME
[Laboratory]

Research Question
How do island size and distance from immigration sources influence the equilibrium between immigration and extinction?

Preparation
In this simulation exercise, egg cartons simulate islands, and ping-pong balls represent species. Remove the hinged lids from cardboard or plastic egg cartons, leaving only the cupped base portion. Dozen-sized egg cartons are needed for the large islands. Half-dozen–sized cartons can be used for the small islands, or you can cut some dozen-sized cartons in half. A hard surface 8' long, either on the floor or on long benches, is required to bounce the balls into the egg cartons to simulate species becoming established on islands.

This simulation works best with four people in each group.

Materials (per laboratory team)
1 dozen-sized egg carton (with lid removed)

1 half-dozen–sized egg carton (with lid removed)

16 ping-pong balls

1 pair of dice, of different colors (e.g., a red die and a green die)

Small plastic cup for shaking dice

Felt-tip markers to match dice colors (e.g., a red and green marker)

Masking tape and meter stick (can be shared with other groups)

Procedure
1. Number the six cups at one end of the large egg carton 1 through 6 by writing numbers with a colored marker on the inside of each depression where an egg would sit. Number the six cups at the other end 1 through 6 with the other colored marker. These numbered cups represent niches for species living on an island (see Figure 18.12). Be aware that calling these cups "niches" is an oversimplification. Since a niche is not just a physical space (Chapter 8), think of each cup as a unit of habitat containing sufficient resources to sustain a population of one species.
2. Number the cups in the small egg carton 1 through 6 with any color.
3. Place the large (dozen-sized) egg carton on a hard floor surface. This represents a large island. Use masking tape to mark the position of the mainland, one meter away. Alternatively, place the large egg carton one meter from the edge of a bench or table.
4. Designate four people on your team as a) colonization simulator, b) extinction simulator, c) ping-pong ball catcher, and d) data recorder.

Egg Carton Simulates Large Island. Balls Represent Resident Species.

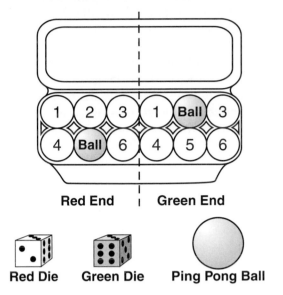

Red End | **Green End**

Red Die   **Green Die**   **Ping Pong Ball**

Figure 18.12 Apparatus for biogeography simulation game. An egg carton represents an island with 12 "niches" available for colonization. Ping-pong balls bounced into the "island" represent colonizing species. Red and green dice are used to decide which species will become locally extinct in each round of play.

5. *Simulate colonization of a large island.* Taking five ping-pong balls in hand, the colonization simulator stands on the mainland (behind the tape line on the floor or at the edge of the table) and tries to bounce ping-pong balls into the niche spaces on the "island." The only rule is that the ball must bounce off the floor at least once before it lands in a cup. The ping-pong ball catcher helps round up the balls that do not go into a cup, returning them to the colonization simulator for the next round. Balls that land on top of the island but do not settle into a "niche" in one of the cups have not become "established," and must be removed.

6. After the five-ball colonization trial, the extinction simulator shakes the dice in the plastic cup, draws one die out at random, and tosses it. The color and number of the die indicates one "niche" on the island. (For example, if the green die comes up with #3, then look at the #3 cup on the green end of the island.) If this "niche" is empty, extinction is zero for this round. If this "niche" is occupied, remove the ball that sits in that place. One "extinction event" has occurred. (A piece of tape wrapped sticky-side-out around your finger can help remove balls from their "niches" more easily.)

7. At the end of one round of colonization and extinction, the recorder writes down the following information in Data Table 1 for Method A: a) how many species existed on the island at the beginning of this round (which will be 0 the first time), b) how many successful immigration events occurred out of the five attempts, c) how many individuals exist on the island after colonization, and d) how many extinction events (zero or one) occurred in this round.

8. Repeat steps 5 through 7 for 30 rounds, simulating five colonization attempts followed by an extinction trial in each round. Record data after each round.

9. *Simulate colonization of a small island.* Replace the large egg carton with the half-dozen–sized egg carton. Remove one of the dice from the extinction simulator's plastic cup. Repeat steps 5 through 7 as before, with the exception that the extinction simulator throws the same die each time, and removes any "species" that exist in the indicated "niche." Continue for 30 iterations, recording your results in Data Table 2 for Method A.

10. *Simulate colonization of a distant island.* If you have time, try to simulate colonization of the large island at a distance of *2 meters* away from the "shoreline" where the colonization simulator stands. Simulate 30 rounds of colonization and extinction as before, recording your results in Data Table 3 for Method A.

11. Pool data from the first (large island) simulation. Begin by counting how many colonization attempts were made on islands containing zero established species. Look down through *Column 1* in your own Data Table 1 to find all cases in which the initial species number is zero. Multiply the number of cases by five to determine how many total colonization attempts were made to colonize islands with zero initial colonists. Write this number in the first line of Column 1 in Data Table 4. Next, for all the cases you have identified, add up all the "successful colonization events on islands with 0 species" from Data Table 1, and record this sum in Column 2 of Data Table 4. The idea here is to calculate a success rate for all attempts made on islands with 0 colonists.

12. Repeat step 11 for all the cases you can find in Data Table 1 in which the island has 1 species at the beginning of a simulation round. Enter the sums for total attempts and total successes in Data Table 4. Repeat for cases in which the island already had 2, 3, 4, 5, 6, 7, 8, 9, 10, and 11 species at the beginning of a round, filling out the first and second columns of Data Table 4. We can assume the rate of immigration onto islands already containing 12 species is 0, since there are no niches left to colonize on a full island.

13. Combine your data with all the other groups in your class, summing total attempts and total successes for each initial species condition for the class as a whole. Record these totals in the third and fourth columns of Data Table 4. Then divide (total successes)/(total attempts) to calculate the rate of immigration onto islands for the class as a whole. When you have completed this step, you should be able to demonstrate that the success of immigration is negatively related to species already on the island.

14. Pool class data for extinction rates in a similar fashion. First, recognize that the numbers in the third column of Data Table 1 are the numbers present when an extinction trial is conducted. We can assume the rate of extinction is zero when no species are on the island. Look down through Column 3 to find all cases in which 1 species is present. Write the total number of these cases in Column 1 of Data Table 5. Then add up the number of extinction events that actually occurred in these cases, from the corresponding lines in Column 4. Record this sum in Column 2 of Data Table 5.

15. Repeat step 14 for cases with 2, 3, 4, 5, 6, 7, 8, 9, 10, and 11 species present at the beginning of the extinction trial. Record your results, filling in the first two columns of Data Table 5.

16. Combine your data with all the other groups in your class, summing total number of extinction trials and total extinction events on islands at each initial species number. Record these totals in the third and fourth columns of Data Table 5. Then calculate the rates of extinction for the whole class by dividing (total number of extinction events)/(total number of trials) at each species density. Record these fractions in the fifth column of Data Table 5. We can assume that the rate of extinction for islands with 12 species present is 1.0, since the die will indicate the removal of one of the 12 balls every time when the island is completely full.

17. *Plot the class immigration rates*, with initial species number on the x-axis and rate of colonization on the y-axis, on the graph labeled Results for Method A: Immigration and Extinction Curves, Based on Class Totals.

18. *Plot the class extinction rates*, with initial species number on the x-axis and rate of extinction on the y-axis, on the same graph with the immigration curve. Examine the graph. The immigration line should have a negative slope, showing decreasing rates of successful colonization as the number of species on the island increases. The extinction curve should have a positive slope, showing increasing rates of extinction as more and more species fill up the island. The x value that lies beneath the point of intersection of the two lines is the predicted equilibrium value for species on the large island. Compare your graph with Figure 18.4.

19 Plot the total species number on the large island, listed in the first column of Data Table 1, against time, measured in trial numbers. Trial number should go on the x-axis and number of species on the island on the y-axis. Compare your graph with Figure 18.7. Make similar plots for your results from the small island experiment and the distant island experiment. Compare the stability of equilibria, and the equilibrium species numbers from the three experiments. Record your thoughts about these three experiments.

20. Analyze your results by answering Questions for Method A.

Data Table (1) for Method A:
Large Island Simulation

TRIAL #	COLUMN 1 INITIAL NUMBER OF RESIDENT SPECIES	COLUMN 2 NUMBER OF SUCCESSFUL COLONIZATION EVENTS	COLUMN 3 # SPECIES AFTER IMMIGRATION (ADD COLUMNS 1 AND 2)	COLUMN 4 NUMBER OF EXTINCTION EVENTS
1	0			
2				
3				
4				
5				
6				
7				
8				
9				
10				
11				
12				
13				
14				
15				
16				
17				
18				
19				
20				
21				
22				
23				
24				
25				
26				
27				
28				
29				
30				

Data Table (2) for Method A:
Small Island Simulation

TRIAL #	COLUMN 1 INITIAL NUMBER OF RESIDENT SPECIES	COLUMN 2 NUMBER OF SUCCESSFUL COLONIZATION EVENTS	COLUMN 3 # SPECIES AFTER IMMIGRATION (ADD COLUMNS 1 AND 2)	COLUMN 4 NUMBER OF EXTINCTION EVENTS
1	0			
2				
3				
4				
5				
6				
7				
8				
9				
10				
11				
12				
13				
14				
15				
16				
17				
18				
19				
20				
21				
22				
23				
24				
25				
26				
27				
28				
29				
30				

Data Table (3) for Method A:
Distant Island Simulation

TRIAL #	COLUMN 1 INITIAL NUMBER OF RESIDENT SPECIES	COLUMN 2 NUMBER OF SUCCESSFUL COLONIZATION EVENTS	COLUMN 3 # SPECIES AFTER IMMIGRATION (ADD COLUMNS 1 AND 2)	COLUMN 4 NUMBER OF EXTINCTION EVENTS
1	0			
2				
3				
4				
5				
6				
7				
8				
9				
10				
11				
12				
13				
14				
15				
16				
17				
18				
19				
20				
21				
22				
23				
24				
25				
26				
27				
28				
29				
30				

Data Table (4) for Method A:
Team & Class Immigration Results

Initial number of resident species	Column 1: Number of colonization attempts by your group	Column 2: Successful colonization events by your group	Column 3: Class total: number of attempts	Column 4: Class total: number of successes	Mean Rate of Immigration: (Divide column 4 by column 3)
0					
1					
2					
3					
4					
5					
6					
7					
8					
9					
10					
11					
12		0		0	

Data Table (5) for Method A:
Team & Class Extinction Results

Initial number of resident species	Column 1: Number of extinction trials by your group	Column 2: Number of extinction events in your group	Column 3: Class total: number of extinction trials	Column 4: Class total: number of extinction events	Mean Rate of Extinction: (Divide column 4 by column 3)
0		0		0	
1					
2					
3					
4					
5					
6					
7					
8					
9					
10					
11					
12					1.00

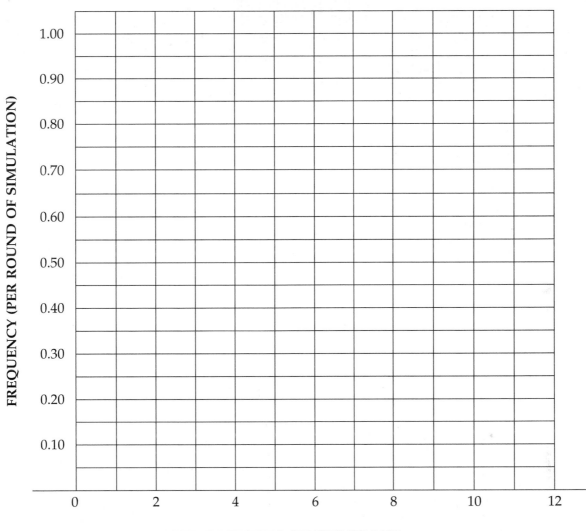

**Results for Method A:
Immigration and Extinction Curves
Based on Class Totals**

FREQUENCY (PER ROUND OF SIMULATION)

NUMBER OF SPECIES ON THE ISLAND

- - - - - - **IMMIGRATION CURVE**
─────── **EXTINCTION CURVE**

Estimate the equilibrium species number for the large island, based on population size at which extinction and immigration lines intersect:

_____

Results for Method A:
Changes in Species Number on Large Island

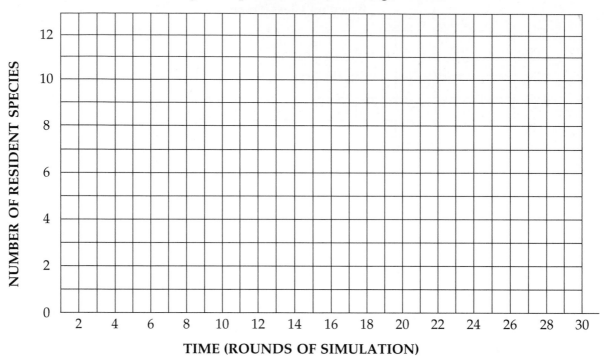

TIME (ROUNDS OF SIMULATION)

Results for Method A:
Changes in Species Number on Small Island

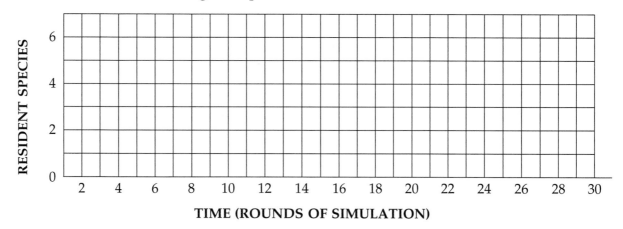

TIME (ROUNDS OF SIMULATION)

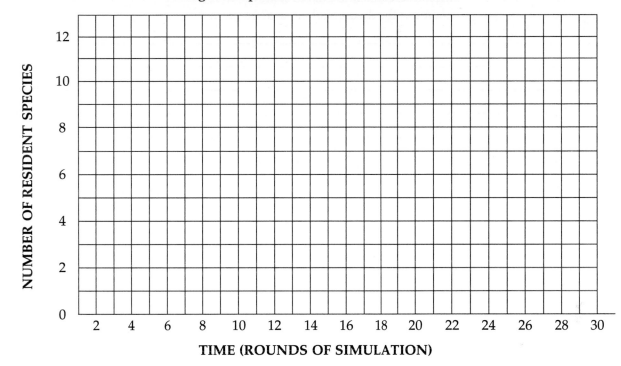

Results for Method A:
Changes in Species Number on Distant Island

Observations:

Questions for Method A

1. Compare your immigration and extinction curves with Figure 18.4. Did your observed immigration and extinction curves resemble the theoretical curves? Explain.

   _____

   _____

   _____

   _____

   _____

2. Compare the hypothetical equilibrium state derived from your immigration and extinction curves with the mean species numbers you actually observed near the end of the large island simulation. Did your large island achieve an equilibrium state near the predicted number? Explain.

   _____

   _____

   _____

   _____

   _____

3. Each group had a different colonization simulator, so the accuracy of tossing balls onto the island could vary from group to group. Is it possible that migration to islands from the mainland might also vary from one situation to another? Propose several ecological factors other than distance that might affect successful colonization.

   _____

   _____

   _____

   _____

   _____

4. How did the small island results compare with the large island results? Could this result have been predicted by the species–area formula? Explain.

   _____

   _____

   _____

   _____

   _____

5. If you had time to conduct a distant island simulation, comment on your findings. If not, what would you expect the results of this simulation to demonstrate?

   _____

   _____

   _____

   _____

   _____

METHOD B: SPECIES–AREA RELATIONSHIP FOR BIRD COMMUNITIES
[Mapping or computer activity]

Research Question
Do resident birds in regional parks and preserves exhibit a species–area curve?

Preparation
This activity requires data collection from species checklists and paper maps or digital maps. A list of 10–15 parks or wildlife preserves near your campus will be useful as a starting point. It is desirable to include a large range of park sizes, from small city parks to large state or national parks. The procedure focuses on forested areas, but undisturbed desert, marsh, or prairie habitats may be more appropriate for some regions. Preselecting parks that are surrounded by impacted habitat will provide clearer results. Maps of local parks and wildlife preserves are usually available on request, and may be used to measure the park's forested area. Alternatively, digital topographic maps are available on-line and in CD-ROM format. Digital mapping software has the advantage of coming with area calculation functions. The U.S. Geological Survey and Topozone (http://www.topozone.com) maintain sites for maps of any location in the United States. Natural Resources Canada (http://maps.nrcan.gc.ca/topo_e.php) offers topographic maps of Canada.

Checklists of resident birds are developed by many parks. On-line checklists can be obtained from state ornithological groups, and from clearinghouse sites such as the Northern Prairie Wildlife Research Center's 2005 *Bird Checklists of the United States,* found at http://www.npwrc.usgs.gov/resource/othrdata/chekbird/bigtoc.htm.

Materials (per laboratory team)
List of regional parks or other forested sites to be included in this survey

Topographic maps (paper or digital) for each site

A checklist of birds found at each site

Ruler (or software) for determining map areas

Calculator with $\log_{10}$ function

Procedure
1. Obtain a list of regional parks or refuges from your instructor. Write site names into the first column of the Data Table for Method B or C. For each site, consult a topographic map and find boundaries of the forested area, which is typically shaded green. Estimate the area of forest habitat, using the map's grid lines and the map scale. Use either km^2 or mi^2, but be consistent. If contiguous forested land outside the park boundary appears free of artificial disturbance, include these areas in the calculation of forest area as well. Record the estimated area of each site in the second column of the data table.
2. Consult a bird checklist for each site, counting the number of resident species. If the checklist identifies rare or transient birds, do not include them in your species count. These species have not succeeded in the establishment phase of colonization. Record the number of resident species in the third column of the data table.

3. Use a calculator to determine $\log_{10}$ of each area and $\log_{10}$ of each species number that you generated in steps 1 and 2. Record these log calculations in the fourth and fifth columns of the data table.

4. In Results for Method B or C, plot $\log_{10}$ of area on the x-axis vs. $\log_{10}$ of species number on the y-axis. If you see a trend line, use a ruler to draw the best straight line through your points. If you have access to regression software, test the significance of a nonzero slope.

5. Calculate the slope of the best line through your points, either with linear regression or by calculating (Δlog species)/(Δlog area). Enter this number in the box at the bottom of the graph. The slope of the log-log plot is equal to the constant z from the species–area equation.

6. Record comments, and analyze your results by answering Questions for Method B or C.

METHOD C: SPECIES–AREA RELATIONSHIP FOR ARTIFICIAL ISLANDS

[Campus activity]

Research Question

Do invertebrates hiding under objects on the ground exhibit a species–area curve?

Preparation

A month or more before the laboratory begins, create habitat islands of five sizes by cutting a plywood sheet as shown in Figure 18.13. Place the "islands" on the ground in a relatively undisturbed part of the campus where killing the grass or other plants under the sheets will not matter. Several replicates are desirable, but the number of plywood sheets you will need depends on the size of the class. The same set of "islands" could be used for more than one section if insects are set free after the laboratory to recolonize the islands. With a waterproof marker or pencil, give each "island" a unique number for site identification.

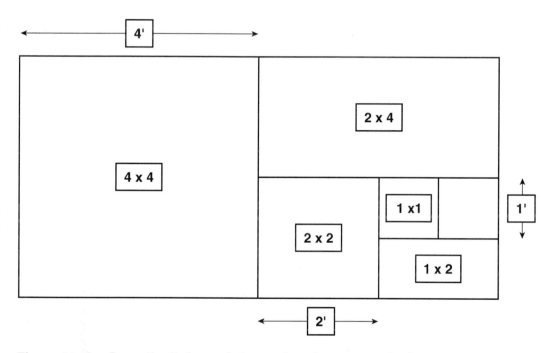

Figure 18.13 Cut a 4' × 8' sheet of plywood as shown to make five sizes of "islands" for Method C.

Pitfall traps consisting of collecting jars buried flush with the ground under the plywood sheets and filled ½ full with isopropyl alcohol can be used to collect colonizing invertebrates, but students may instead collect and count live invertebrates using forceps and small vials. Species may be identified on-site without making a collection if students are given some help with field identification. Instruct students to be careful in collecting scorpions, centipedes, and some kinds of spiders by hand.

Materials (per laboratory team)

Five collecting jars, approximately 50 ml, with isopropyl alcohol preservative

Wax pencil or marker for labeling vials

Meter stick

Forceps

Calculator with $\log_{10}$ function

Procedure

1. Working as a team, you will be assigned one set of five different-sized plywood "islands" as your sampling area.
2. Note the identification number on each of your "islands" and record these site numbers in the first column of the Data Table for Method B or C.
3. Measure the length and width of your "islands" in cm. Calculate area of each "island" in cm^2, and record these numbers in the second column of the data table.
4. Carefully turn over one of the "islands." *Take care when turning over plywood—venomous spiders, scorpions, and snakes often hide under lumber.* It is best to lift the opposite side of the sheet, keeping the plywood between you and any large organism trying to escape. Count the numbers of species you see under the plywood island. Include vertebrates such as mice if they are present, but most of the organisms will be invertebrates such as isopods, centipedes, millipedes, or ground beetles. Using forceps, you may wish to collect one of each species for taxonomic identification. Repeat this procedure for each of the plywood islands in your study area.
5. Use a calculator to determine $\log_{10}$ of each area and $\log_{10}$ of each species number that you generated in step 4. Record these log calculations in the fourth and fifth columns of the data table.
6. For the purpose of replication, pool your data with two other groups to include data from 15 "islands" in your data table.
7. In Results for Method B or C, plot $\log_{10}$ of area on the x-axis vs. $\log_{10}$ of species number on the y-axis. If you see a trend line, use a ruler to draw the best straight line through your points. If you have access to regression software, test the significance of a nonzero slope.
8. Calculate the slope of the best line through your points, either with linear regression or by calculating (Δlog species)/(Δlog area). Enter this number in the box at the bottom of the graph. The slope of the log-log plot is equal to the constant z from the species–area equation. (For further discussion of regression analysis, see Appendix 3.)
9. Record comments, and analyze your results by answering Questions for Method B or C.

Data Table for Method B or C

SITE NAME	AREA (km² *or* cm²)	SPECIES #	LOG$_{10}$ AREA	LOG$_{10}$ SPECIES
1.				
2.				
3.				
4.				
5.				
6.				
7.				
8.				
9.				
10.				
11.				
12.				
13.				
14.				
15.				

Observations:

_____
_____
_____
_____
_____
_____
_____
_____
_____
_____
_____
_____
_____

Results for Method B or C:
Log$_{10}$ Area vs. Log$_{10}$ Species Number

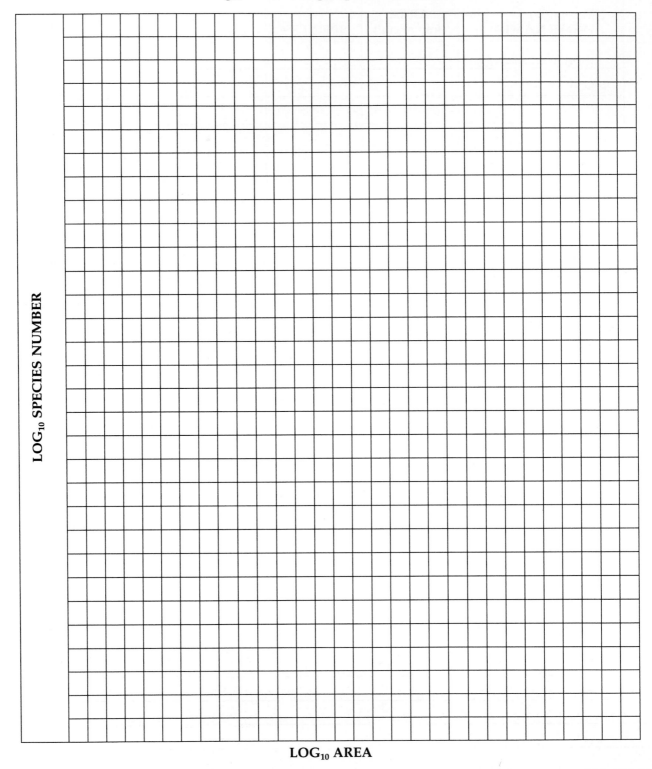

LOG$_{10}$ SPECIES NUMBER

LOG$_{10}$ AREA

Estimate of slope = Δ(log species)/Δ(log area) = z =

Questions for Method B or C

1. How comparable are habitat islands in this study to oceanic islands? Do you think the organisms counted in your study could move easily between habitat islands, or not?

    _____

    _____

    _____

    _____

    _____

2. Are all organisms included in your counts equally constrained by habitat island boundaries? How does this correspond to species on oceanic islands?

    _____

    _____

    _____

    _____

    _____

3. Did your graph show a significant relationship between species and area? Explain.

    _____

    _____

    _____

    _____

    _____

4. Look at the points that fell farthest above or below the line on the graph. On consulting your park maps or the position of the plywood islands, does distance from sources of colonizers play any role in creating these outliers?

    _____

    _____

    _____

    _____

5. If islands promote speciation, can it be argued that habitat fragmentation by humans is speeding up the process of species formation? Will this counterbalance accelerated extinction due to habitat loss? Explain.

    _____

    _____

    _____

    _____

FOR FURTHER INVESTIGATION

1. Calculate an extinction curve for the small island in the biogeography simulation game, and plot this curve on the same graph you prepared for the large island. How do the two curves compare? Can you explain the consequence of this simulation result in biological terms?
2. Measure the area of your own campus and compile a bird checklist, observing resident birds in late spring as well as winter months to include summer migrants. Where does your campus fall on the graph of log (forested area) vs. log (species number) from Method B?
3. Try testing the distance effect by placing plywood islands of the same size at different distances from a source of immigration, such as a patch of woods. Do more distant habitat islands reach equilibrium more slowly? Do they reach a lower equilibrium number of species?

FOR FURTHER READING

Brown, James H. 1971. Mammals on mountaintops: non-equilibrium insular biogeography. *American Naturalist* 105:467–478.

Diamond, Jared M. 2004. *Collapse: how societies choose to fail or succeed*. Viking, N.Y.

MacArthur R. H. and E. O. Wilson. 1967. *The theory of island biogeography*. Princeton University Press, Princeton, N.J.

Northern Prairie Wildlife Research Center. 2005. *Bird Checklists of the United States*. http://www.npwrc.usgs.gov/resource/othrdata/chekbird/bigtoc.htm

Powledge, F. 2003. Island biogeography's lasting impact. *BioScience* 53: 1032–1038.

Quammen, David. 1997. *The song of the dodo: island biogeography in an age of extinctions*. Simon & Schuster, N.Y.

Simberloff, D. S. and E. O. Wilson. 1970. Experimental zoogeography of islands: a two year record of colonization. *Ecology* 51: 934–937.

Simberloff, D. S. 1974. Equilibrium theory of island biogeography and ecology. *Annual Review of Ecology and Systematics* 5:161–182.

Simberloff, D. S. 1976. Experimental zoogeography of islands: effects of island size. *Ecology* 57(4):629–648.

Variance

Variance, symbolized by σ^2, is a measure of data scattering around the mean. Suppose you have two data sets made up of a series of measurements—let's say the wingspans of five ducks hatched from nest A and wingspans of five ducks hatched from nest B.

Each measurement is represented by an "x" on a number line:

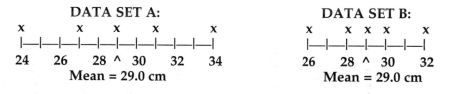

Notice that both have the same mean of 29 cm, but A has a broader "scatter" than B.

To measure the "scatter" we can measure each observation's distance from the mean.

These are called **deviations (d),** and we can mark one for each x on the number line.

DEVIATIONS FOR DATA SET A: **DEVIATIONS FOR DATA SET B:**

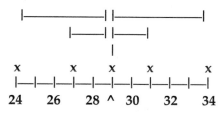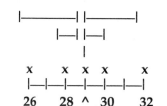

Now, let's square each deviation, and show the resulting areas on our number line:

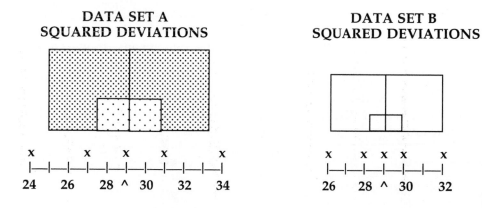

Finally, we can *calculate the average size* of our squared deviations by summing the areas of all the squares and dividing by the sample size*:

Variance A = $\dfrac{5^2+2^2+0^2+2^2+5^2}{5}$ = **11.6 cm²** σ^2 **A**

Variance B = $\dfrac{3^2+1^2+0^2+1^2+3^2}{5}$ = **4.0 cm²** σ^2 **B**

The symbol for variance is σ^2. This is appropriate, because the units on variance are always a square of the units originally measured. *On our number line, variance is an area.* It is a good way to measure data scattering because the farther the average observation is from the mean, the larger our average square becomes.

*If our data set were a sample drawn from a larger population, we would need to calculate sample variance (s^2). This is a similar calculation, but we would divide by (sample size –1), or 4 in this example, to correct for underestimation of variance in small samples. See Chapter 1 for a discussion of populations and samples.

Appendix 2

The t-Test

This is a **parametric** test for comparing means of two sets of **continuous data**. For example, we may wish to know whether lengths of razor clam shells (in cm) collected from two beaches are the same or different. Of course, two shell collections from the same population could have slightly different means, just because of random variation and limited sample size. The question we are asking is, "is the difference in means too large to explain by chance alone?" The **null hypothesis** in a t-test is that the actual means of the two populations are equal.

In addition to the sizes of the samples, there are two things we need to know to determine significance with a t-test. First, of course, is the difference between the two means. If we show the samples as histograms, with one collection of clam shells represented by solid bars and the other by crosshatched bars, we can agree from inspection that the two samples on the left are significantly different, but the two samples on the right are harder to distinguish. The shapes of the distributions are equivalent, but the pair of samples on the left has a greater distance between the means (barbell line), and will generate a more highly significant t-test result.

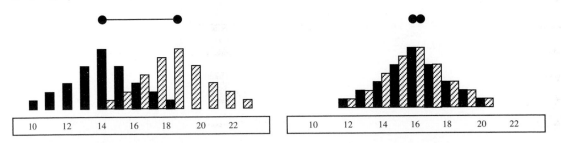

.Secondly, we need to consider the sample variances. (For a discussion of the variance, see Appendix 1.) The following sample pairs have equal differences between their means, but the two samples on the left are more clearly distinguishable because of their smaller variances. Distinctions between means must always be viewed in the context of variances, because more scatter around the population mean provides a greater possibility for large-scale sampling errors.

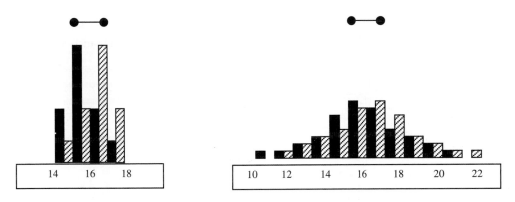

In summary, any difference between two means can be found significant if the sample size is large enough, and the variances small enough, to rule out sampling error as an alternative cause.

To perform a t-test of the null hypothesis that two means are equal, use the following formula. Note that increasing the difference between means or increasing the sample sizes will increase the value of t, but increasing the variance will decrease t.

$$t = \frac{\bar{X}_1 - \bar{X}_2}{\sqrt{(s_1^2/n_1) + (s_2^2/n_2)}}$$

You will also need to calculate degrees of freedom (d.f.), as follows:

$$\mathbf{d.f.} \;=\; n_{1} + n_{2} - 2$$

   t = a statistical measure of the significance of a difference between means
d.f. = degrees of freedom

$\bar{X}_1$ = mean of the first sample
s_1^2 = variance of the first sample
n_1 = sample size of the first sample

$\bar{X}_2$ = mean of the second sample
s_2^2 = variance of the second sample
n_2 = sample size of the second sample

Next you will need to consult a statistical table to determine the significance of this value of t as follows:
1. First look down the left-hand column to find the degrees of freedom that matches the one you calculated. Then reading across that row, compare your t value with the number in the second column. This number is the critical value of t for $p = 0.05$.
2. If your value of t is smaller than the $p = 0.05$ t value, then your two samples are not significantly different. You must *accept the null hypothesis* that the two samples come from the same population. There is no real difference between these means.

3.   If your value of t is larger than the p = 0.05 value, then the difference between your means is *significant*. You can reject the null hypothesis with no more than a 5% rate of error for the conclusion.
4.   Look at the right-hand column to compare your value of t with the p = 0.01 critical value. If your value of t is larger than this number, then the difference between your means is *highly significant*. You can reject the null hypothesis with a 1% rate of error for the conclusion.

Critical Values of the t Distribution

(From Rohlf, F. J. and R. R. Sokal. 1995. *Statistical Tables*, 3rd ed. W.H. Freeman, San Francisco)

d.f.	p = 0.05	p = 0.01
1	12.706	63.657
2	4.303	9.925
3	3.182	5.841
4	2.776	4.604
5	2.571	4.032
6	2.447	3.707
7	2.365	3.499
8	2.306	3.355
9	2.262	3.250
10	2.228	3.169
11	2.201	3.106
12	2.179	3.055
13	2.160	3.012
14	2.145	2.977
15	2.131	2.947
16	2.120	2.921
17	2.110	2.898
18	2.101	2.878
19	2.093	2.861
20	2.086	2.845
21	2.080	2.831
22	2.074	2.819
23	2.069	2.807
24	2.064	2.797
25	2.060	2.787
26	2.056	2.779
27	2.052	2.771
28	2.048	2.763
29	2.045	2.756
30	2.042	2.750
40	2.021	2.704
60	2.000	2.660
120	1.980	2.617
∞	1.960	2.576

Appendix **3**

Correlation and Regression

Both correlation and regression compare values of two continuous variables, such as length and width measurements on a sample of leaves. These are **parametric statistics**, applicable to **normally distributed** data sets. Because correlation and regression can both be calculated from the same data set, they often are, but the purposes and assumptions of the two analyses are not the same.

Correlation analysis measures the extent of association of two variables. They are typically illustrated on axes labeled x and y, but it does not matter which variable is x and which is y. No assumption about cause and effect is made in correlation analysis; you are simply looking for a pattern of co-variation. The **null hypothesis** for correlation analysis is that the two variables are independent. The null hypothesis is rejected if the two variables are positively correlated, or negatively correlated, as shown below.

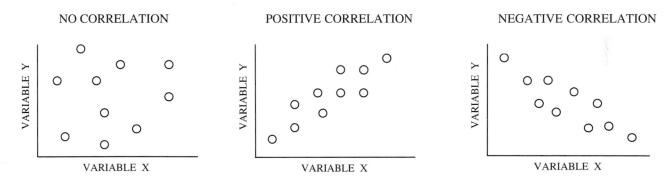

A parameter called the **coefficient of determination**, or r^2, measures how tightly associated the two variables are. If the variables are not independent, their patterns of variation may be linked to a greater or lesser extent. Values of r^2 range from 0 for no correlation to 1.0 for a perfect linear relationship.

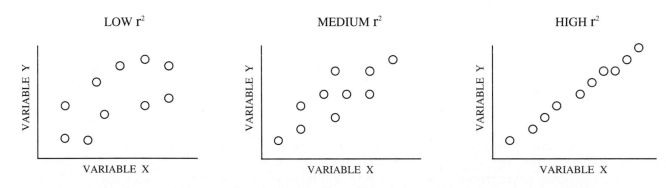

In summary, the three questions addressed by correlation analysis are: 1) *Are the two variables significantly correlated?* If $p > 0.05$, the null hypothesis of "no relationship" must be accepted. If $p < 0.05$, the null hypothesis of "no relationship" can be rejected. 2) *Are the variables positively or negatively correlated?* The sign of the **correlation coefficient (r)** can be positive or negative. Positive correlation means the two variables increase together. Negative correlation means one increases as the other decreases. 3) *How closely are the two variables correlated?* The closer the value of the coefficient of determination (r^2) to 1.0, the more the "cloud of points" on an x, y plot conforms to a straight line.

Regression analysis, unlike correlation, begins with the assumption of an independent variable (x values), and a dependent variable (y values). The x variable is typically under the investigator's control, or measured with greater accuracy as an experimental input. It is appropriate in regression analysis to think of cause and effect, with the independent variable influencing the dependent variable as described by the formula for a straight line, y = a + bx. (Simple analysis of this type is therefore called *linear* regression.)

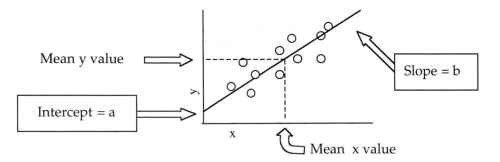

The first task in regression analysis is to determine the values of a and b that produce the best straight line through a set of points. The "best" line in regression analysis always passes through a pivot point representing the joint mean of x and y (indicated by the intersection of dotted lines above). Rotating around this joint mean, the best line minimizes the squared vertical distances from each data point to the regression line. This is where the name "least squares regression" comes from. A close-up view of the regression line, showing the squared distances from three points, is illustrated below. The farther a y value falls from the line, the larger the area of the illustrated square. The sum of all these squares is minimized in calculating the slope of the best line through the points:

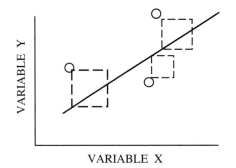

Output from regression programs typically reports a **regression coefficient (b)**, which is equivalent to the slope of the regression line. Regression analysis tests the significance of b vs. a slope of zero. A significant result ($p < 0.05$) means that the line has a nonzero slope. If so, we conclude that x has a significant influence on y.

In summary, the two questions addressed by regression analysis are: 1) *Is the regression coefficient significantly different from zero?* If $p > 0.05$, the null hypothesis "y is independent of x" must be accepted. If $p < 0.05$, we conclude the regression line describes a real relationship between the variables. 2) *What is the equation for the best straight line through these points?* Plug derived values for the slope (b) and the y intercept (a) into the equation y = a + bx to determine the best straight line through your data points.

Appendix  4

Critical Values
for the Mann-Whitney U-Test

(From Rohlf, F. J. and R. R. Sokal. 1995. *Statistical Tables*, 3rd ed. W.H. Freeman, San Francisco)

Tabled numbers are critical values for the Mann-Whitney U statistic in two-tailed tests (the null hypothesis is "no significant difference between the samples"). A procedure for calculating the Mann-Whitney U test is described in Method B, Chapter 15.

To read the table, find the size of the larger of the two samples (n_1) in the first column, then locate the size of the other sample (n_2) in the second column. To the right, in this row are critical values of U for $p = 0.05$ and $p = 0.01$. Reject the null hypothesis at the indicated level of significance if U is greater than the tabled value.

n_1	n_2	$p = 0.05$	$p = 0.01$
4	4	16	
5	3	15	
	4	19	
	5	23	25
6	3	17	
	4	22	24
	5	27	29
	6	31	34
7	3	20	
	4	25	28
	5	30	34
	6	36	39
	7	41	45
8	2	16	
	3	22	
	4	28	31
	5	34	38
	6	40	44
	7	46	50
	8	51	57
9	2	18	
	3	25	27
	4	32	35
	5	38	42
	6	44	49
	7	51	56
	8	57	63
	9	64	70

n_1	n_2	$p = 0.05$	$p = 0.01$
10	2	20	
	3	27	30
	4	35	38
	5	42	46
	6	49	54
	7	56	61
	8	63	69
	9	70	77
	10	77	84
11	2	22	
	3	30	33
	4	38	42
	5	46	50
	6	53	59
	7	61	67
	8	69	75
	9	76	83
	10	84	92
	11	91	100
12	2	23	
	3	32	35
	4	41	45
	5	49	54
	6	58	63
	7	66	72
	8	74	81
	9	82	90
	10	91	99
	11	99	108
	12	107	117

Appendix 5

Critical Values for the Chi-Square Test

(From Rohlf, F. J. and R. R. Sokal. 1995. *Statistical Tables*, 3rd ed. W. H. Freeman, San Francisco)

For a Chi-Square Contingency test procedure, see the Calculation Pages for Chapter 8.

d.f.	p = 0.05	p = 0.01	d.f.	p = 0.05	p = 0.01
1	3.841	6.635	26	38.885	45.642
2	5.991	9.210	27	40.113	46.963
3	7.815	11.345	28	41.337	48.278
4	9.448	13.277	29	42.557	49.588
5	11.070	15.086	30	43.773	50.892
6	12.592	16.812	31	44.985	52.192
7	14.067	18.475	32	46.194	53.486
8	15.507	20.090	33	47.400	54.776
9	16.919	21.666	34	48.602	56.061
10	18.307	23.209	35	49.802	57.342
11	19.675	24.725	36	50.998	58.619
12	21.026	26.217	37	52.192	59.893
13	22.362	27.688	38	53.384	61.162
14	23.685	29.141	39	54.572	62.428
15	24.996	30.578	40	55.758	63.691
16	26.296	32.000	41	56.942	64.950
17	27.587	33.409	42	58.124	66.206
18	28.869	34.805	43	59.304	67.460
19	30.144	36.191	44	60.481	68.710
20	31.410	37.566	45	61.656	69.957
21	32.670	38.932	46	62.830	71.202
22	33.924	40.289	47	64.001	72.443
23	35.172	41.638	48	65.171	73.683
24	36.415	42.980	49	66.339	74.920
25	37.652	44.314	50	67.505	76.154

Appendix 6

A Table of Random Numbers

Random numbers (generated by the random number function in Excel©) are presented in 5-digit blocks. If you need fewer than 5 digits, truncate these values as needed. For example, if you need a number between 0 and 99, use the first two digits in each group. If you need a number between 0 and 500, use the first three digits in each display, and simply skip any numbers that fall outside your range. Random numbers can be used in any order, but you should not use the same numbers twice within a series of experiments.

97002	32380	24930	86089	44434	96895	53514	19792	88693	91689
30267	13952	38773	01469	91771	65642	14677	72862	14186	35589
98480	47508	56055	40090	16671	09012	69899	19502	76540	80683
13218	26308	30506	33521	91601	37020	89333	39881	09377	99279
40235	16429	94569	19006	90723	03270	99074	19210	48827	29603
06536	97522	08101	51851	43209	08886	60500	40863	48283	80427
71719	64667	15564	08256	73911	36511	88072	86674	84730	21899
40325	95826	44708	58775	53729	80013	59119	71250	31287	58072
16472	80308	06838	83767	91579	05224	47172	40347	98847	58367
04086	32811	73049	83575	90374	84195	80603	10915	78346	81403
67870	67684	37395	04598	36778	63588	68997	49298	54070	34954
14083	23917	98870	36928	38485	21720	24061	73999	57343	06571
40427	69913	21886	63482	71720	10911	57493	60998	72563	26633
91978	91064	55064	13112	48835	49337	36791	56254	44250	32241
50900	23152	56039	02957	26384	59053	77316	79999	47556	94596
76269	76801	04321	41444	80753	51440	10363	70150	28064	50324
72270	66602	40049	15740	09444	36610	80857	87382	96923	72471
54837	64474	71355	31723	91298	26907	20588	51289	15525	01621
04604	14715	89240	91549	92728	62811	84641	92234	97012	92499
54130	84493	39289	22800	15610	14480	46570	32224	99367	33303
04328	64461	75309	55687	72553	99854	05488	72345	37795	04328
51239	28455	22931	29154	81782	68955	67274	24254	90479	51239
26655	26081	69995	14378	34226	07985	07297	94351	20664	26655
58653	23834	10784	65724	21324	56468	16310	42419	05135	58653
80445	00947	66211	14568	26273	80680	48318	48581	70108	80445
48427	42777	15825	05455	71838	54243	62385	02401	19267	48427
78436	57621	54183	14166	85878	69309	63769	10101	10206	78436
59728	45210	34811	32283	82517	99775	26559	64688	43278	59728
62459	13091	05657	38666	85939	27416	19443	09364	19560	62459
14153	28680	15726	55865	26954	75620	17609	29812	45233	14153
54393	31300	91770	06356	88442	92840	99362	68548	13784	54393
81816	57962	96720	76343	04712	40337	42862	12622	24582	81816
81805	11828	45913	40816	22954	57785	78235	49655	98578	81805
32888	50980	09160	73373	37208	33996	29363	42699	39530	32888
38847	41182	83533	00482	56442	59181	94087	78238	56521	38847
03673	96054	83389	56969	93257	34529	99853	10195	20761	03673
39785	17748	79242	40437	75129	83509	81212	63312	37003	39785
04639	85078	44833	61974	13067	56194	30034	43507	31404	04639
56767	82807	78048	63186	71236	00136	44304	92648	09587	56767
05168	61895	89006	21129	55074	95200	28531	22987	67580	05168

Appendix **7**

Writing a Laboratory Report

Laboratory reports are written to document your work in lab or field exercises, and to give you practice in the skill of scientific writing. However, their most important function is to clarify your thinking about what you have observed. A well-written report resembles a scientific journal article, both in its format, and in the way it leads us through the scientific process to make sense of a collection of data. Reports contain the following parts, generally set off with subheadings, as shown.

Title

The title should contain a succinct description of the subject of your investigation. Avoid broad general titles, such as "Competition." Instead, provide as much detail about your work as you can express in a few words. For example, a competition study might be entitled, "Competition between corn (*Zea mays*) and oat (*Avena sativa*) seedlings grown in a replacement series under low-light conditions." Your instructor may require the name of the class, section number, date, and other information on the title page, but these do not replace the need for a descriptive title.

Introduction

The introduction demonstrates your understanding of the context of the experiment. Since every study builds on previous work of other scientists, the reader needs to know what these investigations have shown, and how your work follows logically from theirs. Experimental methodology, background information on the organisms in your study, and theory which guided your experimental design are all typical inclusions in an introduction. Most important, your introduction should lead to your own research question, which is typically stated in the final paragraph of this section. Avoid rambling discussions of unrelated material; every statement in the introduction should lead the reader toward an understanding of the question you asked and the reason you asked it.

Cite every study you incorporate, using the authors' names and dates of publication embedded in parentheses just after inclusion of the information, as (Jones, 2005). Citations must show authorship of any facts you did not personally observe, whether directly quoted or not. If the paper has two authors, include both their names, as (Jones and Doe, 2005). If the paper has three or more authors, use the Latin abbreviation meaning "and others," as (Jones *et al.*). Note that *al.* is followed by a period (because it is short for the Latin word *alia*) and that both words are italicized. Finally, make sure bibliographic information on each source you cite is included in the Literature Cited section at the end of your report.

The length of the introduction may be as short as a paragraph, summarizing in your own words the introductory information from the laboratory manual. More formal laboratory reports may involve library research and an introduction several pages in length. Make sure you understand the amount of background reading that your instructor expects you to do before you begin writing.

Procedure

Alternatively known as "Materials and Methods," this section tells the reader exactly what experimental steps you followed, and how your observations were made. Provide sufficient detail for a reader to replicate your work solely from your written description. Specify quantities and types of materials used, chemical formulations, instrument settings, sampling routines, numbers of replicates, and any other details of your procedure which may have influenced the outcome of your experiment. If you tested more than one hypothesis, organize the procedure around questions you addressed rather than the chronological order of your laboratory work. If parts of your procedure are described in the laboratory manual or other source, you need not repeat the description. Simply cite the method, and then add any details in which your procedure differed from the cited protocol. Any problems with instruments, unexpected field conditions, or missing data should be reported here as well. Procedures are traditionally written in passive voice, as in "plants were measured," rather than active voice, as in "I measured the plants."

Results

The results section contains a straightforward display of your measurements or observations. If you noted anything relevant to your topic that you did not expect to see, those observations should be included as well. Unlike journal articles, laboratory reports usually include all the raw data as well as processed data in the form of graphs or statistics. You may be asked to tear out calculation pages or graphs from the manual and include them here. Since this is a learning exercise, it is wise to include your reasoning for the methods of analysis you chose and all calculations involved. If you have made any mistakes, this allows your instructor to help you find them. Avoid discussing the meaning of your results here; those comments belong in the next section.

Discussion

This is your opportunity to make sense of the investigation. Make sure to refer back to the hypotheses your experiment was designed to test, connecting results with arguments supporting or disproving each statement. Be sure to interpret each graph you included and each statistical result you obtained. It is also customary in a discussion section to explain what you still do not know, what questions arose from this project, and briefly, what kinds of experiments might shed light on these questions in future research.

Literature Cited

Every source cited in your paper must be displayed in alphabetical order in a Literature Cited section. Avoid the term "Bibliography," which refers to a comprehensive list of all publications related to your topic. Each citation should include the author's name, the date, the title, and the name of the journal or publisher. Do not use "*et al.*" in the Literature Cited as you did in the citation. Every author's name should be included here. Volume and page data (if applicable) are included at the end of the listing. For examples of literature cited listings, refer to the "For Further Reading" sections at the end of each chapter in this manual. Your instructor may prefer a different style; it is wise to ask in advance.

Glossary

abiotic   the nonliving parts of an ecosystem, including air, water, and soil.

ABO blood groups   human phenotypes, based on alleles coding for the A and B antigens located on the surfaces of red blood cells and capable of triggering an immunological reaction.

acute toxicity   poisonous effects so severe as to cause death a short time after exposure.

adaptation   1) the observed fit of an organism to its environment. Example: the flipper-like wings of penguins represent an adaptation to their aquatic existence. 2) a change in a population in response to environmental challenges. Example: when humans began spraying DDT to control mosquitoes, adaptation to the pesticide resulted in DDT-resistant populations of mosquitoes.

adaptive radiation   rapid development of many species from a few ancestral forms because of a broad range of open niches.

alleles   two versions of the same gene. Example: the P allele codes for purple pea flowers and the p allele codes for white flowers.

allelopathy   depression of the growth or survival of one plant caused by chemicals released by another plant.

allometry   examination of the relationship between the size of a body part vs. the size of the organism. Proportions can change during development of the individual, but evolutionary changes in proportions are also evident from comparisons among species.

allopatric speciation   concept of species formation proposing that populations geographically isolated from others of their kind accumulate unique genetic characteristics eventually leading to reproductive isolation. When reintroduced to the ancestral area, the isolated population no longer interbreeds, and is thus a new species.

ammonification   release of ammonium or ammonia from organic nitrogen sources such as proteins, usually as a result of decomposition by bacteria or fungi.

angle of inclination   the angle formed between the horizon and an object sighted above the horizon.

antigen   a large molecule, such as a protein, that elicits an immune response.

ascospore   a microscopic reproductive cell of any of the sack fungi, generated by sexual recombination. Ascospores are small enough to disperse on the wind, and are highly tolerant of harsh environmental conditions.

assortative mating   nonrandom selection of mates, exhibiting either an excess or a deficiency of reproductive pairings between similar genotypes, in comparison to the number expected by chance.

autotrophic succession   community development driven by producers contributing biomass, which accumulates as the system matures.

benthic   refers to the bottom of a lake, pond, or sea.

bias   a problem with sampling methodology that creates significant differences between the characteristics of sampled individuals and the statistical population they are supposed to represent. Example: netting birds at a feeding station for weight measurements may overrepresent well-fed members of a species, leading to a biased estimate of body size.

biodiversity   the amount of variation of life forms at a specified site. Biodiversity can refer to variation in genetic types, in species, or in ecological communities.

bioindicator   a pollution-sensitive species used to test environmental quality. Example: mayfly larvae, when present in streams, indicate low levels of pollution.

biological concentration   an accumulation of certain water-insoluble toxins in the food chain. Example: DDT concentrations tend to increase with each step in the food chain, with top predators being most severely affected.

biomass   the accumulation of all living matter in an ecosystem or locality, usually measured as dry weight.

biotic   the living parts of an ecosystem, including all organisms found in a given place.

bottom-up control   community interactions in which species at the bottom of the food chain, such as plants, influence the kinds and numbers of species found at higher trophic levels.

carnivores   animals that eat other animals. Also called secondary consumers.

carrying capacity   the maximum population size that can be permanently sustained by natural resources in a designated area of land.

Chi-square test   a statistical method for comparisons of frequency data.

climax community   the final stage in ecological succession, composed of species not easily replaced by others, and characteristic of a given region. Example: beech-maple forest is a climax community encountered in long-undisturbed sites in the northeastern United States.

clonal reproduction   asexual reproduction resulting in genetically identical offspring, common to many perennial plants.

cohort   a group of individuals, all born at or near the same time, followed through their lives in a demography study.

color blindness   a sex-linked trait coded by genes on the X chromosome in humans.

commensal   an organism living on or in another species, but not harming or taking significant resources from the host.

competitive exclusion   a struggle for access to resources resulting in extinction of one species because of the presence of another.

confidence interval   a statement of likelihood that an estimated value falls within a given range of error. Example: a 95% confidence interval around a sample mean implies that the population mean will fall within the stated confidence interval in 95% of similar cases.

congeners   members of the same genus. Example: *Quercus alba* (white oak) and *Quercus stellata* (post oak) are congeners.

critical value   in statistics, the magnitude of a test outcome needed to reject the null hypothesis.

Darwin, Charles   English biologist who published *The Origin of Species* in 1859, giving explanatory power to biological science.

decomposer   organisms such as bacteria and fungi that break down organic material, use the released energy for their own metabolic needs, and return nutrients to the soil or water.

degradative succession   community development on a source of nutrition, such as a dead log or a carcass, characterized by decomposer organisms and other species that feed on them.

degrees of freedom   in statistics, a calculation derived from sample sizes that indicates how many data values could vary independently of one another.

demography   the study of population characteristics such as age distribution, sex ratio, death rates, and birth rates.

density-dependent   describing factors whose influence depends on the numbers of individuals per unit area. Example: mortality is density-dependent if the per capita death rate depends on population size.

detritivore   an animal that feeds on decaying organic matter. Example: an earthworm is a detritivore when it consumes leaf mold in the soil.

dilution series   a method for creating solutions varying over a broad range of concentrations by transferring a fraction of each solution into pure solvent to prepare the next.

diploid   having two sets of chromosomes; one set of genetic instructions inherited from each of two parents.

dispersal   movement of individuals from one place to another.

dispersion   the amount of scattering of a population in space, in comparison to a random pattern. Populations may be uniformly dispersed, randomly dispersed, or show aggregated patterns of dispersion.

dominant allele   the allele having primary influence on the phenotype in heterozygotes. Example: Pp genotype produces purple-flowered peas. Dominance does not imply superiority or likelihood of increase over time, but simply control of expression in heterozygotes.

Drosophila melanogaster   a fruit fly species commonly used in genetics and ecology laboratory studies.

duff   partially decomposed organic material overlying mineral layers of the soil.

ecological density   numbers of individuals per ecological resource unit. Example: the number of weevils per acorn.

ecological efficiency   energy accumulated by organisms in a trophic level, in comparison to the energy available to them. Calculated by dividing productivity at trophic level t by productivity at trophic level t−1.

ectotherm   an animal that regulates its body temperature externally, by sunning or seeking out cooler places in its environment. Example: reptiles and amphibians are ectotherms.

eluviation   the removal of dissolved minerals from upper layers of soil as water moves downward.

emigration   movement of organisms out of a community.

endosymbiotic theory   the concept that small prokaryotic organisms living inside larger cells developed mutualistic associations, and ultimately became organelles we now recognize as parts of the eukaryotic cell.

endotherm   an animal that regulates its body temperature internally, through physiological adaptation. Example: birds and mammals are endotherms.

entropy   a measure of randomness of matter and/or energy in a system. Example: when a mammal consumes chemical energy in food and dissipates energy as heat, it increases the entropy within its ecosystem.

ephemeral   of short duration. Example: Spring ephemeral wildflowers grow and bloom for only a few weeks in early spring.

epilimnion   the upper surface of a stratified body of water, characterized by higher water temperature and high oxygen levels.

eutrophic   describing a body of water with high nutrient levels, commonly exhibiting algae blooms.

evolution   a change in the genetically controlled characteristics of living organisms over time, ultimately leading to speciation when populations accumulate genetic differences sufficient to prevent interbreeding.

exploitative competition   a negative community interaction between two organisms with overlapping resource requirements, resulting from removal of required nutrients rather than from active interference.

exponential population growth   population expansion characterized by an accelerating growth rate, typical of populations experiencing optimal conditions for survival and reproduction.

facilitation (in succession)   a theory that succession proceeds as one seral stage creates conditions allowing the next community type to immigrate and prevail.

facultative   in community interactions, describing a relationship that benefits a species, but which the species does not require in order to survive.

fall turnover   loss of stratification in a body of water in the Northern Hemisphere, due to cooling and sinking of surface water, which results in mixing of upper and lower water layers.

family size   the number of offspring born to the average parent over its lifetime. Also defined as the mean number of female offspring born to each female.

fecundity   the number of offspring born to an individual per unit time, often expressed as a function of maternal age.

field capacity   the amount of water held by the soil against the pull of gravity, expressed as a proportion in relation to the weight of dry soil.

first law of thermodynamics   a physical principle stating that matter and energy are conserved. Example: energy entering an ecosystem, taking stored energy into account, equals energy radiating back out again. Also called the conservation rule.

fitness   an individual's capacity for survival and reproductive success, always measured in comparison to the population as a whole.

fixed allele   a gene frequency that rises to 100%, as other alleles are deleted from the population by drift or natural selection.

foraging strategy   a pattern of behavior in animals seeking food, resulting in a higher rate of return than would result from random searching, and subject to improvement by natural selection or learning.

frequency   in statistics, the number of times an observation occurs within a data set.

functional constraints   limits on evolutionary development due to the basic design of an organism. Example: spiders could not easily evolve chewing mouthparts because their esophagus passes through a circle of neurons in their head.

fundamental niche   the range of environmental conditions supporting a species in the absence of competition.

gene flow   movement of alleles from one area or population to another through dispersal of gametes, spores, fertilized eggs, juveniles, or adults.

gene frequencies   proportional representation of the alleles at a locus in a population. Example: gene frequencies at a locus might be 65% B alleles and 35% b alleles.

genet   a genetically distinct individual. Example: all parts of a patch of violets that were derived from a single seed.

genetic drift   random changes in gene frequencies commonly observed in small populations.

genome   the sum of all genetic information in the DNA of one organism.

genotype   the alleles in an organism's genetic makeup relevant to a trait of interest. Example: PP, Pp, or pp are genotypes for flower color in peas.

genotypic frequencies   proportional representation of all combinations of alleles present in a population. Example: genotypic frequencies for a two-allele system would specify the percentage of BB, Bb, and bb individuals.

gross primary productivity   the rate of energy accumulation by producers in an ecosystem, equal to gross primary productivity minus the energy used (converted to heat) by the photosynthetic organisms themselves.

handling time   the time it takes a predator to process a prey item after capture. Example: the time required by a snake to swallow and digest a bird before it is ready to hunt again.

Hardy-Weinberg rule   a prediction of genotypic frequencies based on gene frequencies in a population with random mating, no selection, no mutation, no immigration or emigration, and no genetic drift.

herbivores   animals that eat plants. Also called primary consumers.

heterosis   superior fitness of the heterozygous genotype. Example: a hybrid corn plant heterozygous for many traits exhibits faster growth, superior disease resistance, and higher seed yield than either of its homozygous parents.

heterozygous   having two dissimilar alleles. Example: Pp genotype in peas.

histogram   a visual display of a frequency distribution. The range of the variable of interest in a data set is divided into a series of equal size classes along the x-axis, and the frequency of observations falling within each class is represented as a vertical bar.

homologous pair   two chromosomes in the genome containing similar genetic instructions, one inherited from each parent.

homozygous   having two similar alleles. Example: PP or pp genotypes for flower color in peas.

hydrometer   an instrument used to measure the density of a solution.

hypolimnion   the lower surface of a stratified body of water, characterized by lower water temperature, and often low oxygen levels.

illuviation   the deposition of minerals in lower layers of soil, particularly clay, as water brings dissolved ions down from the surface.

immigration   movement of organisms into a community.

inclinometer   an instrument for measuring angle of inclination above the horizontal.

independent events two occurrences that have no influence on one another. Example: rolling two dice constitutes two independent events because the face showing on one die is not influenced by the face showing on the other.

index of dispersion a measure of the extent of dispersion of a population in its environment. Example: the variance to mean ratio for individuals counted in each of a series of quadrats measures dispersion.

inhibition (in succession) a theory that some species prevent the establishment of others, so that species found in later stages of succession can thrive only after the earlier inhibitory species are gone.

interference competition a negative community interaction between two organisms in which one competitor actively harms the other. Example: fungi release antibiotics, which reduces competition from bacteria.

interspecific competition a struggle for access to resources involving two or more species.

intraspecific competition overlapping demand for resources by members of the same species.

intrinsic rate of increase (r) the constant that describes growth rate of an exponentially expanding population.

island biogeography a study of community dynamics on islands, or isolated habitat patches.

isocline See zero-growth isocline.

joint abundance a type of graph used to display the Lotka-Volterra models, showing population number of one species on the x-axis, and population numbers of a second species on the y-axis. A point on the graph shows numbers of both populations at the same time.

keystone species a species much more important in the maintenance of its biological community than would be expected based on its abundance.

K-selected species species having life histories adapted for relatively slow population growth in late stages of succession, characterized by small family size, low juvenile mortality, high parental investment, long life span, high tolerance for competition, and relatively poor mechanisms for dispersal.

Latin square an experimental design that spreads replicates among rows and columns to avoid clustering samples with similar treatments.

LD 50 a method for measuring toxicity of a polluting agent by exposing test organisms such as *Daphnia* to a series of chemical concentrations. The concentration of the pollutant causing 50% death rate is the LD 50 level.

legume a family of plants including peas and beans, which tends to form symbiotic relationships with nitrogen-fixing bacteria.

life table a tabulation of survivorship and fecundity values for each age class in the life span of a population. Life tables can be used to calculate generation length and population growth rate.

Lincoln-Peterson Index a simple mark-recapture method for estimating population size.

locus the location of a gene on a particular chromosome.

logarithmic plot a graph with at least one axis scaled as exponential values. Example: exponential population growth can be plotted as a straight line on a graph of log (population size) vs. time.

logistic population growth population expansion characterized by slowed growth as the population approaches carrying capacity.

Lotka-Volterra competition model a simple mathematical description of population dynamics of interspecific competitors, based on population sizes, carrying capacities, and impact of each species on the population growth of the other.

Lotka-Volterra predation model a simple mathematical description of predator-prey population dynamics, based on population sizes, reproductive and death rates of predators, and capture rates leading to deaths of prey.

Malthus, Thomas an economist whose essay on the principles of population emphasized the problem of populations outgrowing environmental limits.

mark-recapture methods estimates of population size based on marking a sample of individuals, letting them mix with the larger population, and counting recaptures in subsequent samples.

mean   the average value of a data set, calculated by dividing the sum of all data by the number of observations.

median   the central value in a data set, calculated by arranging all observations in numerical order and selecting the middle value. In an even-numbered set of observations, the median is the average between the two middle values.

mode   the most commonly observed value in a set of data.

mutualism   a type of interspecific relationship resulting in benefits outweighing costs for both species involved. Example: pollinators and flowers mutually benefit from their interaction.

mutually exclusive events   two outcomes that never occur at the same time. Example: a child can have blood type AB or type O, but not both phenotypes at the same time.

n   symbol representing the size of a sample. Compare with N, the population size.

N   symbol representing the size of the entire population of interest.

natural selection   evolutionary process fostering increase in frequency of genes that provide advantage in survival or reproductive success.

net ecosystem productivity   energy in the form of biomass that accumulates in the ecosystem.

net primary productivity   the rate of energy accumulation in organic matter in an ecosystem, equal to gross primary productivity minus the energy used (converted to heat) by all organisms in all trophic levels.

neutral model   a null hypothesis explaining apparent patterns in terms of random processes, against which other theories of causality can be tested

niche   an opportunity for a species to consume resources and maintain its presence in a community. Niches are defined by characteristics of the environment, characteristics of the organism, and the nature of competition with other species. Example: tapeworms exploit a parasitic niche in the intestines of mammals.

niche overlap   a competitive situation in which two species use at least some of the same resources in similar ways.

niche partitioning   separation of ecological roles in response to competitive pressure. Example: in the presence of closely related species, warblers restrict their foraging to parts of a tree on which their species is specialized.

nitrogen fixation   chemical reactions, usually mediated by enzymes and requiring energy input, converting nitrogen gas to organic forms of nitrogen. This is an important step in the nitrogen cycle.

non-parametric statistics   statistical analyses that do not assume that data are normally distributed.

non-point sources   describing origins of pollution that cannot be ascribed to a single place. Example: nitrogen fertilizer running off lawns and farm fields is a non-point source of water pollution.

normal distribution   a data set producing a bell-shaped histogram with a single peak, with mean, median, and mode at the same point, and 95% of the observations falling within 1.96 standard deviation units of the mean.

null hypothesis   a statement subject to test in an experiment or statistical analysis, generally attributing patterns in the data to random sampling error. Rejection of the null hypothesis provides support for a more interesting alternative explanation. Example: the null hypothesis of a t-test is that means of two samples are not significantly different.

obligate   in community interactions, describing a relationship that the species must have in order to survive. Example: corals have obligate relationships with algal symbionts.

parametric statistics   analyses such as t-tests and linear regression, which should be performed only on normally distributed data sets.

parasite   an organism living in close association with another species, and deriving nutrition from its host.

phenotype   the actual expression of a trait. Example: pea flower color can be white or purple.

phytoplankton   small photosynthetic floating organisms in a body of water. Example: diatoms are important members of the phytoplankton community in the oceans.

plankton a floating community of microscopic life in freshwater or saltwater environments, including both photosynthetic and non-photosynthetic organisms.

point sources describing origins of pollution that can be ascribed to a single place. Example: a smokestack is a point source of air pollution.

polymorphic describing a population with more than one commonly observed allele at a specified locus. Example: if the rarest allele for flower color comprises 12% of all alleles at a flower color locus, then we would say this plant population is polymorphic with respect to flower color.

population 1) in biology, a group of organisms of the same species living in the same locality. 2) in statistics, the set of all items from which a sample is drawn.

population density numbers of individuals per unit area. Example: the number of white-tailed deer per hectare.

primary consumers the second trophic level (after producers), deriving energy from consuming plants. Example: a cow is a primary consumer when it eats grass.

primary productivity the rate of carbon fixation by plants in an ecosystem.

primary succession community development beginning on bare rock, typically taking centuries to reach a climax condition because soil forms very slowly.

producers organisms such as plants that harvest light energy.

productivity energy captured at a particular trophic level. Example: primary productivity is the amount of solar energy captured by plants and stored as chemical energy.

quadrat a standard frame or staked-off area of fixed size, established as a representative sample of a larger area.

ramet a morphological unit of a population, which may be asexually reproduced. Example: a crown of grass leaves.

random event an occurrence whose outcome cannot be predicted based on available information.

range the difference between the highest and lowest value in a data set.

realized niche the range of environmental conditions supporting a species in the presence of competition. The realized niche is typically smaller than the fundamental niche because competitors make life at the margins of the niche more difficult.

recessive allele the allele masked by a dominant gene in heterozygotes. Example: if p is a recessive allele for white flowers, Pp genotypes produce purple flowers. Recessive genes are not necessarily inferior, nor do they necessarily decline in frequency over time.

replacement series an experimental design mixing two species in a number of treatments, with the proportion of the first species varying in a regular progression from 0% to 100%.

Rh factor a genetically coded antigen on red blood cells. The Rh gene is phenotypically expressed as Rh positive blood.

riparian zone the biological community found along the banks of a river or stream.

r-selected species species having life histories adapted for rapid population growth in open habitats, characterized by large family size, high juvenile mortality, low parental investment, short life span, low tolerance for competition, and excellent mechanisms for dispersal.

s the symbol for the standard deviation of a sample.

s^2 the symbol for the variance of a sample.

S.E. a symbol for the standard error of the mean. The standard error is used to calculate confidence limits for a mean.

sample a group of observations or measurements selected to represent a larger statistical population.

sample size the number of observations in a data set.

saturation capacity the maximum amount of water that the soil can hold, expressed as a proportion in relation to the weight of dry soil.

search image   a hypothesis that predators develop a tendency to hunt a particular kind of prey, generally a common type. As a result, this type of prey is captured at a higher rate than would be predicted based on its abundance.

second law of thermodynamics   a principle of energy transformation stating that useful energy declines over time.

secondary consumers   the third trophic level (after herbivores), deriving energy from consuming other consumers. Example: a fox that eats a rabbit is a secondary consumer.

secondary succession   community development beginning on disturbed soil, typically reaching climax conditions within decades, as its rate is not limited by soil formation.

sere   a stage in ecological succession. Example: weedy annual plants make up a sere in old-field succession.

sex chromosomes   X and Y chromosomes which determine sex in humans and fruit flies.

sex-linked trait   a genetic characteristic coded by a gene on a sex chromosome (the X chromosome in humans and fruit flies).

Shannon Diversity Index   a measure of biodiversity which takes both numbers of species and the evenness of their distribution into account. Also called the Shannon-Weaver or Shannon Weiner Index.

sigma (σ)   the symbol for the standard deviation of a statistical population, estimated by s.

significant   in statistics, a departure from results that could reasonably be explained by chance.

size class   an arbitrarily defined grouping of data along an axis. Example: measurements of tree diameters could be grouped in 10-cm size classes as 1–10 cm, 11–20 cm, 21–30 cm, etc.

soil horizon   a layer of soil having characteristic physical and chemical properties. Example: "topsoil" is a common name for the A horizon, and "subsoil" is a common name for the B horizon.

soil profile   a description of the layers found in a vertical sample of soil.

spatial pattern   the way individuals in a population are physically distributed in their environment.

species accumulation curve   a mathematical description of the declining rate of discovery of new species as the same habitat is repeatedly sampled over an extended period of time.

species–area relationship   an empirical observation that species on islands increase with area of the island.

stable age distribution   the predictable fractions of a population found in each age class when survivorship and fecundity values remain constant over several generations.

stable equilibrium   a robust state of balance that tends to restore itself after disturbance.

standard deviation   a measure of scattering in a data set equal to the square root of the variance. In normally distributed data, 95% of all variation falls within 1.96 standard deviations of the mean.

standard error   a measure of the possible discrepancy between sample means and the population mean. In 95% of all samples, the population mean falls within 1.96 standard error units of the sample mean.

standing crop   the amount of biomass currently found in an ecosystem, irrespective of the length of time it took for the ecosystem to accumulate this material.

statistic   a numerical representation of some aspect of a data set such as its mean or its range.

stochastic process   a complex system incorporating both random and nonrandom elements

stratification   formation of temperature layers in a body of water, with warmer water floating on top of cooler water.

stratified sample   a plan for making observations deliberately targeting several subsets of a complex population. Example: soil samples from a hilly site could be stratified so that hilltop, slopes, and valleys are sampled in proportion to the areas they cover on the map.

succession   the replacement of one ecological community by another over time.

super-organism   Frederic Clements' idea that a community functions as a single living thing, with species seen as analogous to body parts and succession analogous to an organism's development. This concept has been rejected by contemporary ecologists, since species function more independently than organs do.

surface/volume   the ratio of an object's surface area (measured in square units) divided by its volume (measured in cubic units). Surface/volume decreases as an object grows in size.

survivorship   the proportion of individuals remaining alive from one age class to the next.

symbionts   organisms commonly found in close association with another species. Example: *Rhizobium* bacteria are symbionts that reside in root nodules of legumes.

systems ecology   the study of ecology at the ecosystem level of organization.

taxon   a group of organisms sharing similar classification at any level of the systematic hierarchy from subspecies to kingdom. Example: beetles are a taxon (called coleoptera) of insects.

thermocline   a boundary between the epilimnion and the hypolimnion in a stratified body of water, characterized by an abrupt temperature difference between water above and below this layer.

tolerance (in succession)   a theory that succession proceeds as species more tolerant of competition replace species less tolerant.

top-down control   community interactions in which species at the top of the food chain, such as predators, influence the kinds and numbers of species found at lower trophic levels.

transect   an ecological sample taken along one straight line through a study site.

transforming data   performing the same mathematical operation on every item in a data set. Example: squaring every number in a series of measurements to facilitate analysis.

transpiration   water moving from the earth's surface into the atmosphere through the stems and leaves of plants.

trophic level   a step in the food chain. Example: producers make up the first trophic level, herbivores make up the second, and carnivores the third.

turbidity   a measure of the amount of light absorbed by water and all materials dissolved or suspended in it.

type I functional response   a linear change in the feeding rate of predators as the density of their prey increases.

type I survivorship   a life history pattern characterized by low juvenile mortality, with most deaths occurring near the maximum age. Example: human beings exhibit a type I survivorship pattern.

type II functional response   a curvilinear change in the feeding rate of predators as the density of their prey increases, characteristic of invertebrate predators. As prey density approaches an upper limit, predation levels off because predators are unable to process prey as quickly as they are encountered.

type II survivorship   a life history pattern characterized by constant risk of mortality throughout the life span. Example: robins exhibit a type II survivorship pattern.

type III functional response   a curvilinear change in the feeding rate of predators as density of their prey increases, characteristic of vertebrate predators. Predation is limited by predators' lack of experience at low prey density, and by satiation at high prey density.

type III survivorship   a life history pattern characterized by high juvenile mortality, with risk of mortality decreasing as organisms age. Example: most trees exhibit a type III survivorship pattern.

unstable equilibrium   a delicate balance, easily disrupted, and not tending to restore itself after disturbance.

variance   a measure of the amount of scattering around the mean of a data set. To calculate the variance, measure the distance from each observation to the mean, square those distances, and then average all the squared values. The variance is the square of the standard deviation.

water potential   a force, measured in pressure units, required to move water from one point to another against its tendency to remain in place.

wilting point   a soil moisture level below which plants are unable to extract water from the soil.

X chromosome   a sex chromosome found in both males and females. In humans and fruit flies, XX individuals are female and XY individuals are male.

Y chromosome   a sex chromosome normally found only in males. In humans and fruit flies, XX individuals are female and XY individuals are male.

zero-growth isocline   in ecological models, a line connecting all points on a joint abundance graph in which the indicated species neither increases nor decreases in number.

zooplankton   very small non-photosynthetic organisms floating in freshwater or saltwater habitats.

Index

Photo Credits

1.CO Rob & Ann Simpson/Visuals Unlimited; **2.CO** Herman Eisenbeiss/Photo Researchers, Inc.; **2.2b** Dr. Kari Lounatmaa/Photo Researchers, Inc.; **2.6a** Tony Heald/naturepl.com; **2.6b** Michael Nichols/National Geographic Image Collection; **2.6c** Miles Barton/naturepl.com; **2.6d** blickwinkel/Alamy; **2.7** Rob Kingsolver; **2.8** Harry Taylor/Dorling Kindersley Media Library; **2.9** Rob Kingsolver; **3.CO** Ted Nichols/New Jersey Division of Fish and Wildlife; **4.CO** Anup Shah/naturepl.com; **4.6** Jan Baks/Alamy; **4.7** SciMAT/Photo Researchers, Inc; **5.CO** Joseph Van Os/Getty Images; **5.5** Laboratory of Tree-Ring Research, University of Arizona; **5.6** Dan Suzio/Photo Researchers, Inc.; **5.7** Dominique Braud/Animals Animals - Earth Scenes; **6.CO** Priscilla Connell/Indexstock/Photolibrary; **6.2** Biological Photo Service; **6.3** W. Atlee Burpee & Co.; **7.CO** Fritz Polking/Bruce Coleman Inc; **7.5** George H. H. Huey/CORBIS; **7.6** Gareth Jones; **7.7** Fred Habegger/Grant Heilman Photography; **7.10a** Dr. Lee W. Wilcox; **7.10b** Astrid & Hanns-Frieder Michler/Photo Researchers, Inc.; **7.11** Jerry Young/Dorling Kindersley Media Library; **8.CO** Dorling Kindersley Media Library; **8.2** Royalty-Free/Corbis; **8.5** Adrienne Hart-Davis/PhotoResearchers, Inc.; **8.6a** Custom Medical Stock Photo; **8.6b** Dorling Kindersley Media Library; **8.7a** TNT MAGAZINE/Alamy; **8.8b** Alan J. Southward; **8.8c** London Scientific Films/OSF/Photolibrary.com; **8.8** Rob Kingsolver; **9.CO** Robert Holmes/Alamy; **9.6** Rob Kingsolver; **9.7** Rob Kingsolver; **10.CO** Alexander Wild; **10.2** R J Erwin/DRK Photo; **10.5** Inga Spence/Visuals Unlimited; **11.CO** PH School/Photo Researchers, Inc.; **11.3** Andre Seale/age footstock; **11.4** D. PARER & E. PARER-COOK/AUSCAPE/Minden Pictures; **12.CO** Ronald L. Day; **12.3** Doug Backlund; **12.4** George Grall/National Geographic Image Collection; **12.5** Michael Abbey/Visuals Unlimited; **12.6a** Stephen Sharnoff/National Geographic Image Collection; **12.6b** Jack Dermid; **12.6c** Duncan McEwan/naturepl.com; **12.6d** Inga Spence/Visuals Unlimited; **12.8** Stephen Sharnoff; **13.CO** Jeff Hunter/Getty Images; **13.3** John Meyer; **13.5** Ted Mead/Getty Images; **13.6** Toby Smith; **13.7** Lyntha Scott Eiler/Library of Congress; **14.CO** National Park Service; **14.2** Theo Allofs/zefa/Corbis; **14.3** AP Photos/Courtesy USDA Forest Service; **14.4** Ron Buskirk/Alamy; **14.5** Phil Schermeister/CORBIS; **14.6** Gregory K. Scott/Photo Researchers, Inc; **14.7** Mark Turner; **14.8** Rob Kingsolver; **14.9** Scott Camazine/Photo Researchers, Inc; **14.10** Kevin Schafer/Alamy; **15.CO** Bret Robinson/USGS; **15.2** NASA/John F. Kennedy Space Center; **15.3** USDA/NRCS/NCGC/National Cartography and Geospatial Center; **15.4** Ted Spiegel/CORBIS; **15.5** Mauro Fermariello/Photo Researchers, Inc; **15.9** David M Dennis/Photolibrary; **16.CO** Gerry Ellis/Digital Vision/Photolibrary; **16.2** Wolfgang Kaehler/CORBIS; **16.3** Victoria McCormick/Animals Animals - Earth Scenes; **16.4** Ken Hammond/USDA; **16.7** Edward Kinsman/Photo Researchers, Inc.; **16.9** Bryan Smith; **17.CO** Jerry D. Greer; **17.2** Fritz Polking/Peter Arnold, Inc.; **17.6** Connecticut Agricultural Experiment Station; **17.9** Science VU/Visuals Unlimited; **17.9** Roland Birke/Peter Arnold, Inc.; **17.10** Wally Eberhart/Visuals Unlimited; **18.CO** J P Nacivet/Getty Images; **18.6** Larry Korhnak; **18.9** Bettmann/CORBIS; **18.10** Landsat Images/USGS; **18.10** Landsat Images/USGS; **18.11** Victoria McCormick/Animals Animals - Earth Scenes.

Figure Credits

Fig 2.4 C. Edward Stevens and Ian D. Hume. 1998. "Contributions of Microbes in Vertebrate Gastrointestinal Tract to Production and Conservation of Nutrients." *PHYSIOLOGICAL REVIEWS* 78(2): 393-427, fig. 3. Copyright © 1998 by the American Physiological Society. Used with permission.; **Fig 3.3** Material copyright 1996-2004, Enchanted Learning Software (**www.enchantedlearning.com**).; **Fig 3.4** Material used with kind permission of HortNET, a product of The Horticulture and Food Research Institute of New Zealand.; **Fig 7.8** Image courtesy of Missouri Botanical Garden.; **Fig 9.2** Hardy, AC (1924). "The herring in relation to its animate environment. Part I. The food and feeding habits of the herring with special reference to the east coast of England." *Fishery Investigations*, Series 27 (3): 53.; **Fig 9.3** EPA Kuchler Potential Natural Vegetation of the Conterminous United States, Kuchler, A.W. 1964. "Potential Natural Vegetation of the Conterminous United States," American Geographical Society, Special Publication No. 36. Used by permission.; **Fig 10.3** George F. Gause, Images of Paramecium caudatum and Paramecium Aurelia, *The Struggle for Existence*, Chap. 5. **www.ggause.com**. **Fig 10.4** George F. Gause, Results for Paramecium Aurelia and Paramecium caudatum, *The Struggle for Existence*, Chap. 5. **www.ggause.com**.; **Fig 11.2** *The American Heritage ® Dictionary of the English Language*, Fourth Edition. Copyright © 2000 by Houghton Mifflin Company. Published by the Houghton Mifflin Company. All rights reserved.; **Fig 11.5**